Industrial Oil Plant

Changzhu Li · Zhihong Xiao ·
Liangnian He · Martino Di Serio ·
Xinfeng Xie
Editors

Industrial Oil Plant

Application Principles and Green Technologies

Springer

Editors
Changzhu Li
Hunan Academy of Forestry
Changsha, Hunan, China

Liangnian He
Institute of Elemento-Organic Chemistry
Nankai University
Tianjin, China

Xinfeng Xie
School of Forest Resources and
Environmental Science
Michigan Technological University
Houghton, USA

Zhihong Xiao
Institute of Biomass Energy
Hunan Academy of Forestry
Changsha, Hunan, China

Martino Di Serio
Department of Chemical Science
University of Naples Federico II
Centralino, Napoli, Italy

It was financially supported by the State Key Laboratory of Utilization of Woody Oil Resource of China (Grant No. 2019XK2002).

ISBN 978-981-15-4919-9 ISBN 978-981-15-4920-5 (eBook)
https://doi.org/10.1007/978-981-15-4920-5

This Springer imprint is published by the registered company Springer Nature Singapore Pte Ltd.
The registered company address is: 152 Beach Road, #21-01/04 Gateway East, Singapore 189721, Singapore

Foreword

Industrial Oil Plant: Application Principles and Green Technologies

With the advance of human society and civilization, nonrenewable resources including petroleum will be depleted, and at the same time environmental and ecological burdens are created by the increased use of fossil fuels. In order to achieve sustainable development, a right balance in economic development, social development, ecological conservation and environmental protection must be established. Renewable resources have a great potential to displace nonrenewable resources for ecological and environmental protection. Since the new century, plant oil as a renewable resource has attracted a great attention for research and development in several industrial sectors. Varieties of intermediate products such as natural fats and oils, proteins, starch and cellulose have been derived from oil plants, which can be further converted to biodiesel, bio-jet fuel, bio-lubricants, oil-based materials, polymers, surfactants and emulsifiers.

The book *Industrial Oil Plant: Application Principles and Green Technologies* systematically introduces the underlying principles of industrial oil plant and related green technologies for cultivation, harvesting, extraction and applications of oil plants. First, various types of industrial oil plant sources, their cultivation and product properties are introduced and examined comprehensively, with areas of further research and development being identified and highlighted. Second, a solid theoretical foundation is laid out on how to efficiently and effectively extract oil from industrial oil plant resources, which is a critical step in the global development of industrialized modern oil plant business. Third, various technologies and processes are introduced for converting extracted plant oil to different oil products, materials and chemicals, and the applications of those green products. Finally, potential environmental and ecological problems/concerns in developing a sustainable oil plant industry, including the protection of biodiversity and the balance of the forest ecosystem during the plant cultivation and harvesting, are addressed. The processes and technologies for green conversion of oil plant resources to oil-based materials, biofuels and high-value chemical products will be of particular interest to oil plant researchers, producers and end-users.

It is impossible to achieve sustainable development without deploying green products and technologies in all human activities. Industrial oil plant is becoming an important feedstock for bioenergy, chemicals, oil-based materials and pharmaceuticals. With the continuing improvement of the human living standard, it is urgent to develop renewable energy and chemical products which are of low eco-toxicity and environmentally benign. Actively and creatively addressing global problems especially on climate change, energy shortage and ecological crisis has become a global consensus, which calls for green and innovation-driven development of green products and clean technologies. Industrial oil plant is certainly one of the important renewable resources for displacing the chemical products derived from petroleum, natural gas and coal. The book *Industrial Oil Plant: Application Principles and Green Technologies* clearly demonstrates that researchers are ready to tackle those challenges.

Xiaotao Bi, Ph.D., PEng, FCAE
Professor in Chemical
and Biological Engineering
The University of British Columbia
Vancouver, Canada

Contents

Chapter 1
Introduction

Jingzhen Chen, Changzhu Li, and Peiyi Yu

Abstract The problem of energy and environment is a great challenge for mankind in the twenty-first century; with the increase of energy crisis and environmental pollution, the exploitation and utilization of new energy and renewable resources has become a global hot point. Plant energy has many advantages, such as environmental protection and safety, rich reserves, convenient, and transportation, and its development and utilization prospect is second only to coal, oil and natural gas. The evaluation of industrial oil plant germplasm resources is to describe and compare the characteristics of various oil plant germplasm samples under suitable environments, and to comprehensively evaluate the status of various oil plant germplasm resources in industrial oil plant and their potential for development and utilization. The ideal evaluation of germplasm resources includes not only genes that determine phenotypic traits, but also the interaction between genetic traits and the environment. Nowadays, we are in an energy-driven society, and relying solely on petrochemical resources no longer supports the sustainable development of the social economy. The industrial application of oilseed plants has ushered in historic development opportunities and it will provide new avenues for solving energy and environmental problems. The main use of oil plants is the consumption of their oil resources and industrial products. It is an important subject to strengthen the research on the development and utilization of energy plants to implement the energy strategy of sustainable development in China. Industrial oil plant resources are used as an important raw material resource of energy, chemical and material. Large-scale utilization of industrial oil plant resources is of great significance for saving cultivated land, improving ecological environment, increasing farmers' income and cultivating new economic growth points.

J. Chen · C. Li (✉)
Hunan Academy of Forestry, Changsha 410004, Hunan, China
e-mail: lichangzhu2013@aliyun.com

P. Yu
Central South University of Forestry and Technology, Changsha 410004, Hunan, China

C. Li et al. (eds.), *Industrial Oil Plant*,
https://doi.org/10.1007/978-981-15-4920-5_1

1.1 Industrial Oil Plant Resources

1.1.1 Distribution

Industrial oilseed plants refer to those plants that produce oils, aromatic oils or similar alkanes and have the characteristics of replacing petroleum, chemical and material resources [8]. They are mainly used in the production of energy and chemical products. Industrial oil plants have high efficiency in capturing solar energy, absorbing greenhouse gas carbon dioxide, synthesizing of biomass energy raw materials and releasing oxygen, improving the ecological environment with renewable and environmentally friendly [9].

Industrial oil plant resources are an important source of energy, chemical and material raw materials. The large-scale utilization of industrial oil plant resources has great significance in saving cultivated land, improving the ecological environment, increasing farmers' income and cultivating new economic growth points. Industrial oil plant utilization technology mainly includes high-efficiency oil preparation and grease clean conversion technology. The leading products are biodiesel, bio-lubricants and other energy products, such as ink resins and bitumen materials.

Industrial oil plant seeds or raw materials of fruit oil and fat can be indirectly converted into products including biodiesel, bio-lubricants, bio-aviation fuel oils and oleochemical [10]. The trimmed stems can also be directly prepared or converted into solid and gaseous fuels. Mainly include (1) plants rich in oil-like: there are many varieties of this kind of plants, such as jojoba, Caper Spurge, Indiantree spurge, rubber tree and then. They can be used as diesel oil through simple degreasing treatment, which has low production cost and high utilization rate, and is the best source for plant energy; (2) oil rich in plants: an organ of these type of plants (such as seed or fruit) has highly oil content and we can extract the oil from them and convert it to biodiesel. The traditional industrial oil plants, such as *Arecaceae*, *Aleurites fordii Hemsl*, *Linum usitatissimum* L. are also important raw material for paint, and as a woody oil tree with high economic value and rich in unsaturated fatty acids, these plants are mainly included in the *Euphorbiaceae*, *Lauraceae*, *Myrtaceae*, *Apocynaceae*, *Compositae*, *Leguminosae*, *Comaceae*, *Flacourtiaceae* and *Asclepiadaceae*. The types of plant fuel oils that have been used for production include eucalyptus oil, *Jatropha carcas*, *Lindera flavinervia Allen*, *Litsea cubeba*, *Xanthium sibiricuma* and then, most of these plants are closely related to human life and are an important part of people's life and industrial production. Some industrial oil plants are shown as follow:

Genus	Species	Function	Distribution
Theaceae	*Camellia oleifera Abel*	*Camellia oleifera Abel* is an important industrial raw material. It is a raw material for soap, petroleum jelly, mechanical lubricating oil and mechanical anti-rust oil	It is 18°28′–34°34′ north latitude and 100°0′–122°0′ east longitude. It is mainly distributed in 16 provinces and regions such as Fujian, Guangdong, Guangxi and Hunan in China
Euphorbiaceae	*Jatropha curcas* L	*Jatropha curcas L* is China's key development of green energy tree species. It can extract sulfur-free, non-polluting biodiesel and be used to produce plant protein, bio-pesticide and other chemical products	It is native to the tropics of the Americas, later introduced to Southeast Asian countries. In China, it is distributed in Guangdong, Guangxi, Sichuan, Guizhou, Yunnan and other provinces
	Ricinus communis L	One of the world's top ten oil crops, the seed oil content is about 50%. *Ricinus communis* L is an important industrial oil, It can be used as surfactant, fatty acid glycerin, stabilizer for polymerization, polyfoam and elastic rubber	It is widely distributed in tropical regions of the world or cultivated in tropical to warm regions. It is the most in North China and Northeast China, followed by Northwest China and East China
	Sapium sebiferum (L.) Roxb	It is an important industrial oil tree species in southern China. The wax of the seed is called "wax of Chinese tallow", which can be used to make "skin oil" for premium soap, wax paper, candles and then; the oil extracted by the seed is used for paints, inks and then	It is distributed in south of the Yellow River in China, and reaches Shaanxi and Gansu. Japan, Vietnam, and India also have; in addition, It is also cultivated in Europe, America, and Africa

(continued)

(continued)

Genus	Species	Function	Distribution
	Euphorbia lathyris L.	Euphorbia lathyris L. is an ideal energy plant. Seed oil can be used as biodiesel crude oil; one hectare of land can produce 25–125 barrels of oil	It is widely distributed or cultivated in Europe, North Africa, Central Asia, East Asia and North and South America, produced in Jilin, Liaoning, Inner Mongolia, Hubei, Hunan, Guangxi, Sichuan, Guizhou, Yunnan, Tibet and then
	Euphorbia tirucalli	The *Euphorbia tirucalli* is similar to the composition of petroleum. It is rich in 12 kinds of hydrocarbons, and the milk is rich in hydrocarbons such as olefins, hydrazines, sterols and the like. It can be directly or mixed with other substances to form crude oil	It is native to the Mediterranean coast of Africa, and distributed in Hong Kong, Taiwan's Penghu Islands, Hainan, the United States, Malaysia, India, the United Kingdom and France now
	Hevea brasiliensis (Willd. ex A. Juss.) Muell. Arg	Natural rubber is widely used in industrial, national defense, transportation, medical health fields and daily life because of its strong elasticity and good insulation. Seed oil is the raw material for making paints and soaps. The rubber husk can produce activated carbon, furfural and then	It is mainly distributed in Southeast Asian countries, including Malaysia, Indonesia, Thailand, Sri Lanka and India, the rubber production area and rubber production accounted for 90% of the world. China's gum-coating areas are mainly distributed in Hainan, Guangdong, Guangxi, Fujian and Yunnan
Xanthium	*Fructus Xanthii*	Seeds can be pressed for oil, *Fructus Xanthii* and *Camellia oleifera Abel* are similar. They can be mixed with tung oil paint or used as raw materials for ink, soap, linoleum, hardened oil and lubricating oil	It is widely distributed in the plains, hills, low mountains, wilderness roads and fields on the northeastern, north, east, south, northwest and southwestern provinces of China

(continued)

(continued)

Genus	Species	Function	Distribution
Comaceae	*Swida wilsoniana*	The fatty acid composition of the *Swida wilsoniana* oil is mainly composed of oleic acid and linoleic acid, and the content is above 70%. It can be used as raw material for biodiesel or processed into edible oil. *Swida wilsoniana* oil is similar to 0# petrochemical diesel. It is a safe and clean biomass fuel oil	It is distributed in Shaanxi, Gansu, Zhejiang, Jiangxi, Fujian, Henan, Hubei, Hunan, Guangdong, Guangdi, Sichuan, Guizhou and other provinces
Symplocaceae	*Santalum album Linn.*	*Santalum album Linn.* has a wide range of uses in industry. Such as lubricating oil in the machinery industry, softening and cleaning of wool in the textile industry, used as a blending agent in the ink industry	It is a native tree species in China with a wide distribution range from Liaoning to South of Sichuan, Yunnan, Fujian and Taiwan
Sapindaceae	*Xanthoceras sorbifolium Bunge*	It can produce industrial raw materials such as advanced lubricants, advanced paints, plasticizers and cosmetics. The content of biodiesel-related hydrocarbon lipids prepared by *Xanthoceras sorbifolium Bunge* seed oil is high, and the hydrocarbons containing 18C account for 93.4%, which are in line with the ideal biodiesel index	It is distributed in China's north latitude 28°30′–46°, east diameter 100°–127°, and naturally distributed in mountains and hilly areas with an altitude of 400–2000 m. South from Anhui Province, north to Liaoning and southwest of Jilin, east to Shandong, west to Gansu, Ningxia, Xinjiang

(continued)

(continued)

Genus	Species	Function	Distribution
Anacardiaceae	*Pistacia chinensis Bunge*	*Pistacia chinensis Bunge* has strong adaptability, fruits and seeds have higher oil content. As a biofuel oil tree species, it has good development and application prospects. In August 2007, in 《the Medium and Long Term Development Plan for Renewable Energy》, it was specifically proposed to develop Chinese pistache as one of the important biofuel plants	It is distributed in the Mediterranean coast, Afghanistan, central Asia, eastern and southeast, the Philippines to Central America, Mexico and South America, Guatemala. There are three species in China, with the exception of the Northeast and Inner Mongolia

1.1.2 Evaluation of the Germ-Plasm

There are about 1 million species of animals, 300,000 species of plants and many microorganisms on the earth, which contain rich and diverse genetic resources. As long as a small part of them is explored and utilized, it is enough to open up a broad world for the cultivation of new varieties of agriculture and forestry. Plant germplasm resources, also known as genetic resources and gene resources, are all kinds of plant types and varieties that can be used for breeding, cultivation or other biological research, such as cultivated species, wild species, semi-wild species, wild near-edge species and the newly created new germplasm materials [11–13]. It is in the form of plants, seeds, organs, tissues, pollen, cells and DNA fragments.

Plant germplasm resources are an important natural resource formed by natural evolution and artificial creation in a long historical process. It carries all kinds of germplasm, is the material basis of plant evolution, scientific and technological innovation, crop breeding, biotechnology industry, and is an important source of human food, drugs and industrial raw materials. It plays a stable role in the agricultural ecosystem and provides the key raw materials for the rise and development of modern plant breeding. Its richness is directly related to the screening of excellent varieties and genes. In a generalized sense, industrial oil plant resources refer to all biological genetic resources that can be used for industrial oil development, which is the sum of all industrial oil species. As far as a specific species is concerned, the narrow sense of genetic resources is usually the carrier of all available genetic materials, including cultivated varieties, wild species, related species and special genetic materials.

With the development of modern science, scientists have collected most of the useful genes of plants in the world and stored them in a "warehouse" called "gene pool", commonly known as "seed bank" [14]. It is used to preserve the germplasm resources, and there are advanced heat-preservation and moisture-insulation structures and air-conditioning instruments in the storehouse. It maintains a low temperature and dry environment all the year round, slows down the metabolism of seeds, and prolongs the life span of seeds, so that the seeds in a few or even the last 100 years still do not lose the original heredity and germination ability. In nature, all organisms show their own genetic phenomenon, which is the guarantee of life continuation and racial reproduction of germplasm resources. Farming proverb says that "as a man sows, so he shall reap" which is a vivid description of genetic phenomena. The reproduction of beans and melons is determined by genetic material.

The evaluation of industrial oil plant germplasm resources is to describe and compare the characteristics of various oil plant germplasm samples collected for industrial purposes under suitable environment and specific conditions, and to comprehensively evaluate the status of various oil plant germplasm resources in industrial oil plant and their potential for development and utilization. The ideal evaluation of germplasm resources includes not only the description of phenotypic traits, but also the description of genes that control phenotypic traits, as well as the description of phenotypic traits, the interaction between genetic traits and environment. The evaluation of industrial oil plant germplasm resources is the focus of research and development of industrial oil plant germplasm resources, and it is the basis of the protection, utilization and deep development of industrial oil plant germplasm resources, and the comprehensive and accurate evaluation of industrial oil plant resources. The analysis of its position and the potential of its development and utilization is of great significance to carry out directional breeding, expand the genetic basis, improve the breeding level, cultivate the resources with characteristics, and realize the sustainable and multilevel development and utilization of resources.

1.1.2.1 Evaluation of Plant Germplasm Resources for Industrial Oil

Industrial oil plants play an important role in industrial plant resources, which are rich in resources and varied in variety. According to the traditional methods, they can be divided into herbaceous industrial oil plants, woody industrial oil plants and aquatic industrial oil plants [15]. According to the current distribution, domestication degree, utilization status and source of species, industrial oil plants can be divided into natural wild industrial oil plants and wild cultivated industrial oil plants. There are four categories of cultivated industrial oil plants and foreign oil plant resources. According to the oil content of plant oil, it can be divided into two categories: high oil content oil (rape seed, cotton seed, peanut, sesame with oil content more than 30%) and low oil content oil (soybean, rice bran and other oil content that is about 20%). At present, industrial oil is mainly used as an alternative to petroleum production energy, materials and oil-based chemical products.

1.1.2.2 Evaluation of Basic Characters of Industrial Oil Plant Germ Resources

It mainly includes the description and development of the botanical characters such as the root, the stem, the leaf, the flower, the fruit, the seed and the like, the yield character (fruit or seed, grease, etc.), the quality character and the like.

The composition of the root system: taproot and lateral root. Types of roots: fixed roots and adventitious roots. Type of root system: straight root system and fibrous root system. Root metamorphosis: storage root, supporting root, air root, climbing root, water root and parasitic root. The storage root is divided into fleshy root and tuber root.

Section shape of stem: cylindrical, square, triangular and flat. According to texture, stems can be divided into wooden stems, herbaceous stems and fleshy stems. Wooden stems are divided into trees, shrubs and subshrubs. Herbaceous stems are divided into annual herbs, biennial herbs, perennial herbs and evergreen herbs. According to the growth habit, the stem can be divided into vertical stem and rattan stem. The latter can be divided into twining stems, climbing stems and stolons. The perversion of the stem: there are two types of stem metamorphosis, aerial stem and underground stem metamorphosis. Among them, the metamorphosis of aerial stem can be divided into leafy stem, spiny stem, hook stem, stem tendrils, small tuber and small bulb. The metamorphosis of underground stem can be divided into root stem, tuber, bulb and bulb. Branch pattern of stem are divided into uniaxial branch, combined axis branch, binary branch and pseudo-binary branch.

The morphology and size of the leaves are very different, but their components are basically the same and can be divided into three parts: leaf, petiole and bracket. Leaves with all three parts, called complete leaves, lacking petioles or stipules are called incomplete leaves. The morphology of leaves is manifested in the whole shape of leaves, leaf apex, phyllopodium, leaf margin, leaf vein, texture and surface appendages. Leaf morphology: aciculiform, strip-shaped, lanceolate, oblanceolate, rectangular circle, ellipse, oval, obovate, circular, cordate, obcordate, cochlear, fan-shaped, reniform, panduriform, rhomboid, cuneate, triangular, scaly, shield-shaped, arrow shape, Euphorbia pulcherrima, tubular, etc.

The composition of flowers consists of pedicels, receptacles, calyx, corolla, stamens and pistil groups. Flowers with calyx, corolla, stamens and pistil, called complete flowers, lack part or parts of flowers called incomplete flowers. According to the presence or absence of calyx and corolla, flowers can be divided into Multiple perianth, Single perianth, no perianth and multiplicate flower. According to the presence and distribution of stamens and pistil, flowers are divided into bisexual flowers, unisexual flowers and asexual flowers. Corolla also has a variety of colors, including white, red, purple, yellow and so on. Inflorescence is divided into infinite inflorescence and limited inflorescence. Infinite inflorescences are divided into racemes, spikes, catkin, spadix, corymb, capitulum and hypanthium. Limited inflorescences can be divided into single inflorescences, dichotomous cymes, polydichotomous cymes and cymes.

There are many types of fruit, which can be divided into true fruit and pseudocarp, according to the parts involved in fruit formation. According to the source, structure and properties of the fruit, it can be divided into three categories: single fruit, polymeric fruit and aggregated fruit.

The shape, size, color and surface texture of seeds vary with the variety of medicinal plants. The shape of the seed is divided into circle, ellipse, kidney shape, egg shape, conical shape, polygon, etc. There are differences in size. The seed color includes green, white, red, black and so on. Some seed surface are smooth, glossy, and some are rough.

Growth and development habits are the basic characteristics reflecting the adaptation of industrial oil plants to the environment [16]. According to the different requirements of plants for environmental temperature, they can be divided into four types: cold-tolerant plants, semi-cold-tolerant plants, warm-loving plants and heat-resistant plants. The different varieties of the same plant are divided into winter and spring, which are generally divided into five grades: strong winter, weak winter, partial spring and spring. According to the different needs of plants for light intensity, they can be divided into positive plants, shade plants and intermediate plants. According to the response of plants to photoperiod, they can be divided into long day plants, short day plants and medium sunshine plants. According to the adaptability and adaptation mode of plants to water, they can be divided into xerophytes, wet plants, mesophytes and aquatic plants.

The inter-specific or intra-species differences of plants were mainly evaluated from the aspects of phenological stage, flowering habit, fruiting habit, seed setting habit, propagation mode and so on [17]. Phenological stage, seed sowing stage, transplanting stage, growing stage, budding stage, initial flowering stage, unflowering stage, fruiting stage, mature stage and harvest stage. According to the maturity period can be divided into: early maturity, medium maturity and late maturity. Flowering habits include the number of flowers, single flower life, flowering uniformity and so on. Result habits include fruit setting rate, number of results, weight of single fruit, etc. Seed setting habits include seed fullness, 1000-grain weight, etc. The mode of reproduction is divided into sexual reproduction and asexual reproduction.

Yield traits refer to those factors related to yield composition [18]. Industrial oil plant yields are usually divided into biological yield and economic yield. Biological yield is the total amount of all substances produced and accumulated by industrial oil plants through photosynthesis and absorption in a growth period. Economic output refers to the part of industrial oil plants that can be used for industrial use, and it is the harvest part of biological production. The proportion of economic output to biological output is called economic coefficient or harvest index. The relationship among the three is as follows: economic yield = biological output * economic coefficient.

The evaluation of industrial oil plant quality mainly refers to the test, identification and gene analysis of the industrial value of oil plants. The internal quality of oilseeds is an important symbol of fruit commercialization and practicability. The internal quality of fruit is mainly composed of texture, aroma, nutrition and other factors, including fat, protein, carbohydrates, vitamins, minerals and other structure and nutrition, as well as glycosides. Secondary metabolites such as tannins affect the formation of fruit intrinsic quality. The kind and content of fatty acids are the important indexes of fruit quality of industrial oil plants.

1.1.2.3 Evaluation of Stress Resistance of Industrial Oil Plant Germ Resources

It mainly includes the adaptability and resistance of industrial oil plants to diseases, pests, low temperature, high temperature, drought, waterlogging, saline-alkali soil, heavy metal stress and so on.

The vast territory of our country, diverse ecological environment have created rich industrial oil energy plant resources [19]. In the long process of coevolution with the environment, each kind of industrial oil plant has formed a unique suitable environment, but the industrial oil energy plants in nature do not always live under suitable environmental conditions, and often encounter diseases and insects. Cold, drought, salt and other adverse environmental factors, this industrial oil energy plant survival or growth of the environment is called adversity, also known as stress. The stress of industrial oil plants can be divided into biological stress and abiotic stress. Biological stress mainly includes disease, pest, grass damage and so on. Abiotic stress includes low temperature, high temperature, drought, waterlogging, saline and alkali soil, heavy metal and so on. Some industrial oil and energy plants cannot adapt to these bad environments. Some plants can adapt to these environments, and this adaptability and resistance to bad environment is called the resistance (tolerance) of industrial oil plants to the adverse environmental resistance, which leads to the reduction of production or quality, or even the inability to survive, but some plants can adapt to these environments.

The evaluation of stress resistance mainly refers to the systematic identification and gene analysis of adverse environmental factors, such as low temperature, high temperature, waterlogging, drought, saline-alkali soil, acid soil, excess or lack of individual elements in soil, resistance to diseases and insect pests, tolerance and gene analysis.

Disease identification: there are many methods for identification of plant disease resistance, which can be divided into population level, individual level, tissue level and molecular level according to the different levels of non-identification [20]. According to the place of identification, it can be divided into field identification and laboratory identification, and plant materials can be divided into adult identification, seedling identification and retirement identification. In order to evaluate the disease resistance of industrial oil plant germplasm resources quickly and comprehensively, the disease resistance should be identified according to the different requirements of disease, infection type, epidemiological standard and yield, and various methods should be used flexibly to give full play to the advantages of various identification methods.

Field identification: it is the most basic identification method and the main basis for evaluating the identification results of other methods. Field identification needs to be tested in an ad hoc disease resistance identification nursery, according to the different sources of infection. Disease nurseries can be divided into natural disease nurseries and artificial disease nurseries.

Indoor identification: it is done in greenhouse, artificial climate room, plant growth box or other artificial facilities through artificial vaccination to identify plant disease resistance method which is called indoor identification. The indoor identification should strictly control the plant cultivation, pathogen vaccination method and environmental conditions after vaccination, which can identify the disease resistance of plants at each growth stage, but due to the limitation of space it is mainly used for seedling stage identification.

In vitro organ identification: in vitro organs, tissues or cells are used as materials, inoculated with pathogens or toxin treatment to identify disease resistance, so it is only suitable for identifying disease resistance that can be expressed at the organ, tissue and cell level. In vitro leaves, branches, stems, spikes and so on are the most important in vitro materials. In vitro materials need to be cultured in water or culture medium and supplemented with plant hormones to maintain their normal physiological state and disease resistance. In vitro identification, as a laboratory-aided identification method for disease resistance, not only has the advantages of rapid operation and small space occupation. Its advantages are also reflected in the fact that the resistance of the same plant material to different pathogens or different physiological races can be measured at the same time, and that the whole plant cannot be infected, and several physiological races of several diseases or the same disease can be determined at the same time for individual plants. It can also identify the contemporary disease resistance of any single plant in the field and hinder its fruiting. It is convenient to select resistant individual plants among the progenies of key hybrid combinations.

Identification of toxin activity: many crop experiments showed that the sensitivity of hosts to pathogenic bacteria was positively correlated with the sensitivity to toxins, and the resistance to toxins at the cell level was consistent with that in the field. The resistance of plants to pathogenic bacteria and toxins was also positively correlated with the resistance of their cell membrane systems. Bioassay or resistance methods were often used to identify and screen cells and plant resistance.

Cytological identification and screening: this method is designed according to the theory of partial inhibition of cell division by Trichoderma in resistant hosts, and is suitable for infectious wilting diseases caused by Fusarium and various species of Trichoderma.

Identification and screening of cell tissue culture techniques: in the process of tissue culture, disease resistant mutations could be induced by physical and chemical factors, and further isolated, selected and purified, and the resistant mutants could be screened out, which could be selected in vitro under artificial control. The physiological, biochemical and molecular genetic mechanisms of disease resistance were explored by isotopic tracer and semi-microanalysis at the cell, tissue and whole plant levels, and the resistant mutant plants were identified and screened out.

Biochemical identification and screening: isozyme analysis, resistance component analysis, serological methods and other biochemical techniques were used to identify and screen resistant plants. It has been found that potato resistance to late blight is related to catalase activity in leaves. There was also a certain correlation between the activity of polyphenol oxidase and its disease resistance in the field. Whether the enzymes in other plants are also related to their resistance needs further biochemical analysis. At present, the differences of isozymes (such as peroxide isozymes and esterase isozymes) between resistant plants and controls were studied.

Molecular identification and screening: The resistant genes of resistant plants were cloned and identified by PCR, RAPD and RFLP molecular biological techniques, and the differences between them and susceptible plants were analyzed to determine whether the resistance identification and screening of resistant plants could be carried out by gene molecular method. At present, it is still in the stage of exploration.

Disease resistance classification: the main basis for evaluating plant disease resistance is the performance of the disease. The disease investigation can adopt the established index and standard, so the disease resistance can also be graded or quantitatively graded.

Qualitative classification: to the disease resistance of allergic necrosis reaction, or to the disease resistance which is mainly resistant to expansion, the method of grading according to the reaction type is generally adopted. The classification of the reaction type is mainly based on the strength of the infection point necrosis reaction, the size of the disease spot, the shape and color of the disease. The degree of development of fruiting body and the thickness of mycelium layer on the lesion spot were divided into two parts: the degree of development of fruiting body and the thickness of mycelium layer.

Quantitative grading: quantitative classification is the basis of most resistance identification, and the most commonly used quantitative survey indexes are incidence, severity and disease index. The incidence rate refers to the diseased plant rate, diseased leaf rate, diseased stem rate, diseased fruit rate and so on. The number of infected individuals accounts for the percentage of the total number of patients investigated, which is used to indicate the degree of prevalence of the disease. The severity indicates the severity of the disease, and the severity grade can be divided according to the incidence degree of a single leaf, plant or whole field, or can be divided into comprehensive consideration, and a grading standard index table or a grading standard chart can be made. According to the visual estimation, the severity of leaf disease can be graded according to the percentage of spot area in the total area, and the distance between severity and severity can be equal distance or equal ratio sequence. The disease index is the comprehensive value of incidence and severity, when the plant survey unit? The formula is as follows: disease index (%) = disease index = 100% $\times \sum$ (number of leaves at all levels $\times$ severity of all generations)/(total number of leaves investigated $\times$ maximum severity).

The disease grade itself is not disease resistant grade, but the disease data can be used to divide the disease resistance grade. The division method must be based on the disease data of the susceptible control varieties, that is, the establishment of a reference system, because the incidence of the same susceptible variety can be as high as 100% or as low as 20% under different induced intensities. Therefore, it is necessary to set up a reference system, which is based on the disease index of susceptible varieties, and the standard of induction intensity is less than 100%. When the difference between resistance and susceptible varieties is the most obvious, after transforming the disease number into relative disease resistance index (rrd), the effect is better, and the formula is as follows:

$$\mathrm{RRI} = \ln \frac{X}{1 - X} \ln \frac{Y}{1 - Y}$$

In formula x is the disease index of the susceptible control variety and y is the disease index of the tested variety. According to the value of resistance index, the grade of disease resistance was divided.

Evaluation of insect resistance: the insect resistance of industrial oil plants refers to its ability to prevent pests from infestation, growth, development and damage. It is a heritable characteristic formed in the process of long-term coevolution of industrial oil plants, and widely exists in the varieties (lines) of industrial oil plants. In wild species and related species, their performance is related to plant genetic characteristics, genetic characteristics of pest damage, environmental conditions and other factors. The evaluation of insect resistance is the premise and basis of breeding and utilization of plant insect-resistant varieties, and its main task is under the condition of natural or artificial vaccination. Identify the resistance types of plant genetic resources to pests and evaluate the degree of resistance.

Pest identification: laboratory biological identification: the materials needed for indoor biological identification are insect source (commonly used larva of different ages), plant tissue to be identified (official, leaf, stem, root, etc.). Other equipment and tools (petri dish, gauze, wormkeeping tank, brush, etc.) are required for identification. It is common practice to take the insect source of the same age, according to the designed experimental scheme, purposefully connect to the plant tissue to be identified, observe the number of larval deaths and the number of live insects at a fixed time after exposure. According to the purpose, some experiments also investigated the damage degree of plant tissues and organs, the growth and development of larvae (such as pupation rate, emergence rate and reproductive rate, etc.). Strictly speaking, the observation time can be divided into continuous observation and stage observation. Continuous observation means that the changes of larvae are recorded and investigated every day by observing the fixed time period (such as one day) after receiving insects for several days. Although the workload of this kind of survey is very large, it can get more experimental data. The results of identification are more persuasive: stage observation, that is, a fixed period of time (such as two or three days), is investigated for a longer period of time. It can be said that continuous observation includes stage observation, and the observation time of investigation should be determined according to the specific situation.

When carrying out indoor biological identification, it is very important to keep the identified plant tissues and organs fresh and the humidity of the identified environment, otherwise it will make the identification results produce great errors and make the identification futile. The preservation of plant tissues and organs can be often replaced by fresh tissues and organs, or fixed in fresh medium (Ma Lihua 1998).

Moisturizing is usually done by placing moist filter paper under identified plant tissues and organs, or by artificially controlling humidity in the air by receiving insects in an insect culture room. In addition, in the process of receiving insects, we should try our best to avoid the damage to the insect body, otherwise it will have a great impact on the results of the experiment.

Identification of outdoor cage insects: the identification of outdoor cage insects is to artificially cover the yarn net for the whole plant or a part of the plant that grows in potted plants or in the field. In order to isolate the natural enemies of identified pests and other pests, and to create an environment for the feeding of identified pests. You can also use a fixed network room or in the greenhouse for identification, can achieve the same effect. The number and mode of receiving insects should be determined according to different crops, different pests and different experimental purposes, and there is no fixed model. In the process of identification, it is also necessary to investigate the victimization of plants, the death and development of insects, and so on. At the same time, there should be controls that can be combined in a single way or in two ways. For a single way, indoor biological identification is generally used.

Identification of naturally susceptible insects in the field: plants must grow under natural conditions and identify insect resistance under natural conditions in order to be meaningful. The identification of naturally susceptible insects in the field, also known as the field control investigation method, is the identification method that can achieve this goal, but this identification method is restricted by the population size of transgenic plants, and the small population cannot be identified in the field. Only if the population is large enough to identify naturally susceptible insects in the field can it be persuasive. It should be noted that because the field is an open environment, it is vulnerable to natural factors such as weather conditions, insect population, natural enemies and so on. This requires us to do field insect identification in a variety of areas, long time identification, to spend a lot of manpower, material and financial resources. Therefore, field identification requires not only enough population size, but also enough strength to ensure.

Insect resistance classification: according to the harm characteristics of pests, determine the location of the investigation, such as the number of leaves, stems, fruit or pest caused by the fruit drop rate. According to the characteristics of different industrial oil plant pests, the damage degree was divided into several grades, usually divided into $0 \leq 9$ grades, which were resistant to (mr), and (hs) in (r), with high resistance to (hr), and so on. The number of plants at all levels was counted, and the pest index was calculated according to the following formula, and then the insect resistance of germline materials was judged according to the pest index. Pest index $= 100\% \times \sum$ (representative grade value $\times$ number of plants or leaves)/(number of highest values $\times$ number of total plants or leaves).

Cold resistance evaluation: temperature is the dominant factor affecting the geographical distribution of industrial oil plants in climatic factors. Under the premise of the limitation of genetic background of industrial oil plants, it plays a decisive role in some growth and development processes of industrial oil plants. Low temperature stress is a common disaster in the cultivation of industrial oil plants. The cold resistance of industrial oil plants determines its growing season and planting range.

Cold resistance of industrial oil plants refers to the ability of industrial oil energy plants to endure and resist low temperature, and the harm of low temperature to industrial oil plants, according to the degree of low temperature and damage. It can be divided into cold injury (zero upper and lower temperature) and freezing injury (subzero low temperature). Although there is no freezing phenomenon above zero temperature, it can cause physiological obstacles of warm industrial oil plants and cause them to be injured or even killed. The adaptability of industrial oil plants to low temperatures above zero is called cold resistance. When the temperature drops below zero, the industrial oil plant freezes, so it is injured or even death occurs. Its adaptability to low temperature below zero is called freezing resistance.

Field identification: the plant species exposed to natural conditions are used as the research object, and after the low temperature test in winter, the plant species can be observed from the external character, or some index of the plant can be determined by means of the instrument, so as to investigate the frost damage and the survival condition of the plant, thereby identifying the cold resistance of the plant.

Indoor identification method: by means of artificial refrigeration equipment, the identified plant material is subjected to freezing treatment under a certain freezing condition, and the survival and freezing conditions of the plant material are observed and detected by a certain method. According to its detection principle, it can be divided into physiological and biochemical research, molecular biology research and physical research. The physiological and biochemical changes, such as the change of photosynthesis and respiration, soluble sugar, soluble protein, free amino acid and other organic acid levels and the change of the hormone level. The cold resistance of different plants can be compared by the determination of these indexes, which mainly include gas chromatography, spectrophotometry and isoenzyme method. The physical method mainly includes the recovery of the growth method, the electric conductivity method and the microscopic examination method. The biological research mainly includes the dna molecular marker and the cold-resistant gene clone.

Cold resistance evaluation index and classification: according to the criteria of the internal column standard, the cold resistance force of the industrial oilseed plants with different species and varieties is divided into five grades, and the cold resistance force of the plant is the strongest. The whole plant has no freezing damage, the victim rate of the bud is below 101%, or the length of the branch of the branch is less than 10%, and the cold resistance of the second grade is strong. The tree trunk has no freezing damage, the victim rate of the bud is about 30%, or the length of the branch is less than 50%, and the cold resistance of the third grade is moderate. The frost damage of the trunk is less than 50%, the injury rate of the bud is less than 70% and the frostbite length of the branch is below 50%. The cold resistance of the 4-stage cold resistance is weak. The frost damage of the trunk is more than 50%, the rate of the bud is 70% or the branch is all frostbite, but the next year can still germinate. All or a large part of the above-ground part of the ground is frozen, the victim rate of the bud is above 90%, or the root can still germinate in the next year.

Drought resistance evaluation: plant drought resistance refers to the adaptability of plants to drought, that is to say, the ability of plants to grow, survive and reproduce in arid environment, as well as the ability of plant yield recovery after drought relief, but scholars of different disciplines. There are different understandings of early resistance, so different definitions of drought resistance are put forward from different disciplines. (1) Molecular biologist: drought resistance is caused by water stress caused by salt or epg. Survival ability of a single cell or cell mass (tissue). (2) Biochemists: important biochemical reactions, such as the protection of protein synthesis mrna against water deficit. (3) Plant physiologists: continue to grow under water stress, The ability to recover rapidly when water stress is lifted. (4) Ecologists: crops can not only survive under atmospheric or soil drought conditions, And the ability to stabilize the yield at a certain level. (5) Agronomists define drought resistance as the yield stability of a crop or a variety under the condition of water shortage.

Drought is the main environmental factor affecting species richness, distribution and survival, development and growth of individual plants. Plants adopt drought avoidance, drought resistance and drought tolerance strategies to deal with drought stress. There is a shortage of water resources in most marginal lands in China. It is of great significance to evaluate the drought resistance of industrial oil plants in order to screen and utilize the industrial oil plants with strong drought resistance and improve the water use efficiency of marginal land. It is the basic work to promote the rapid development of industrial oil plants in marginal land. Plant drought resistance is a comprehensive response involving the morphological and anatomical structure of plants, water physiological and ecological characteristics and physiological and biochemical reactions to tissue cells, photosynthetic organs and even protoplast structures. It is a complex trait controlled by multiple genes. Using a single index to evaluate plant drought resistance is very limited, and the evaluation results are often very inconsistent.

Drought resistance identification: at present, the main methods used to study the drought resistance of crops are field identification method, drought shed or artificial climate box method and laboratory method. The field identification method is to plant the tested materials directly in the field, and to control the soil moisture by natural precipitation or irrigation to affect the external morphology or yield of the plant, in order to evaluate the drought resistance of the varieties. This method is greatly affected by environmental conditions, takes a long time and has a large workload, but the method is simple, does not need special equipment and takes the output as the evaluation index. Therefore, it is easy for breeders to accept it; the drought shed or artificial simulated climate box method is to sow the tested material on the soil with controllable moisture, which can study the drought resistance of different growth stages and the effect of soil moisture on growth and development. Under the influence of physiological process or yield, this method needs certain equipment and limited experimental quantity, but it is reliable and reproducible. The laboratory method is divided into pot method and hyperosmotic solution method. Pot cultivation (soil culture, sand culture and water culture) is to cultivate plants of different seedling ages by pot cultivation method, and then transfer the normal growing seedlings to osmotic solution for dehydration treatment. Dehydration can be sudden dehydration or dynamic dehydration, but also repeated dehydration. In this process, some physiological indexes can be determined to reflect the drought resistance at seedling stage. The method is simple and feasible, and can be carried out in large quantities, but it is inconvenient to identify the drought resistance in the later stage. The hyperosmotic solution method is to germinate the seeds in hyperosmotic solution. The drought resistance of varieties at seedling stage was evaluated by the percentage of seed germination or germination percentage. Because of the different standards, there are two views at present. One is that the germination rate in hyperosmotic solution cannot represent the drought resistance at seedling stage, but there is little relationship between them. On the other hand, it is considered that the germination and extension period of seeds in hyperosmotic solution is related to drought resistance at seedling stage.

Drought resistance evaluation index: drought resistance is a complex biological character, which is reflected in a series of physiological and morphological changes, and has a certain impact on yield. Therefore, it is difficult to identify drought resistance. Scholars at home and abroad have done a lot of work in the identification of drought resistance of corn, and put forward a variety of drought resistance identification indicators. Morphological indexes such as root number, root dry weight, maximum root length, root–shoot ratio, radicle number, leaf size, shape, angle, leaf curl degree and male panicle size can be used as identification indexes of crop drought resistance. Physiological and biochemical indexes include leaf water potential, leaf relative water content, exosmosis conductivity, aab content, sod activity, mda content, osmotic regulation ability and so on. The common indexes for identifying drought tolerance of maize by CIMMTY were leaf extension index, time between male heading and silking, leaf necrosis grade and yield, etc. The leaf extension index was measured before anthesis, and the formula was as follows: leaf extension index = (leaf length under irrigation condition-leaf length under drought condition)/leaf length $\times$ 100% under irrigation condition. The small leaf extension index indicates that drought has little effect on leaf growth, and such varieties are more drought-resistant than drought-resistant varieties, on the contrary, drought resistance is poor.

After drought, the drought resistance of the varieties with short drawing delay time and short drawing interval time was stronger, on the contrary, the drought resistance of the varieties with short drawing delay time and short drawing interval time was stronger, but on the contrary, the drought resistance of the varieties was poor. The grade of leaf tissue withering was investigated at filling stage. According to the degree of leaf senescence and death during drought, where the cell water holding capacity was strong, the leaf senescence was slow after drought, and the drought resistance of the varieties with long leaf functional period was stronger, on the contrary, the drought resistance was poor. Drought resistance refers to the adaptability and resistance of crops to drought. For crop production, drought resistance and drought resistance are mainly reflected in yield. Therefore, many scholars believe that the evaluation of crop drought resistance should be based on whether it can stabilize the yield and high yield under the condition of drought, among which the following are the most important ones. Although the drought resistance coefficient (dryland yield)-water field yield proposed by Chinoy [21] has been used by many researchers to measure the drought resistance of crops, this index can only indicate the stability of dryland crop varieties. However, it cannot explain the plasticity of high yield or high yield potential, and it is difficult to provide the basis for breeders to select high yield and drought resistance genotypes. In 1978, Fisher put forward the concept of drought sensitivity index: drought sensitivity index = 1 − (dryland yield/water-land yield)/(total dryland yield/tested variety water-land yield summation).

The index still fails to provide information on high and low genotypic yields. In order to make up for the deficiency of drought resistance coefficient proposed by Chinoy [21] and drought sensitivity index proposed by Fisher (1978), Lan Jusheng et al. (1990) put forward drought resistance index. The yield index of crop drought resistance identification has been substantially improved in biological sense. Drought resistance index = (dryland yield × drought resistance coefficient)/mean dryland yield of reference genotype. Hu Fushun (1997) suggested that the drought resistance genotype must have the double standards of high yield and high drought resistance coefficient in dryland, and set up the control at the same time. Based on this, the modified drought resistance index was proposed as an index to measure drought resistance: drought resistance index = (dryland yield/dryland yield of control varieties) × (drought resistance coefficient/drought resistance coefficient of control varieties). In the identification of crop drought resistance? Taking the anti-early index as the index, good results have been obtained.

Salt resistance evaluation: the adaptability of plants to salt stress is called salt resistance. Salt stress is one of the main environmental factors to inhibit plant growth and reduce plant yield. According to the adaptability of plants to salt damage, it can be divided into salt avoidance and salt tolerance. Salt avoidance refers to plants avoiding salt stress or living in high-salt habitats in various ways. In addition, halophytes can also take various forms and physiological ways to avoid excessive salt injury. Many halophyte protoplasts and enzymes related to metabolism can withstand high concentration of salt stress without harm, that is, so-called salt tolerance; salt tolerance is an important mechanism of halophyte resistance to salt damage. Plant salt tolerance is divided into biological salt tolerance and agricultural salt tolerance.

The evaluation of plant salt tolerance is mainly to study the adaptability of plants to excessive salt content, is to study the mechanism of salt tolerance and the basis of salt tolerance ability, and is also the key to the breeding, introduction and screening of salt tolerant plants. Plant salt resistance involves many factors of physiology and biochemistry. It is a very complex reaction process controlled by multiple genes and a manifestation of comprehensive traits. Because of the different salt tolerance mode and salt tolerance mechanism of different plants, the physiological metabolism and biochemical changes of their tissues or cells are also different, or under different salt concentration and environmental conditions, plants may resist salt toxicity through different ways or mechanisms. Therefore, in the screening and evaluation of salt-tolerant resources, different methods and various ways should be used to comprehensively evaluate the salt tolerance of plants.

Identification of salt resistance: the methods of salt tolerance identification vary according to plant and saline-alkali environment, and can be divided into two categories: direct identification and indirect identification. The direct identification method is to place the tested materials in the natural or artificial saline-alkali environment, taking the saline-alkali free condition as the control, taking the germination ability, the growth of each organ of the organism, the yield, the injured symptom and so on as the index. The differences were recorded in different growth stages. For traits that cannot be quantified, the damage index is calculated by population visual classification or individual classification, and the relative damage rate or salt (alkali) tolerance coefficient can be calculated for the measured index. It mainly includes germination comparison method, morphological comparison method and yield comparison method. The yield comparison method is more accurate, but this method has long cycle, high cost and the change of year and climate has great influence, so it can only be used when there are a few materials. The morphological comparison method is based on the visual fuzzy form standard as the classification standard, and can only be qualitative and quantitative, and the error is large. The indirect identification method is based on some physiological and biochemical characteristics, and indirectly determines the salt tolerance of the material to be tested. Such as proline, betaine, albumin, boron content determination, electrolyte extravasation electro-foreign method, mass wall separation method, water potential method, leaf green protein damage method, electro-dissolution method and the like.

Evaluation of resistance to heavy metals: excessive heavy metals in soil are a kind of stress factor to plants, which will limit the normal growth, development and reproduction of plants, so as to change the structure of plant communities. The resistance of plants to heavy metals means that plants can survive in a certain heavy metal environment with high content without toxic symptoms such as decline in growth rate or death, and can still survive and reproduce in an environment under heavy metal pressure, and pass on this ability to the next generation of traits. Plant resistance to heavy metals can be obtained through two ways: avoidance and tolerance. Some plants can protect themselves from poisoning by absorbing high levels of heavy metals in the environment through some external mechanism. Called avoidance in this case, the concentration of heavy metals in plants is not high, and tolerance means that plants have certain physiological mechanisms so that plants can survive in a high content of heavy metals without damage. At this time, there are high concentrations of heavy metals in plants.

Identification of resistance to heavy metals: because of the complexity of soil conditions, plants are also affected by other factors in addition to heavy metal factors, so it is difficult to achieve the consistency of identification results between different experiments and sites in field experiments. Most of them are indirect experiments, including hematoxylin staining, nutrient solution culture and artificial heavy metal toxic soil culture. The roots were treated with heavy metal solution without nutrient elements for a certain time, and then the reaction of roots was observed by staining method to judge the weight resistance of plants. Plant seedlings were cultured in nutrient solution containing heavy metals, and then the response of root growth to heavy metals was observed. The germinated seeds were seeded in heavy metal toxic soil for one week, and the root length was measured to judge the resistance of different genotypes to heavy metal.

1.1.2.4 Evaluation of Genetic Diversity of Industrial Oil Plant Germ Resources

This paper mainly evaluates the characteristics and influencing factors of genetic diversity of industrial oil plants by using genetic, cytological and physiological methods and knowledge from the aspects of adaptability, uniformity, stability and specificity [22], and the relationship between the law of action and the factors.

Genetic diversity, also known as genetic diversity, is an important part of biodiversity, which refers to intra-species gene changes, including genetic variation between populations and within the same population. It covers all the genetic differences among individuals of the same species, is the most important source of biodiversity at all levels, determines the potential of species evolution, and is also the material basis for the survival and development of human society. As an important part of biodiversity, genetic diversity is the basis of ecosystem diversity and species diversity. Any species has its own unique gene pool and genetic organization, and the diversity of species shows genetic diversity.

The study of transmission diversity is of great significance both for the protection of biodiversity and for the sustainable utilization of biological resources. It can (1) identify areas with high genetic diversity, (2) develop strategies for priority collection and sampling, (3) guide the delineation of conservation areas, (4) monitor genetic loss or genetic vulnerability, (5) guide relocation nurseries. Management (6) maximization of genetic Diversity of Core Germ Resources (7) comparison of useful areas of agronomic traits in different crop genomes (8) defining the characteristics of improved varieties and other plant genetic resources (9) Monitoring plant heredity the flow of resources.

Genetic diversity is the basic characteristic of life system and the genetic basis of species adaptation to nature and evolution. The variation of tree population genetic structure and the resulting population genetic diversity are important fields of genetic research. The degree and distribution of genetic diversity of natural population are affected by genetic fading, migration, mutation and selection, and its gene frequency will fluctuate at a certain level, that is to say, genetic diversity will be reflected at dna level. Thus, the detection of polymorphism was realized, and the phenotypic difference and isoenzyme analysis were developed to the present molecular marker method. The same genotype can develop different morphological or physiological characteristics under different environmental conditions, and the same morphology may involve different genotypes. Phenotypic variation cannot be strictly called genetic diversity if it is not confirmed by genetic analysis. The protein polymorphism of isozymes is limited by the limited analysis sites due to the great influence of the environment. Molecular markers are the analysis of genomic polymorphism of species, thus avoiding the limitation of environment and analysis sites.

Genetic markers can clearly reflect the biological characteristics of the genetic polymorphism, and can help people to better study the law of the genetic and variation of the organism. It mainly includes morphological markers, cytological markers, biochemical markers and molecular markers.

1.1.2.5 Development Potential and Environmental Assessment of Industrial Oil Plants

It mainly includes the potential of oil plant development and utilization and environmental impact assessment.

The "potential", that is, the potential ability, refers to the ability of a certain benefit index to increase or a certain loss index may be reduced under a certain period of time, certain productivity development level and certain established application condition, and the potential belongs to the relative concept. The resource potential is the difference between the theoretical stock and the actual resource stock that can be achieved by some kind of resource in a certain period of the future under the common action of the nature and the human. The resource potential is estimated by a certain scientific method and the assumed condition, and also belongs to the relative concept. The resource potential of the industrial oil energy tree species is the theoretical supply scale of the industrial oil resource which can be improved through the improvement of the space expansion and the output capacity of the oil forest planting area in the future for a certain period of time under the established technical condition.

Evaluation method of industrial oil plant resources potential: with the continuous in-depth study of the development and utilization technology of oil plant resources, in order to make more reasonable utilization of this kind of resources, the calculation methods and evaluation criteria of energy and oil plant resource potential have also been studied in various countries. Hiromi, begging mamotoak. From the University of Tokyo It was suggested that the final potential of energy, oil and plant resources was determined by the available land area and the output level of energy plants. The supply capacity of raw materials is the land area of energy oil plants multiplied by the unit output of energy plants on the corresponding land. At the same time, they used glue model to evaluate the potential of biomass resources in developed and developing countries. The European Union has taken three main steps in evaluating biomass resources, including woody oil resources: (1) Assessment of the total output of biomass resources at the regional level of each country, which could serve as the upper limit of biomass resources; (2) to evaluate the availability of all kinds of biomass energy, that is, to consider the actual availability of various biomass resources under the constraints of afforestation, economy, technology and environment; (3) the energy potential of different types of biomass resources is expressed by calorific value, that is, the potential to provide calorific value, so as to compare the average calorific value of various types of biomass resources.

There are a few calculation methods and evaluation criteria on the potential of energy and oil plant resources in China. Yin Tianyou and others of Jilin University found that the technology of various types of biomass energy utilization and conversion technology should be taken into account in order to evaluate the availability of biomass resources. Whether economic and other conditions are feasible. His assessment model for the availability of biomass resources mainly takes into account the following three aspects. The main contents are as follows: (1) the amount of theoretical resources. That is, the theoretical upper limit of the amount of resources a region may have in a year. (2) quantity available. That is, at the existing economic level, technical conditions can be converted into useful biomass resources. (3) availability? That is, the amount of biomass energy that can actually be used for energy production = the amount available × the available coefficient. The availability coefficient is a series of comprehensive expressions of the development potential and industrial development of industrial oil resources in China, which is a series of non-technical constraints on biomass energy production. Generally, it includes the share of biomass resources as energy use and the restriction factors of ecological environment, such as in order to protect the ecological environment, the harvesting volume of trees should be moderate, and the sewage charge can be used as external income and so on. Combined with the above theoretical research and methods, the calculation methods of resource potential of industrial oil energy tree species can be obtained as follows:

$$\mathrm{Pt} = \mathrm{Sot} \times \mathrm{Qot} - \mathrm{Srt} \times \mathrm{Qrt}$$

Pt The resource potential of oil tree species in t period;
Sot The theoretical upper limit of planting land area of industrial oil tree species in t period;
Qot The theoretical upper limit of unit yield of industrial oil tree species in t period;
Srt During the t period, the actual land area of industrial oil tree species that was planted;
Qrt The actual unit production of the first-stage industrial oil tree species at the time of t.

1.1.3 Relationship Between Cultivation and Ecological Environment

Fossil fuels were the main source of energy in the past, but in addition to effective combustibles, fossil fuels often contain a lot of impurities, which may produce toxic gases and put them into the atmosphere after combustion [23]. The non-combustible waste is discharged into the ecological environment, which contains toxic substances, which seriously destroy the ecological environment. And fossil fuel mining process will also cause environmental pollution; fossil fuels are not inexhaustible; fossil fuels are heading for the road of energy depletion in recent years.

Interest in alternative energy is rising due to the depletion of oil resources and the adverse impact of fossil fuel use on the environment. As the limited reserves of fossil fuels and raw materials in the world, we have aroused active research interest in non-oil, renewable and pollution-free fuels. Biofuel is the only viable source of energy. In the future, we can form the foundation of sustainable development on social, economic and environmental issues. The emergence of biodiesel and bioethanol is a promising energy source in the future. Renewable energy is energy obtained from renewable resources, which cannot be exhausted for all practical purposes.

Therefore, renewable energy is completely different. Fossil fuels do not produce as many greenhouse gases and other pollutants as fossil fuels burn [24]. Renewable energy such as wind, solar, geothermal, hydrogen and biomass plays an important role.

There are plants in all parts of the world, and there are unique oil plants in different regions. Their growth is different. These oil plants, such as *Swida wilsoniana* (*Wanger.*), *Xanthoceras sorbifolium Bunge*, *Pistacia chinensis Bunge*, *Jatropha curcas*, leprosy, are mainly distributed in China. Most oil plants are barren, drought-tolerant and saline-alkali tolerant, which can be used not only as oil output, but also as soil and water conservation projects, especially in the transformation of mountains, and the greening benefit is more obvious. Biofuels are in the process of acquiring worship status. It does not provide an opportunity to address energy security and climate change. It is simple to use, biodegradable, nontoxic and basically does not contain sulfur and aromatic compounds. Biofuels such as bioethanol themselves or as mixtures can help reduce oil imports and carbon emissions. There is need to develop the right biofuel model based on raw materials and technology, and develop a long-term action plan to ensure availability. Efforts to determine the correct variety of raw materials and the establishment of R&D and production facilities throughout the country are not reliable. Shrubs and trees that can survive in difficult climatic and soil conditions are often considered to be a good source of biodiesel in crop selection.

Swida wilsoniana (Wanger.) Sojak

Swida wilsoniana (*Wanger.*) *Sojak*, a perennial deciduous tree of Cornus officinalis, is an ideal wild woody oil tree species, which mainly grows in central and southern China. It is resistant to barren land and drought, and is suitable for limestone mountain growth. The oil content of the fruit is $33\% \leq 36\%$, and the average oil production per plant is more than 15 kg. The refined oil of *Swida wilsoniana* (*Wanger.*) *Sojak* contains 77.68% unsaturated fatty acids, including 38.3% oleic acid and 38.85% linoleic acid. The refined oil was identified as the first-grade edible oil by the Ministry of National Food. The oil refining of *Swida wilsoniana* (*Wanger.*) *Sojak* is a kind of biodiesel prepared by transesterification. Its physicochemical properties are similar to those of 0# diesel oil, good exercise, stable operation, safety and cleanliness, an ideal alternative to 0 # diesel [17 # 19]. The ecological environment of *Swida wilsoniana* (*Wanger.*) *Sojak* is perennial deciduous shrubs or trees of Cornus officinalis. The trees are tall and grow rapidly. They are widely distributed in the south of the Yellow River in China, and are mainly concentrated in the central and southwest regions of the Yangtze River Basin. Especially in Hunan, Jiangxi, Hubei, Sichuan and other provinces, Hunan, Jiangxi, Hubei, Sichuan and other provinces have the most, wide environmental adaptability, resistance to diseases and insect pests, early flowering and fruiting, fast wood production, high fruit yield, and a long income period. The results showed that the growth, development and fruiting of the light bark tree in the edge of the forest were significantly better than those in the open area. The growth and development of forest edge wood is significantly better than that of forest edge wood, and it is an ideal tree species for soil and water conservation, especially in barren limestone mountains.

Pistacia chinensis Bunge
Pistacia chinensis Bunge is a woody oil and timber forest tree species. *Pistacia chinensis Bunge* is widely distributed, drought tolerant, saline-alkali, barren and adaptable. It is not only an excellent tree species for afforestation in barren mountains and beaches, but also an oil tree species with important economic value. The oil content of *Pistacia chinensis Bunge* seed is 42.5%, and the oil content of seed kernel is 56.7%. It is a kind of dry oil, which can be used as industrial raw material or edible oil. Biomass fuel oil produced by Zhenghe Bioenergy Co., Ltd., which uses *Pistacia chinensis Bunge* seeds as raw material, has been tested by the Petroleum products Inspection Laboratory of the Institute of Petroleum and Chemical Sciences of China Petrochemical Corporation. It is proved that the main physical and chemical products of the oil products are physical and chemical. The standard meets the standards of biomass fuel oil in the United States and light fuel oil in China. *Pistacia chinensis Bunge* ecological environment has strong adaptability and can grow in a variety of soils, and its distribution is also very wide, covering most of China, Hubei, Shanxi, Shandong, Henan and other places in the Yellow River Basin to the north. To the south of the Pearl River Basin, Guangxi, Guangdong, east to Fujian, Taiwan, west to Gansu, Sichuan and other places, most of which are sporadic distribution, only a very small number of areas are pure forest or mixed forest *Pistacia chinensis Bunge* as a kind of adaptability. It can grow in a variety of neutral, slightly acidic or even slightly alkaline soils, and *Pistacia chinensis Bunge* can be used in desertification areas and to carry out large-scale afforestation activities.

Xanthoceras sorbifolium Bunge
Any deciduous shrubs or small trees of the genus Elaeagnus of the unaffected family, also known as Wenxiangguo, is a unique excellent woody oil tree species in China. Elaeagnus angustifolia has strong adaptability, barren tolerance, salt and alkali tolerance, and can grow in abandoned land, sand wasteland, sticky land and bare rock land. It has strong cold resistance and drought resistance, and can grow wherever it can fall under rain. The oil content of seed was $30\% \leq 36\%$, and that of seed kernel was $55\% \leq 67\%$. Among them, oleic acid and linoleic acid accounted for 52.8%–53.3% and 37.8%–39.4%, respectively. Elaeagnus angustifolia oil contains iodine value 125.8, diene value 0.45, is a semi-dry oil, is the manufacture of paint, machine A fine raw material for oil, lubricating oil, and soap.

Jatropha carcas L.
Jatropha curcas, also known as *Jatropha curcas*, irrigates deciduous trees of the genus Euphorbia. Wood or small trees. *Jatropha curcas* is bright and warm, resistant to drought and barren; it can grow on barren land such as barren mountains and wasteland, and can grow in gravel soil, coarse soil and limestone bare land. *Jatropha curcas* develops roots and is easy to survive. It is a good greening tree species for soil and water conservation, wind and sand fixation and ecological environment improvement. *Jatropha curcas* is also a high-quality woody oil plant; its seed oil

content is more than 40%, and is the main raw material tree species for extracting bio-friendly clean diesel oil. The biodiesel produced by *Jatropha curcas* seeds can be used in all kinds of diesel engines, and the key techniques such as flash point, freezing point, sulfur content, carbon monoxide emission, particle value and so on are suitable for all kinds of diesel engines. Compared with the traditional diesel oil, *Jatropha curcas* biodiesel has the advantages of cleaner and higher efficiency, and its performance is obviously better than that of ordinary diesel oil and other plant diesel oil.

Oil tree species are resistant to barren, good planting and have certain economic value. While not affecting the normal local agricultural production activities, many areas have carried out "energy farms", the so-called "energy farms", that is, countries according to their specific national conditions. In order to develop energy and solve the energy crisis, the biomass production base is mainly based on planting plants with high photosynthetic efficiency and for the purpose of obtaining energy. Through the rational planning of the land, the "energy farm" makes the best use of mountains, uncultivated wasteland and water area, selects the biomass varieties suitable for the local growth conditions, and breeds and propagates biomass on a large scale in order to obtain a sufficient number of high-yielding plants, and is processed into available energy.

At present, there is no clear stipulation on the classification of "energy farm" in the theoretical circle. I think, according to the region where energy plants are cultivated, "energy farms" can be divided into "energy forest farms" that use mountainous areas to grow woody plants and "energy fishing grounds" that use water area to cultivate energy crops. Since chemist Calvin succeeded in growing large areas of petroleum plants in California in 1986, there has been a rapid wave of development and research on petroleum plants around the world. The oil "energy forest farm" is divided into two kinds: one is that the planted woody plants produce hydrocarbons, which can be extracted into fuel; the other is made into fuel. Woody plants directly produce almost petroleum fuel. At present, the research on energy farms at home and abroad is mainly based on the study of energy plants.

Energy plants usually refer to plants that have the ability to synthesize higher reductive hydrocarbons and can produce products close to oil composition and alternative petroleum use, as well as plants rich in oil. In a broad sense, energy plants contain all land and marine plants. Energy plants in a narrow sense refer to plants (not grains) that contain oils or similar substances.

From the domestic point of view, the research on energy plants in China started a little late, mainly concentrated on for more than 10 years after 1990. Although the development speed is relatively fast, there is still a big gap with the advanced level of the world. At present, the research mainly focuses on oil-rich energy plants such as *Camellia oleifera Abel*, *Elaeagnus angustifolia* and other carbohydrate-rich energy plants. The necessity of planting energy plants in a large area is as follows:

(1) In poor rural mountainous areas, especially in rural areas where there are large numbers of barren mountains and wasteland, because of the poor ecological environment, they cannot be used to grow food, resulting in a small area of

arable land, and a limited grain harvest for farmers in these areas, living in poverty, is quite poor. If we make rational use of these natural resources, grow energy plants and develop "energy farms", we can not only restore vegetation construction, improve their ecological environment, but also increase rural employment opportunities and improve the living standards of rural people and revitalize the rural economy.

(2) Woody oil plants have the characteristics of planting once and benefiting for many years, and the economic benefits are considerable. They play an important role in adjusting the industrial structure of rural China and improving the living standards of the people. Combined with the ecological project of returning cropland to forest which is being carried out in our country, the large area construction of biodiesel raw material forest can change the inferiority of barren mountain into its advantage. Under the premise of the existing economic level and resource level, we should make full use of mountain resources, cultivate woody oil plants on a large scale and develop the road of bioenergy with Chinese characteristics.

(3) China is a large country of energy production and consumption, and it is also a country with coal as the main energy consumption. Coal consumption accounts for more than 75% of the total primary energy consumption in China, which is close to the average value of that of the same kind of countries in the world. 80% of such large coal consumption is directly burned, resulting in serious environmental pollution. In the absence of effective methods to control the ecological environment pollution caused by the use of fossil fuels, the only way to reduce the use of fossil fuels and to develop clean and alternative energy is to reduce the use of fossil fuels. And the development of "energy farms" is to grow energy plants on a large scale to produce a large number of clean and alternative energy sources, which is one of the best ways.

There are the following reasons why energy plants can be grown in large areas in China:

(1) Land resources

China has a vast land area. In addition to the existing cultivated land, forest land and grassland as traditional agriculture, there are nearly 300 million mu of barren mountain wasteland in the south and 1.5 billion mu of saline-alkali land in the north. Sand can be used to grow energy plants. Because the arable land area in rural areas of our country is limited, there is no free cultivated land to grow energy crops, and woody oil plants have the characteristics of wild, drought-tolerant and barren, so we make use of the large area of barren mountains in rural areas. The development of energy crops by "energy farms", which do not compete with grain production, will become a potential development trend in the future.

(2) Energy and plant resources.

China has tropical, subtropical, warm temperate, cold temperate and plateau alpine regions. The flora of plants is complex and diverse, among which there are only a few species of energy plants in the world. Up to now, there are 151 families and 1553 species of oily plants, of which there are 154 species of plants with oil content of more than 40%. There are nearly 30 kinds of shrubs which can be used to establish large-scale biomass fuel oil raw material base. There are 10 species of biomass fuel oil woody plants which can be used as raw material base and can be used for afforestation in barren mountains, sandy land and other suitable forests to establish a large-scale supply base of biomass fuel oil. From this, Among the energy plants, some species have been screened out, such as *Pinus tabulaeformis*, *Sapium sebiferum*, *Camellia oleifera Abel*, *Cinnamomum cassia* and so on.

The development and utilization of energy plants in foreign countries began as early as 1986. In order to solve the problem of oil shortage faced by the United States at that time, Dr. Calvin selected and cultivated oil plants, and planted a large area of oil plants in California. His success in harvesting 120 barrels of oil per 1 hm^2 has set off a wave of development and research on oil plants around the world. Since then, scientists have discovered some "diesel trees", "alcohol trees" and "wax trees" around the world. So far, more than 40 species of "petroleum" plants have been found in the world, mainly in the family Euphorbia and Euphorbia pulcherriaceae, in Lamiaceae, Compositae, Peach and Leguminosae.

India is rich in biomass resources that can be converted into renewable energy. The Indian government planning commission has launched an ambitious national biodiesel program. It is composed of a number of government agencies and coordinated by the Ministry of Rural Development. The focus of the visit was on planting Jatropha, a shrubby plant of the genus Castor: Ameli. The seed contains 30% less than 40% oil and can be mixed with diesel oil after transesterification.

Because of the large number of energy plants, it can make great contributions to the development of the economy, the solution of energy shortage, the protection of the environment and so on. Therefore, many countries in the world have carried out the selection of energy plant species, the introduction and cultivation of oil-rich species, genetic improvement and the establishment of "energy farm" and so on. Tens of thousands of hm^2 of fast-growing oil forests are planted in the United States; the Philippines has 12,000 hm^2 of acacia trees and can collect 10 million barrels of oil in 6 years. Scientists have also found two perennial weeds, eucalyptus vines and horned melons, that can extract oil in northern Australia. These weeds grow very fast, 30 per week cm, can be harvested several times a year if it is cultivated artificially. In recent years, a number of initiatives have been taken to link poor and marginal farmers in different parts of the world, especially in developing countries, to global biofuels. Without prejudice to food security. The main motivation is to benefit these farmers. The rise in fuel prices may provide an opportunity to achieve this goal. Innovative research on bioethanol from sweet sorghum and biodiesel extracted from Jatropha curcas and Pongamia ensure energy, livelihood and food security for dryland farmers and reduce the use of fossil fuels, which helps reduce food supplies, climate change. These crops It doesn't need too much water, can withstand pressure, and is not expensive to cultivate, so it is a suitable crop for dryland farmers to cultivate. There is a need to develop partnerships between the public and private sectors in such initiatives to make them economically beneficial to poor farmers.

1.2 Economic Value and Environmental Benefits of Industrial Oil Plant

1.2.1 Integrated Utilization of Industrial Oil Plant

With the development of modern society and the progress of civilization, people have become more and more aware that the depletion of petrochemical resources is an indisputable fact, and the ecological and environmental problems caused by the massive use of petrochemical resources are also becoming increasingly prominent, such as incomplete burning of fossil fuels. The resulting smog weather caused by PM2.5 has been shocking. To achieve sustainable socioeconomic development, we must seek an environmentally friendly, renewable alternative to petrochemical resources. China has abundant oil and plant resources, and more than 150 kinds of plants whose oil content are more than 40% have been identified. There are more than 54 million hectares of land in Yilin, and there are a large number of marginal lands such as saline-alkali land, sandy land, mountainous areas, and mines and oilfield reclamation sites. About 100 million hectares of land are suitable for planting specific oil plants. It has very important strategic significance to make full use of existing marginal land resources to develop industrial oil and plant industries which can not only effectively deal with the crisis of existing petrochemical resources exhaustion, but also promote rural ecological construction. It can also solve the unemployment of some farmers and increase their income. However, for a long time, vegetable oil resources have been mainly consumed, and their industrial use has not been properly developed due to the exclusion of coal and petrochemical industries.

Today's society has entered an energy-driven society, and relying solely on petrochemical resources can no longer support the sustainable development of the social economy. The industrial application of oilseed plants has ushered in historic development opportunities and it will provide new avenues for solving energy and environmental problems. The main use of oil plants is the consumption of their oil resources and the use of industrial products. The use of plant oils in the early stage was mainly based on edible oils and fats, and formed a relatively mature basic theoretical technology system. Due to the different application targets, there are certain differences in the production processes of oil plants. Although the industrial application of vegetable oils and fats has a long history, due to the long-term lack of effective manpower, material resources and funds to invest in industrial oil plants, the use of industrial oil plants mainly includes the products transformed with biomass energy conversion technology, such as biodiesel, bio-lubricant energy products, ink resins and asphalt materials. As a renewable biomass energy tree, oil plants are an important alternative resource for solving future energy problems. The use of non-cultivated land for planting oil plants is a feasible way to achieve ecological integration and large-scale production of industrial oil resources. It can promote the development of renewable energy and ecological construction such as soil and water conservation.

1.2.1.1 Application of Industrial Oils in the Market of Basic Oleochemicals

① Fatty Acid Market

Yield: Fatty acids are the most abundant of the basic oleochemicals. According to incomplete statistics, global fatty acid production has been above 3.3 million tons in recent years (including about 10,000 tons of tall oil fatty acids). Uses: A considerable part of fatty acids are used directly, such as for the production of candles, cosmetic bases and surface treatment agents for fillers. About 40% of the fatty acids produced in Europe and the United States are used for soap making, and Japan accounts for about 20%. Malaysia also has nearly 200,000 tons of fatty acids used in the production of soap. Therefore, it is estimated that the global fatty acid used in the soap making industry is more than 1 million tons.

Processed products of fatty acids include fatty acid esters, aliphatic nitrogen compounds, metal soaps and dimer acids [25]. Fatty acid sugar is produced by transesterification of fats and oils, with a global production of 500,000 to 1 million tons. Most of the methyl esters are currently converted to fatty alcohols by hydrogenation. Methacetate (MES) has been used as a detergent raw material with an annual output of about 50,000 tons. One potential area of application for methyl vinegar is biofuels. The relevant technical issues have been resolved, mainly depending on the competition with petroleum fuels and environmental legislation. Fatty acid esters such as higher alcohols, polyhydric alcohols and sorbitol are widely used as emulsifiers, greases, plasticizers and so on. Merely, the annual production of food emulsifiers is around 300,000 tons.

Aliphatic nitrogen-containing compounds primarily reflect derivatives of amides and amines; their main raw materials for cationic and amphoteric surfactants. In the early 1990s, the production of fat nitrogenous materials was estimated to be around 700,000 tons. Later, because of the constraints of environmental conditions, the demand for bis-tallow dimethyl ammonium chloride was sharply reduced in Europe, and the esters were replaced. The amine is still a fatty acid derivative. According to reports, the production of fatty amines in Europe in 1998 was 175,000 tons, which is 18.3% of fatty acid production. North American production is around 200,000 tons. The production of fatty amines in China has also been around 15,000 tons. In addition to alkanolamides, imidazolines, etc., the global production is estimated to be around 700,000 tons.

Metal soaps (fatty acid heavy metal salts such as lead salts, barium salts and zinc salts) are used as heat stabilizers in plastic processing. According to reports, there are 7% of fatty acids in Europe and the United States used to reconstitute metal soaps. In Japan, due to the development of the plastics industry, the consumption of fatty acids in metal soaps is as high as 21%. It is estimated that developed countries consume 150,000 tons of fatty acids in this area. However, due to the toxicity of heavy metal salts and environmental safety issues, metal soaps are being gradually replaced. However, in developing countries, they are still the largest heat stabilizers.

For example, in China, metal soaps accounted for three-quarters of the 45,000 tons of heat stabilizers produced in 1996.

Dimer acid is an additional product of unsaturated fatty acids such as linoleic acid and linolenic acid under thermal or catalytic conditions [26]. The polymerization product of fatty acids has been found for nearly 80 years. After the Second World War, the United States used tall oil fatty acid as raw material to realize the industrial production of dimer acid. At present, the yield of dimer acid is about 40,000 to 50,000 tons. The primary use of dimer acid is to produce reactive polyamide resins, which are excellent epoxy resin curing agents. Another important use is in the production of non-reactive polyamide resins. They have good adhesion to cellophane and polyolefin films, Therefore, it is widely used in such special printing inks. Other uses for dimer acid are metal preservatives, gasoline additives, synthetic lubricating ester additives and coatings. China has only begun to develop dimer acid since the 1970s. There are problems of the small scale of production and unstable quality, and it still mainly relies on imports. According to China's economic development, the demand for dimer acid should be optimistic. The key issue is to improve the quality and reduce costs.

Market Trends: After the analysis above, we can see that the traditional products account for the majority in the fatty acid market; they are being replaced by new products; the rough processed products are the majority while the technical content is low, and the added value of the products is also low. Therefore, the prospect of the market is not optimistic. The further development of the fatty acid industry depends on the development of high value-added deep processing products and the development of new application areas.

② Fatty alcohol market

Yield: Fatty alcohol is the second largest variety of basic oleochemicals, and is also the main breed that drives the growth of the oleochemical industry. According to reports, in 1995 the global production of fatty alcohol was 1,244,500 tons, and increased by 4.65%. Among them, synthetic alcohol was 649,000 tons, with a growth rate of 1.6%; natural alcohol was 575,500 tons, with a growth rate of 9.76%. If estimated at a 4% growth rate, the production of fatty alcohol in 2000 should be about 1.5 million tons, and the synthetic alcohol and natural alcohol will be equally divided.

Uses: Most fatty alcohols are used to produce daily chemical products. 70% fatty alcohol is used to produce anionic, nonionic surfactants such as AS, AE, AES, and APG. 4% is used as a raw material for cosmetics, 8% is used as a lubricant, and the rest is processed into an emulsifier, a demulsifier, a petroleum additive, a fabric treatment agent, etc.

Market Trend: As one of the main raw materials for anionic and nonionic surfactants, the fatty alcohol industry will continue to grow with the further development of the surfactant industry.

The speed of development depends on two aspects of competition: the competition between fatty alcohol and LAB. As the main active of detergents. LAS has a slight advantage in performance/price index. On the other hand, fatty alcohol derivatives are in a favorable position in terms of environmental adaptability, resource renewability and liquid detergent development. From the perspective of development speed of competition between synthetic alcohols and natural alcohols, oil-based fatty alcohols have a clear advantage. As a raw material for nonionic surfactants and other deep-processed products, further development will be made with the expansion of the application fields of these products.

1.2.1.2 Application of Industrial Oil in the Field of Oleochemicals

① Biological liquid fuel

There are two most representative products in bio-liquid fuels: one is traditional biodiesel and the other is bio-aviation fuel.

Biodiesel can completely replace petrochemical diesel, and has been approved as an alternative fuel in the United States, Europe and other countries. At present, due to its high cost, the United States is listed on the B20 (that is, 20% of biodiesel mixed with 80% of petrochemical diesel) to reduce the consumption burden of users. Even so, B20 has proven to be satisfactory for improving air quality. The United States has already listed biodiesel additives, adding 1% of additives to petrochemical diesel to improve the lubricity of sulfur-containing petrochemical diesel. It is also an economically viable method for environmental protection work. In terms of biodiesel, biodiesel is usually produced from edible vegetable oils such as soybean (US) and rapeseed (Germany, Italy, France, etc.) at a cost of 34–59 cents/kg. In order to reduce costs, some countries began to use diesel oil and specialized woody oil plants to produce diesel, and their production costs dropped to 20 cents/kg and 41 cents/kg, respectively. At present, biodiesel has been widely used in the EU and has entered a stage of steady development of commercialization. In 2005, the EU produced more than 3 million tons of biodiesel, more than half of which was produced in Germany. At present, there are 23 bio-oil production plants and 1,800 gas stations in Germany, and the industrial standard (EDIN51606) was issued.

Although the research and development of biodiesel in China started late, it has developed rapidly. The research content involves the selection, cultivation, clean conversion and comprehensive utilization of new varieties of nonedible oil plants. Some of the scientific research results have reached an international advanced level. During the "Eighth Five-Year Plan" period, the Liaoning Provincial Energy Research Institute carried out the project "Improving Agricultural Mechanization and Rural Electrification with Vegetable Oil as Power" (China-European Cooperation Research Project). During the research, the equipment of Italy and Germany was used to esterify the light-skin tree oil produced in Hunan Province. The ester fuel obtained in the golf diesel car (German Volkswagen) and the 40-horsepower wheeled tractor (Italian company)), 30 KVA diesel generator sets (Tessari, Italy) and other equipments confirmed the applicability of the light-skinned oil. After the esterification of the light-skin tree oil, it was burned in the engine of the golf diesel car, and compared with the power performance of burning 0# diesel, the driving speed of the car was 120 km/h. The test results show that the diesel engine fuel is esterified and the diesel engine fuel is basically the same as the diesel fuel. The starting performance of the internal combustion engine is good and the operation is stable; the exhaust emissions are close to each other.

The life cycle assessment (LCA) method is applied to study the biodiesel project from soybean, including soybean planting, bean power generation, soybean oil refining, biodiesel production, various transportation and biodiesel combustion emissions. The six subprocesses were analyzed in a list and their energy consumption and environmental impact were calculated separately. The results showed that the total impact load on biodiesel per kg of soybean oil was 9.69 milli-equivalents; the impact of biodiesel production on environmental impact was mainly CO_2 emissions, and the impact of global warming took the lead; it absorbs 22.264 kg of CO_2 and releases 22.527 kg of CO_2 to the environment. Biodiesel projects can play an active role in reducing greenhouse gas emissions, and biodiesel is an environmentally friendly project compared to diesel. Yao et al. and others conducted a net energy life cycle assessment of biodiesel from waste oil. The study showed that the energy output of the biodiesel production system has an energy surplus compared with the fossil energy input, and the NaOH catalyst is 20.7 g MJ kg^{-1}. The concentrated H_2SO_4 is 24.71 g MJ kg^{-1}. Hu et al. established a life cycle energy consumption and emission assessment model for biodiesel from soybean, rapeseed, light-skinned tree and jatropha, and evaluated life cycle energy consumption and emissions. Compared with petrochemical diesel, the overall energy consumption of soybean and rapeseed biodiesel life cycle is basically the same as that of petrochemical diesel; the overall energy consumption of the life cycle of light skin tree and jatropha biodiesel is about 10% lower than that of petrochemical diesel; biodiesel life cycle fossil energy consumption is significantly reduced, life cycle of HC, CO, PM10, SO_2 and CO_2 emissions are reduced, and NO_2 emissions are increasing. Xing et al. applied the life cycle assessment method to the process of preparing biodiesel from rapeseed oil, jatropha oil and waste oil as raw materials, and applied subprocesses such as raw material planting, collection and transportation, raw material pretreatment, biodiesel production and product distribution. Resource occupancy, water resources and energy consumption were calculated, and parameter sensitivity analysis was performed on energy consumption. The results show that the land resources of the three raw materials for the production of 1 t biodiesel are 13132 m^2, 3333 m^2 and 5 m^2, respectively. The water consumption is 9063.55 m^3, 12306.62 m^3 and 1.97 m^3, respectively. The fossil energy consumption is 0.9 MJ, 0.67 MJ and 0.25 MJ. Since water consumption and land occupation are mainly from planting links, energy consumption mainly occurs in planting and transformation. In China, it is suitable to produce biodiesel from waste oil and jatropha oil. The development of drought-tolerant, high-yield, high-oil-bearing oil plant varieties and new high-efficiency transesterification catalysts and optimized reaction processes are effective measures to reduce the biodiesel life cycle resource consumption and energy consumption.

China's "Medium and Long-term Science and Technology Development Plan" states that biodiesel production in 2010 was 2 million tons, and production in 2020 was 12 million tons. Under the guidance of this plan, the base of biodiesel raw materials based on woody oils such as Jatropha curcas, Pistacia chinensis, Wenguan fruit and Guangpishu has also been greatly developed. Inspired by the country's huge demand for biodiesel and policy support, a large number of domestic enterprises have begun to invest in the biodiesel industry, and the development momentum is relatively fast. According to incomplete statistics, there are currently more than 40 biodiesel producers in China.

In terms of bio-aviation fuels, due to high oil prices, coupled with rising environmental standards, many large airlines are looking for alternative fuels to achieve green flight [27, 28]. The carbon dioxide emission reduction of the aviation industry has become the focus of global response to climate change. The development of bio-aviation fuel is the fundamental way to reduce emissions in the aviation industry. Research and industrialization of diesel or aviation kerosene for the deoxidation of animal and vegetable oils at home and abroad has developed rapidly. Finland is the first country to carry out basic research and industrial production of second generation biodiesel. The Finnish energy company (Fortum OYJ) proposed a method for preparing C12-C24 alkanes by fatty acid hydrodeoxygenation and hydroisomerization (NEXBTL, Next Generation Biomass) to liquid process. Petrobras has developed a process for the mixing and blending of vegetable oils and petrochemical diesel called H-BIO to convert triglycerides into alkanes. In terms of pilot and industrial scale, Sinopec in 2011, organized and designed to transform the original equipment of the Hangzhou refinery into an industrial plant with an annual output of 6,000 tons of bio-aviation kerosene, which can be blended with the appropriate proportion of aviation kerosene blending products as required. Compared with foreign technology, the oil hydrodeoxygenation—the cracking heterogeneous technology—has a small gap, and it is necessary to carry out the industrialized scale production process and equipment amplification design based on the pilot test. Compared with the traditional fossil aviation fuel, the aviation biofuel emits greenhouse gas emissions during the life cycle. The amount has been reduced by 50–80%. So far, nearly 30 aviation biofuel test flights have been carried out around the world. Materials are animal and plant oil or microalgae oil as raw materials, using production hydrogenation process.

According to the latest report released by the International Air Transport Association (IATA), aviation biofuels will account for 1% of aviation kerosene by 2015 and will increase year by year. By 2040, the proportion will reach 40–50% of the total demand.

② Lubricant industry

At present, most of the lubricating oil products are based on mineral oil. Animal and vegetable oils were first used as lubricants, but because they are highly oxidizable, they have a short service life and are gradually replaced by mineral oil. In recent years, due to the need for environmental protection, considering that vegetable oil is easily biodegraded, people have focused their attention on the use of vegetable oils to prepare biodegradable lubricants.

Vegetable oil-based lubricants are environmentally friendly, renewable, less toxic and easily biodegradable, and are a direction for the development of environmentally friendly lubricants in the future. The main component of vegetable oil is triglyceride, which is formed by linking glycerol molecules and three long-chain fatty acids via ester bonds, and has ideal boundary lubrication properties. Compared with mineral oil, vegetable oil has the disadvantages of low thermal stability, low oxidative stability, high freezing point and poor corrosion resistance.

The disadvantages of vegetable oils can be improved by means of additives, chemical modification of vegetable oils and genetic improvement of oils. For example, benzotriazole-containing phosphate esters can increase rapeseed oil load capacity, abrasion resistance and friction properties; amine phosphate can increase the abrasion resistance of vegetable oils; synthetic compounds, S-[2-(acetylamino)thiazolyl-L-yl]dialkyldithiocarbamic acid, added to rapeseed oil at a 1% addition significantly reduces wear and has good no-bite loading capacity; functional additives such as oleic acid, triethanolamine and Triethanolamine oleate significantly improved the friction properties and thermal stability of rapeseed oil. The chemical modification of vegetable oil includes carboxyl functional group modification, fatty acid chain modification, etc., and the modification of the base oil can greatly improve the performance of the base oil. In the genetic modification of oilseeds, various methods of genetic engineering are mainly used to develop new oil plant varieties, such as high oleic soybean oil and sunflower oil, which are currently developed, which have high load-carrying capacity and heat compared with traditional vegetable oils. Oxidative stability.

③ Plastic Industry

Among the three major synthetic materials, the production of synthetic resin is the largest. It is made of plastic and has the widest application range. According to reports, the global production of synthetic resin in 1997 has exceeded 13,000 tons. Among them, the United States produced 42.5 million tons, Japan 15.21 million tons and South Korea 8.17 million tons. The output of Taiwan Province of China also reached 4.64 million tons.

Among the five general-purpose resins (PE, PP, PVC, PS, ABS), PE produced the most, PP increased fastest, PVC ranked third and PE added PP accounted for about one-half of the total resin.

In the 1990s, China's synthetic materials industry gained rapid development. According to statistics, between 1990 and 1996, China's synthetic resin increased at an annual rate of 23.3%, and the growth rates of synthetic fiber and synthetic rubber were 19.7% and 14.2%, respectively. In 1998, the output of synthetic resin reached 6.67 million tons, an increase of 16.0%. Despite this, it is still far from meeting the needs of China's economic growth. In 1998, the performance consumption of China's five general-purpose resins was 14.44 million tons, while the output was 6.43 million tons, and the self-sufficiency rate was only 45%. This shows the potential of China's plastics industry.

Plastic additives: In 1996, the global plastic additives consumed were about 7.9 million tons, accounting for about 6% of the total resin. In 1997, China's additive production was about 700,000 tons, but the varieties were monotonous, of which only plasticizers accounted for two-thirds of the total output. According to the performance classification, plastic additives can be classified into three categories: stabilizers, processing aids and modifiers.

With the development of the plastics industry, new types of additives are constantly emerging, such as anti-reflection, infrared blockers, light-transfer agents (for agricultural film anti-fog). In addition, the voice of nontoxic and harmless additives is getting higher and higher. However, China's plastics auxiliaries industry is relatively backward, the varieties are monotonous and there are many old products. The special materials market is almost completely occupied by imported products. At present, the additives produced by using oils and fats as raw materials are limited to heat stabilizers, filler modifiers and lubricants. The output is also less than one-tenth of the total output of the additive. Obviously, China's auxiliaries industry has a long way to go compared with the fast-growing plastics industry. This also means that the oleochemical industry can add brilliance to the plastics auxiliaries industry and make new contributions.

④ Rubber Industry

In 1997, the world's rubber consumption reached 17 million tons (63% synthetic rubber, 37% natural rubber), with an annual growth rate of 2–3%. The same year tire production reached 10.7 billion, with 17 million pairs of rubber shoes. In 1999, China's rubber consumption was 2.5 million tons, according to the second place in the world. But the output is only 1.15 million tons. The self-sufficiency rate is only 45%. Due to the rapid development of the automobile, home appliance, construction and other industries, the Chinese rubber industry is still facing excellent development opportunities. It is estimated that the annual growth rate will remain at 5–7% in the coming period.

Rubber auxiliaries: According to the purpose of use, rubber auxiliaries are divided into four categories: vulcanization accelerators, anti-aging agents, filling and reinforcing agents, and processing auxiliaries. The current demand for rubber auxiliaries is around 70,000 tons, with an average growth of nearly 3.6% in 5 years. In 1997, the output of auxiliaries in China was 90,100 tons, including 51,300 tons of accelerators and 30,400 tons of anti-aging agents. The bulk additives have formed an oversupply situation.

In the rubber industry, the application of oleochemicals includes vulcanization accelerators, reinforcing agents, dispersing agents and plasticizers. They are used much less than plastic auxiliaries. At present, as the application fields of rubber products continue to expand, the research and development of new products including additives are very active. For example, nano-grade rubber particles, rubber and inorganic ceramic composite wear-resistant lining, safety tires that can still operate under deflation and development of special rubber capable of absorbing light, electricity, magnetism and heat. Requirements are opening up new opportunities for the application of oil deep processing products.

⑤ Inorganic filler industry

Inorganic fillers are widely used in the plastics, rubber, paper, paint, ink and other industries. In addition to reducing production costs, they also give the product a variety of properties. According to reports, the average growth rate of fillers for plastic products in the world in the 1990s was 9%, and demand is expected to rise after 20,000 years.

As a filler, calcium carbonate and talc are used the most. The total consumption of fillers for plastics is about 10% of the amount of resin used, especially in PVC and PP. For example, in the PP packing tape, the $CaCO_3$ filler can be added in an amount of more than 50% of the total weight, which greatly reduces the production cost.

In order to improve the compatibility between inorganic fillers and synthetic resins, and to improve the processing technology and product quality, the surface modification technology of fillers has been widely used. Commonly used surface agents are two types of coupling agents and surfactants. The coupling agent is bonded to the surface of the filler by chemical bonds, and the effect is good, but the cost is high, sensitive to moisture and the operation is inconvenient; the surfactant is modified by adsorption, the operation is convenient and the cost is low. The amount of the surface treatment agent is about 1–2% of the amount of the filler, and it is estimated that the surface treatment agent consumption for the plastics industry alone is between 150,000 and 250,000 tons per year. The oleochemicals currently used as surface treatment agents are mainly stearic acid and its salts. In addition, long-chain quaternary ammonium-treated organic clay has long been used as an additive to coatings.

⑥ Coatings Industry

The use of unsaturated oils such as tung oil to prepare oils can be seen as the earliest examples of oleochemicals used in the coatings industry. After nearly 10 years of efforts, China's coating production has ranked fourth in the world, and the product structure has entered the era of synthetic coatings. According to statistics from the chemical industry, in 1997 China's coating output was 1.37 million tons, of which synthetic resin coatings accounted for 65%.

The development direction of the coatings industry is water-based, solvent-free and powdered. It is predicted that the twenty-first century will be the century of functional coatings, and leaving the additives and their applications, this prediction is just empty talk. Among the main additives of class 19, there may be emulsifiers, pigment-wetting agents, dispersants, defoamers, leveling agents, thickeners, antistatic agents, adhesion promoters and the like.

⑦ Food Industry

China is a country with a large population. The output value of the food industry is above 600 million yuan, accounting for 11.5% of the total industrial output value. The average growth rate in the past 10 years is around 12%. However, from the perspective of development speed, processing depth and product structure, there are still many problems. One of the important issues is that the food additive industry can't meet the needs: less varieties, lower quality and higher prices.

Food emulsifiers are one of the most demanding additives, and the current annual consumption in the world is between 250,000 and 300,000 tons. Due to the backwardness of food processing technology in China, the dosage of emulsifier is only about 11,000 tons, accounting for only 4% of the world total. The annual output of food-grade monoglyceride in China is about 50 million tons, and it is mainly mixed ester containing 50% monoglyceride. Secondly, there are food-grade sorbitol esters (Span, Tween), sucrose esters and other small quantities. Polyglycerol esters and monoglyceride citrate are still in the development stage.

With the development of China's economy, the demand for food emulsifiers will further increase. However, there is still a lot of work to be done in the development of new varieties and in improving product quality.

⑧ Paper Industry

Although the paper industry is not a traditional application area for oleochemicals, the development of the paper industry today offers good opportunities. In 1998, the total demand for paper and paperboard in China was 32.57 million tons, but the annual output was 27.5 million tons, and the self-sufficiency rate was 84.4%. China's per capita consumption is around 25 kg/a, which is less than half of the global average. Poor product quality, low grades, low wood pulp ratio (about 10%) and backward technical equipment are the main problems in the paper industry. However, the backwardness of the papermaking auxiliary industry is also a key reason. For example, it is reported that the annual output of paper chemicals in Western Europe is above 1.3 million tons, with an output value of 3.7 billion US dollars; China's output is only 50,000 tons, and the output value is 36 million US dollars. Among the additives in the paper industry may be related to oleochemicals including deinking agents, defoamers, lubricants, water repellents and the like. The secondary utilization of waste paper and three uses are one of the development trends of the current paper industry. The utilization rate of waste paper in developed countries has exceeded 50%, and China currently only accounts for 1/3. The deinking agent used for waste paper recycling is mainly a surfactant, and the current consumption is only 200 tons, which is expected to increase exponentially. Defoamers mainly use kerosene, while higher education defoamers such as ethylene bis-stearic acid and polyethylene glycol esters are not widely accepted for economic reasons.

1.2.2 The Prospect of Oil Plant Industry Development

1.2.2.1 Industrial Development and Applications of Plant Oils and Their Bio-Based Oleochemicals

In the concepts for new products, performance, product safety and product economy criteria are equally important. They are taken into account already when the raw materials base for a new industrial product development is defined. Here, renewable resources gain again after the earlier "green trend" in the 1980s—increasing attention as an alternative raw materials source compared to fossil feedstock. The industrial use of carbohydrates, proteins and plant oils aligns perfectly with the principles of Responsible Care and is an important part of green chemistry and sustainability in general. Since the 1950s, oleochemistry has grown to a major research and technology area in several institutions and industries. A large variety of products based on fats and oils have been developed since then for different uses, such as specialties for polymer applications, biodiesel, surfactants, emollients for home and personal care industries, pesticides and biodegradable mineral oil replacements for lubricants. However, at present it seems that the use of renewable resources, especially plant oils, have to compete more and more with the increasing demand for bioenergy, which could cause an unbalanced supply and demand in the future or even a threat for the increasing demand for food in certain areas of the world.

The industrial use of agricultural commodities has been an issue generating a significant amount of interest. As the cost of petroleum-derived products increases, the need to change to a more bio-based economy can be clearly seen. In addition to food uses, plant oils have found their way into industrial products in the plastics, pharmaceutical, inks, adhesives, coatings, and many other industries. The advantages of plant oil-derived industrial products can be illustrated by several of the 12 principles of green chemistry (Schwartz et al. 2008) including the call for the use of renewable feedstock, the minimization of hazards and the generation of substances with as little toxicity as possible. Because of their bio-based nature, products formed from plant oil are often biodegradable, and because the CO_2 generated from their degradation can be incorporated into the next year's crop, they can be nearly CO_2 neutral.

1.2.2.2 Examples of Products

Oils and fats are triglycerides with different compositions of the alkyl chains depending on their origin. In industrial processing, they are transferred into fatty acid methyl esters, fatty acids, glycerol, and, as hydrogenation products of the fatty acid methyl esters, fatty alcohols by applying standard manufacturing technologies [29]. Further chemical processes lead to the desired specialty chemicals.

1.2.2.3 Oleochemicals for Polymer Applications

Oleochemicals as polymer materials represent a relatively small market, but are well established. We have to keep in mind that before crude oil was explored, the only possibility to do chemistry was by using renewable resources. One example is linseed oil, which is used to produce linoleum. Here, the demand has increased from 10,000 tons in 1975 to 50,000 tons in 1998 (coming from 120,000 tons in 1960!). Another example: epoxidized soybean oil (ESO) as a plastic and coating additive has a relatively stable market of approximately 100,000 tons/year. It is worth mentioning that the dicarboxylic acids are industrially produced either via ozonolysis of oleic acid to produce azelaic acid, (one of the few examples of large-scale industrial ozonolysis) or by dimerization of linoleic acid and oleic acid to obtain complex mixtures of high molecular weight diacids (e.g., EM-POL types), originally introduced in the 1950s by General Mills Chemicals and Emery (both now Cognis Corp.).

In a variety of polymer applications, such as coatings, a large amount of solvents has to be used, including hydrocarbon and chlorinated solvents. Here, a clear trend is seen toward the so-called "green solvents", such as ester solvents, typically produced from naturally based fatty acids and/or fatty alcohols, representing the largest group.

1.2.2.4 Plant Oils and Their Oleochemicals as Alternative Diesel Fuels

The use of plant oils in diesel engines is nearly as old as the diesel engine itself. The inventor of the diesel engine, Rudolf Diesel, reportedly used groundnut (peanut) oil as a fuel for demonstration purposes in 1900 [30]. Some other work was carried out on the use of plant oils in diesel engines in the 1930s and 1940s. The fuel and energy crises of the late 1970s and early 1980s as well as the accompanying concerns about the depletion of the world's nonrenewable resources provided the incentives to seek alternatives to conventional, petroleum-based fuels. In this context, plant oils as fuel for diesel engines were remembered. They now occupy a prominent position in the development of alternative fuels. Hundreds of scientific articles and various other reports from around the world dealing with plant oil-based alternative diesel fuels ("biodiesel") have appeared in print. They have advanced from being purely experimental fuels to initial stages of commercialization. Nevertheless, various technical and economic aspects require further improvement of these fuels.

Numerous different plant oils have been tested as biodiesel. Often the plant oils investigated for their suitability as biodiesel are those which occur abundantly in the country of testing. Therefore, soybean oil is of primary interest as biodiesel source in the United States while many European countries are concerned with rapeseed oil, and countries with tropical climate prefer to utilize coconut oil or palm oil. Other plant oils, including sunflower, and safflower, have also been investigated. Furthermore, other sources of biodiesel studied include animal fats and used or waste cooking oils. Sources of biodiesel with some emphasis on developing countries have been discussed [31].

1.2.2.5 The Development and Prospect of Palm Oil Industry

In 2013, the world palm oil production reached 55.7 million tons. Indonesia and Malaysia shared 85% of it with their production of 26.70 and 21.70 million tons, respectively. The aims of this study were to identify and analyze the development and prospect of the Indonesian palm oil industry and its derivative products. Data were analyzed by using a descriptive approach. The area of oil palm plantation increased to 9,074,621 ha in 2012. Indonesia's palm oil production in 2012 was 23.5 million tons. There was also an increased Indonesia's palm oil export to 18.15 million tons in 2012. In 2010, the installed capacity of the cooking oil industry in Indonesia was 15.4 million tons. For the fatty alcohol industry, the installed capacity was 320,000 tons/year. In 2007, with real production of 300,000 tons, the 100% real capacity was almost reached. The installed capacities of biodiesel and glycerin industries in 2011 were 3.4 million kiloliters/year and 142,700 tons/year, respectively. In 2010, glycerin production was about 204,394 tons. Glycerin production was 121,640 kiloliters from fatty acid and fatty alcohol industries and 61,694 kiloliters from the biodiesel industry.

The world demand for plant oil is increasing. In 2013, the world demand for this oil was 162.8 million tons and was projected to reach 315.2 million tons in 2030 as a result of the world population growth and the shift in demand from fossil fuel to biofuel (Indonesian Palm Oil Board, 2014). Today, the fulfillment of global need of biofuel comes from palm oil (36.1%), soybean oil (27.4%), rapeseed oil (15.2%) and the other nine kinds of biofuel (21.4%).

In 2013, the world palm oil production was 55.7 million tons. Indonesia and Malaysia shared 85% of it with their production of 26.70 and 21.70 million tons, respectively (Fig. 1.1). In the same year, Indonesia had the export volume of palm oil and its derivative products of 21.2 million tons with a value of US $19.1 billion (47% of international palm oil trade) while Malaysia had 19.8 million tons (44% of international palm oil trade) (Indonesian Palm Oil Board 2014).

As the world's biggest palm oil producer, the export share of Indonesian palm oil and its derivative products is less competitive than that of Malaysia. Of the total national crude palm oil (CPO) production in 2013, only 4 million tons was consumed as cooking oil, 7.7 million tons was used for oleochemicals and biodiesel, and the remaining was exported in the form of CPO. On the other hand, Malaysia exported more oil palm derivative products with higher added values. According to Malaysian Palm Oil Board (MPOB), in 2013, Malaysia exported only 3.8 million tons of CPO

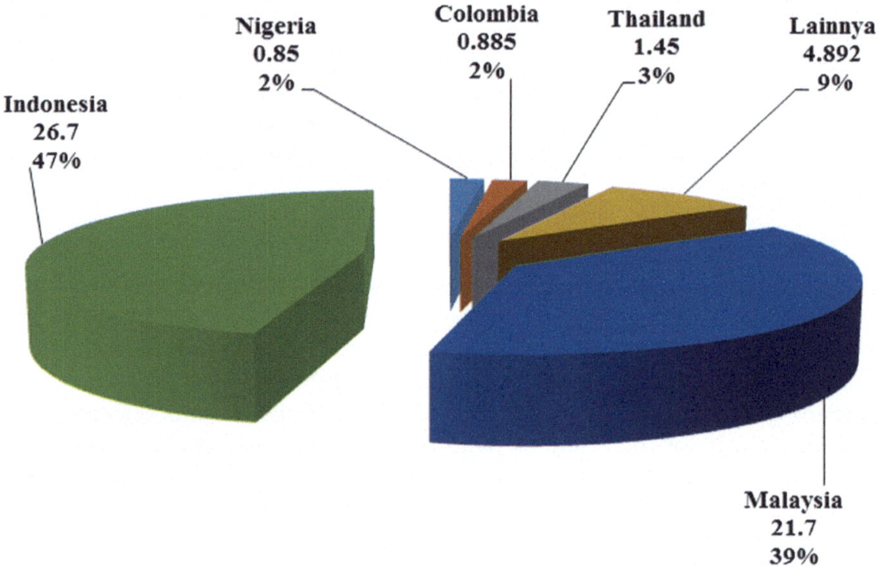

Fig. 1.1 The percentage of world palm oil

and 17.9 million tons of CPO derivatives produced by refinery industries. This condition was very different from that of Indonesia with 40.34% export in the form of CPO and only 59.38% in the form of downstream products.

With the production target of 40 million tons in 2020, it is necessary that Indonesia develop its palm oil downstream industries, improve its export and create new markets for its palm oil and palm oil derivatives in order to avoid excess supply of palm oil. The development of palm oil downstream industry plays a significant role in the national economy, especially as foreign currency generator which can contribute to national GDP by 4.5%, provide employment for about 3.3 million families or 13.2 million people. In the future, the need for food, feed and fuel is increasing which, consequently, will make the need for raw materials to produce them increase and oil palm has big potential to provide those materials.

1.2.2.6 Consumption Development of Palm Oil Derivatives

In the last decade, the development of the palm cooking oil industry was found to occur in line with the shift of public consumption patterns from coconut cooking oil to palm cooking oil. Indonesian Association of Plant Oil Industry (GIMNI), with the assumption of 240 million populations, reported a total domestic consumption of cooking oil of 3.24 million tons in 2010 with the average consumption of 13.5 kg/capita/year. This consumption was estimated to slightly increase to 13.6 kg/year in 2011 making up the total domestic consumption of cooking oil of 3.26 million tons. This might be caused by the lower buying power of Indonesian people.

Cooking oil consumption in the USA and Europe was about 55 kg/capita/year and in China and India was 20 kg/capita/year. If cooking oil consumption is 20 kg/capita/year, the total domestic cooking oil consumption can reach 4.8 million tons. As cooking oil is a normal consumer good, this figure will be easily obtained if there is an increase in people's average income.

Consumers of oleochemicals in the form of fatty acid are producers of soap and detergent, intermediates, plastic, rubber, paper, lubricant, coating, personal care products, food and feed, candles, etc., with their shares as seen in Fig. 1.2.

Fatty alcohol markets include producers of soap and detergent, personal care products, lubricants, amines, etc., with their shares as can be seen in Fig. 1.3. More than two-thirds or about 80% of fatty acid was used as feedstock for surfactant production. As the processing industry of fatty alcohol into surfactants in Indonesia is still limited, most fatty alcohol produced in this country is exported to the global market.

Market opportunity for biodiesel is domestic consumption and export. Domestic consumption is estimated to keep increasing after the issuance of the Regulation of Ministry of Energy and Mineral Resources Number 32 Year 2008. According to this regulation, the utilization of plant oil by 20% for biodiesel, 15% for bioethanol and 10% for PPO is mandatory in 2025. This regulation is valid for the utilization of biodiesel in transportation as part of Public Service Obligation (PSO), non-PSO

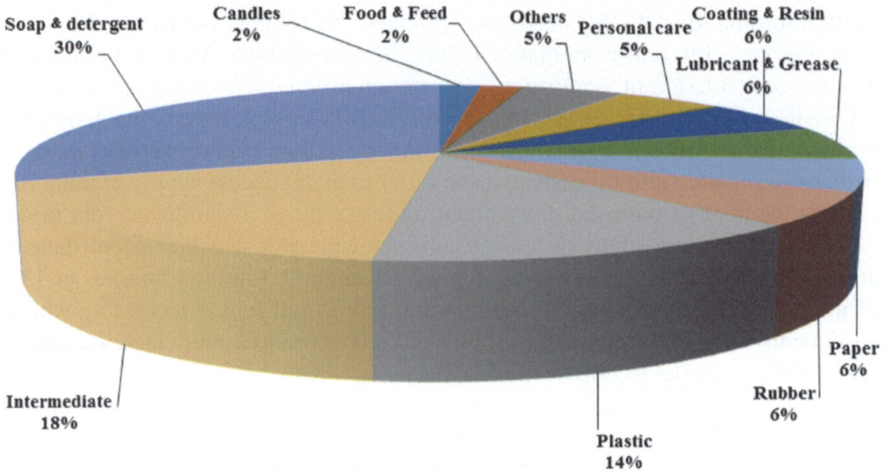

Fig. 1.2 Market of oleochemicals in forms of fatty acid products

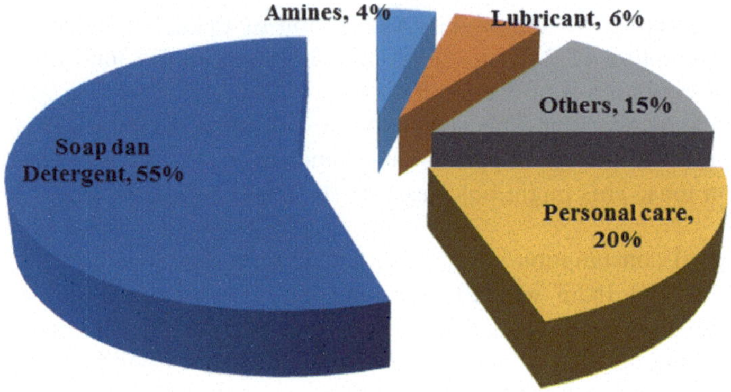

Fig. 1.3 Market of oleochemicals in forms of fatty alcohol

transportation, industry, commercial and power plant. This regulation stipulates a mandatory use of other types of fuel in a gradual and sustainable way by commercial bodies which hold a business license and are direct users of fuel by utilizing and prioritizing locally produced biofuel.

Based on the mandatory use of biodiesel, in 2010, biodiesel production was 460,180 kiloliters which was used to fulfill the need of transportation sector (236,502 kl), non-transportation sector (88,650 kl), industrial and commercial sectors (48,495 kl) and power plant sector (86,533 kl). It is estimated that the use of biodiesel is increasing and reaches 6,866,983 in 2025.

The Indonesian palm oil industry is developing. However, the development only occurs in its upstream industry. The development of the palm oil derivative industry occurs so slowly that Indonesia's palm oil export is still dominated by CPO. This condition is in contrast with that in Malaysia whose palm oil export is dominated by the downstream products.

Therefore, in order to improve Indonesia's export of palm oil and its derivative products and to avoid excess supply, palm oil derivative industry should be developed. This is to improve the added value of oil palm for the most benefit of Indonesian people. With limited domestic palm oil consumption, the government needs to make policies that stimulate the improvement of export by stipulating more flexible export duty tariff and lower export duty tariff for oil palm derivative products than that for CPO. It also recommended that the government provide more budgets for promoting Indonesia's oil palm products overseas by forming a special institution filled in with oil palm experts.

1.3 Utilization Status of Industrial Oil Plant

1.3.1 Industrial Oil Plant in Europe and America

In the United States, the soybean industry has concentrated production areas, large production scale, reasonable farming system, high mechanization level, many varieties without chaos, large yield potential, good quality consistency, simple cultivation technology, relatively low material input and soybean. Utilizing a single bean oil, soybean meal processing and diversified product development, the United States has become the world's largest soybean producer and exporter. In 2004, soybean production reached 79.67 million tons, accounting for 52% of the world's total production. It reached the highest level in history; the total soybean production in 2007 was 79 million tons, the output per unit area was 2.79 t/hm^2, and the planting area was 25.94 million hm^2. About 55% of the soybeans produced in the United States are used for domestic processing of oil, 10% for reserves and seed feed, and 35% for export. The annual export volume is between 24 million and 28 million tons, accounting for about half of the world total export volume. However, in recent years, the proportion of US soybean production in the world has shown a downward trend. From 2000 to 2007, the proportion has dropped by 7 percentage points.

The United States adopts diversified subsidy tools for the oil industry, which is the remarkable feature of the oil policy of the United States. According to the new American agricultural law of 2002, the subsidies provided by the US government for the production of oil crops are mainly divided into two categories: one is direct subsidies, namely subsidies directly provided to oil producers, mainly including marketing loan subsidies, loan spread subsidies, direct income subsidies and countercyclical subsidies. The other is the general service support provided by the government, including the research, technology promotion, pest control, testing and inspection, infrastructure construction, environmental protection and other support. In the case of soybeans, US subsidies for soybean production are on the rise, according to estimates of producer support from the organization for economic cooperation and development. After the implementation of the new agricultural law in 2002, the soybean subsidy was further increased. The government subsidy for soybean production increased from 15.2 dollars in the early 1990s to 59.1 dollars in 2006. The soybean subsidy rate increased from 6.5 to 24%. The nominal protection rate for bean farmers (the ratio between the total value of agricultural products at domestic prices and the value of agricultural products at border prices) was increased from 1.1 to 1.3; At the same time, the United States also implemented export promotion policies for oil products, such as providing support for oil products export in the form of projects (three export promotion programs, namely the new agricultural act of 2002 and the market promotion program, the overseas market development program and the quality sample program in 2004), For increased investment in the export of oil products, mainly in domestic market circulation and overseas market development; timely adjustment of export strategies to shift export regions to developing countries and middle-income countries; efforts to force other countries to reduce import tariffs, reduce agricultural subsidies and so on. The above policies have effectively raised the target price level of oil products and the production income of farmers. However, there are many disadvantages of the above agricultural policies, such as heavy financial burden of the government, bad circulation of agricultural subsidies, international trade friction and unequal distribution pattern of subsidy income.

Robert Pater, a botanist at the center for agricultural research in Bettsville, has been studying the use of an African plant to make industrial oils since 1984. Native to East Africa, this plant is called the Vernoniagolamenit. It has no thorns, lilac flowers, and achenes that are brown in color. As we all know, there is a bad smell when you refresh the paint, because the paint contains volatile solvents. These volatile organic solvents have been considered as air pollutants in recent years and have been banned in many countries around the world. The chamomile seed oil prepared by Patou et al. has a low viscosity and is nonvolatile, and has achieved satisfactory results in diluting dry paint with it instead of solvent. In addition, there are other industrial uses for the oil, such as a raw material in the production of plastics instead of petroleum. Chamomile grows well in arid and semi-arid tropical and subtropical regions with rainfall less than 200 mm. Patou is working with agronomists in Zimbabwe and Kenya to study the best way to grow the wild plant. In its first year, Zimbabwe produced 165 kg of seeds per acre. By improving management, the agronomists increased the yield to 410 kilos the following year. Pai Zhu predicted that through the cultivation of new varieties, yield can be further improved. Patou had tried to introduce galemennis to the United States, but it was a typical African short-day plant. But a variety recently picked up at 11 degrees north latitude in Nigeria, flowering 6 weeks earlier than previously discovered chamomile, it promises to be a late summer and early fall crop in the southeastern United States.

In September 2012, thanks to QE3 (the third round of quantitative easing) in the United States, the oil price in New York approached $100 at one time. As the macro news tended to fade, the trend of crude oil gradually returned to the fundamentals, especially many investors chose to sell at a high level, which also aggravated the decline of oil price. At the same time, US crude oil inventories exceeded expectations and were at historically high levels, weighing on the market. Beijing believes in the medium term that with the global economic downturn, the market is particularly concerned about future demand and may be the main factor to suppress the trend of oil prices in the future, and with the US presidential election approaching, the plan to release strategic petroleum reserves is still under consideration. Industrial products in the crude oil price slump led non-ferrous metals, chemicals, steel and others into the correction shock. And the agricultural products also turned the trend, the early rising soybean meal and other oil and grease varieties showed a significant decline. Monetary easing in Europe and the US saved the sustained decline in industrial production in the short term, but the subsequent rebound in industrial production will still lack momentum if it fails to stimulate the improvement of the real economy. Copper, aluminum, lead, zinc and other non-ferrous products' futures decline in about 1%; Rebar chong high fell back, down nearly 2.4%.

Business club chief researcher Liu Xintian believes that a slump in crude oil can be regarded as the third function of signal fading only relying on the third can bring the commodity market "bull" is obviously unrealistic, afternoon there will be other varieties also can follow the role of the third fade and "flashy tun", this will give the commodity market "the gold nine" pose a challenge.

It is worth noting that in this round of QE3-driven market, the reaction of agricultural products is relatively exceptional, especially the soybean and soybean meal, which had a strong rally in the early stage. QE3 has limited effect to boost the market, and has shown an obvious decline this week. "Crazy" soybean meal starts to calm down. US soybean and soybean meal futures were down 6.6% and 7.3%, respectively, while domestic soybean and soybean meal were down by 1.5% and 5.6%, respectively. The drop in the price of beans and crude oil also led to a sharp drop in soybean oil, palm oil and rapeseed oil, with the three domestic oils dropping by 4.7%, 5.4% and 1.7%, respectively.

Analysts believe that the United States soybean harvest into the month, dry weather bad news gradually faded; the United States began to pull back beans. Domestic, temporary storage soybean auction continued, the spot market supply has increased, easing the upward momentum of soybean. In the early stage of the new beans market, and no demand can bring new good news to the market, the trend of oil and oil is still inclined to adjust, but supported by the global supply shortage background, before the confirmation of the South American harvest is still high volatility market.

According to the investigation to Europe of personage of economic and trade bound, the breed of small oilseed that can offer an export and small miscellaneous grain is very much, wait for seed of millet, pine, sunflower to wait for a dozen kinds. The end use of these varieties in the European grain and oil markets is for the production and processing of animal and bird feed and fish feed. Our small oil and small grain color, taste can meet their needs. European merchants offer more reasonable prices, as long as the quality of the products is stable. Concerned expert thinks, develop the production of small oilseed and small miscellaneous grain energetically and export, should be a content that development earns foreign exchange agriculture henceforth.

Bio-liquid fuel

Biodiesel can completely replace petrochemical diesel, in the United States, Europe and other countries have been approved as alternative fuel. At present, because of its high cost, the United States uses B20 (20% biodiesel and 80% petrochemical diesel) mixture to market, in order to reduce the consumption burden of users. Even so, B20 has been shown to improve air quality satisfactorily. The United States has been listed in the biodiesel additives, adding 1% of the additives in the petrochemical diesel oil, can improve the lubricity of sulfur content of petrochemical diesel oil, for environmental protection, but also do not lose an economic and feasible method. In terms of biodiesel, edible vegetable oils such as soybean (USA) and rapeseed (Germany, Italy, France, etc.) are usually used to produce biodiesel in foreign countries. The cost is as high as 34–59 cents/kg. In order to reduce costs, some countries have started to use waste cooking oil and special woody oil plants to produce diesel, with the production cost reducing to about 20 cents/kg and 40 cents/kg, respectively. At present, biodiesel has been widely used in the EU and has entered the stage of steady commercial development. In 2005 the EU produced more than 3 m tonnes of biodiesel, more than half of which was produced in Germany. Germany now has 23 bio-oil production plants and 1,800 gas stations and has set industrial standards (EDIN51606).

The United States was the first country to study biodiesel. At present, the United States has a number of biodiesel production plants and suppliers; the production of raw materials are mainly soybean oil; the annual production of biodiesel is more than 300,000 tons, and the biodiesel tax rate is zero. The United States actively explores other ways to produce biodiesel while producing soybean biodiesel, and the national laboratory of renewable resources in the United States makes "engineered microalgae" through modern biotechnology. Laboratory conditions can make the lipid content of 40–60% and is expected to produce per acre "engineering microalgae" from 6400 to 16000 of biodiesel, for biodiesel production has opened up a new way. Before April 2005, including the preparation of the factory, a total of 60 biodiesel production plants in the United States, and plans to, biodiesel production 1.15 million tons in 2011, 3.3 million tonnes in 2016.

Pure form of biodiesel, also known as net biodiesel, has been formally listed by the US energy policy act as an alternative fuel for vehicles. Depending on the feedstock and the producer, net US biodiesel prices are currently less than $0.515–$0.793 per liter. The market price for a blend of 80% biodiesel is 7.93–1,057 cents a liter more expensive than conventional diesel.

Finland is a country that has earlier carried out the basic research and industrial production of the second generation biodiesel. Fortum OYJ, a Finnish energy company, proposed a method to prepare C_{12}–C_{14} alkanes by hydrodeoxidation of fatty acids and hydrogenation (NExBTL Next Generation Biomass to Liquid process). Petrobras has developed a process called h-bio, a blend of vegetable oil and petrodiesel that converts triglycerides into alkanes. At present, some countries and regions in the United States, Europe and Asia have begun to establish commercial biodiesel production bases and widely use biodiesel as alternative fuel.

1.3.2 Industrial Oil Plant in Asia

Oil plants are an important source of edible vegetable oil and protein. Oil production and supply are closely related to people's life, agricultural products processing, animal husbandry and grain rotation system. They occupy an important position in national, economic and social development. Japan and South Korea have been the world's major importers of oil crops and vegetable oil for a long time. In recent years, imports of China and India have continued to increase, which has shaken the development of their vegetable oil industry. Japan mainly imports soybeans and rapeseeds. South Korea has changed from a major imported oil crops to a fully imported oil crops and vegetable oil. Over the past two decades, import markets of the world oil crops and vegetable oil have moved rapidly from countries of organization for economic cooperation and development to developing countries. The majority of import growth has occurred in Asia [32].

As two major palm oil exporters in the world, Malaysia and Indonesia impose higher export taxes on crude palm oil than refined palm oil to promote the export of high value-added products. About 80% of Malaysia's palm oil production is exported, which determines the vegetable oil market in India, China and Pakistan. Indonesia is also continuing to expand its export market. But unlike Malaysia, the main goal of Indonesia is to ensure enough edible oil supplies for its 220 million people [33].

Since the mid-1990s, India's edible oil imports have grown rapidly and became the world's largest importer of vegetable oils in a few years. The huge population and stable economic growth are the basis for India's edible oil consumption and import growth, but policy also plays an important role.

Japanese vegetable oil market is dominated by rapeseed oil, soybean oil and palm oil. Palm oil is all dependent on imports. Most soybean oil and rapeseed oil are squeezed by domestic companies. The Japanese oil pressing industry mainly relies on imported oil crops and all domestically produced soybeans and peanuts are almost used for food processing.

South Korea's oil pressing department also relies on imported raw materials. Like Japan, South Korea does not restrict oil imports, but imposes certain tariffs on vegetable oil imports. Geographical location of South Korea is adjacent to China and Japan. The transportation costs of imported oil crops and vegetable oil from the United States, Argentina, Brazil and Canada are relatively close.

Chinese oil crops mainly include rapeseed, peanuts, soybeans, sesame, sunflower, flax, etc. The sum of rapeseed, peanut, soybean acreage and total output accounts for more than 90% of oil crops, which is the main part of oil production and consumption. Among the domestic field crops, planting area, production, output value and employment of farmers of oil crops are second only to rice, wheat and corn. Since the reform and opening up, Chinese oil crop production has generally developed rapidly, reaching the highest level during the Eleventh Five-Year Plan period.

According to the statistics released by the Ministry of Agriculture, the annual planting area of the national oil crops, including rapeseed, peanut, sunflower, sesame and flax, during the "Eleventh Five-Year Plan" period was 13.855 million hm^2, 13.93

million hm^2, 140.23 million hm^2, 14.043 million hm^2 and 1.403 million hm^2; the five-year average was 13.777 million hm^2, the general situation remained stable. Total annual output of oil crops was 33.68 million t, 34.368 million t, 351.6999 million t, 35.074 million t and 35.57 million t; the five-year average was 3461.0 million t, the overall to assume the growth tendency. Total production in 2015 increased by 9.5% in 2010, while total production growth was entirely dependent on higher yields. In the past 5 years, Chinese rapeseed planting area, production and total output have shown an increasing trend. The total output has continuous record-setting in 2013–2015. Planting area of peanuts has remained at 4.6 million hm^2 and the total output in 2013 has set a new record. Planting area of sunflower was 94.7 million hm^2, but the total output continued to grow. The area of sesame and flax planting showed a slight downward trend, but the total output remained basically stable.

Although Chinese oil production has generally increased since the reform and opening up, the market demand for oil products has risen linearly and the growth of domestic production is far from meeting the growth of consumption. After China joined the WTO in 2001, oil crops imports increased rapidly and the products with the largest increase in imports were soybeans, palm oil and rapeseed. In the past 5 years, Chinese imports of oilseed materials were 63.52 million t, 75.58 million t, 83.15 million t, 87.93 million t and 90.48 million t, maintaining a continuous growth trend, with an average annual import volume of 80.15 million tons, accounting for about 57% of the global oilseed trade volume during the same period. China is the largest oilseed importer of the world. In addition to the import of bulk oil, the import of small yield oil such as sesame and flax has also been rising in recent years. China is a traditional peanut exporting country. However, in recent years, due to changes in market demand, price and quality competitiveness factors, peanut exports have been declining and imports have risen rapidly.

In the past two decades, biodiesel has received widespread attention from countries around the world because of issues such as climate warming, environmental pollution, employment and ensuring energy supply security. Governments, scientific research institutions and enterprises of the EU, the United States, Canada, Australia and Asia work together to continuously increase the production and promotion of biodiesel and many countries have entered the practical stage.

Some countries in Asia are also positively developing the biodiesel industry. Japan was an early country to study biodiesel. In 1999, it established an industrialized experimental base for the production of biodiesel from frying oil. At present, the annual output of biodiesel in Japan has reached 40 * 104 t. Thailand's first biodiesel equipment has been put into operation and petroleum companies in Thailand are committed to acquiring palm oil of 7 * 104 t and coconut oil of 2 * 104 t per year to implement a tax reduction policy. South Korea and so on are also promoting the use of biodiesel throughout the country.

The main material of biodiesel is animal or vegetable fats and there are many types of oils and fats. In order to pursue low-cost biodiesel materials, chemical companies around the world use different raw materials to try and explore. China has a wealth of woody oil plant resources and the identified energy oil plant species account for 5% of the country's seed plants. There are about 8,000 kinds of woody plants in

China, about 1,000 species are introduced and about 300 species of oils content of more than 20%, such as *Camellia oleifera*, *Xanthoceras sorbifolia*, *Vernicia fordii*, *Armeniaca sibirica*, *Sapium sebiferum* and so on. Although China has carried out many investigations and research work in recent years, there are still many problems in producing biodiesel from woody plants. As the world's second largest oil consumer, China's energy imports are nearly 50% and annual oil demand is growing. In addition, air pollution in China is more serious and tends to deteriorate. Therefore, energy conservation and emission reduction and energy and environmental safety are China's long-term problems. The development of environmental protection and renewable biofuels have become the strategic needs of the country. China introduced the Renewable Energy Law in 2005, so the production of biodiesel has become an important way to replace traditional diesel.

Current biodiesel activities are mainly concentrated on research and development and demonstration projects. The first biodiesel conference was held in New Delhi on December 12, 2002, and was initiated by the Ministry of Rural Development and the Petroleum Conservation Research Association. As a tropical country, many plants in India can produce a sufficient number of oily materials, such as *Shorea robusta*, Neem Tree and *Jatropha*, which can be used as raw materials for biodiesel.

As the first palm power, Malaysia has gradually let the research and development of biodiesel as an important research direction. It was reported that preliminary test of biodiesel produced by the experimental equipment of the Malaysian Palm Oil Research Institute as a diesel engine fuel was successfully completed at the International Biofuels Symposium organized by Malaysia Palm Oil Research Institute in 1998. The test included detailed testing of the bus fleet (tested by Daimler-Benz since 1987). Due to the limited availability of raw materials in Europe, the export of palm oil has become the best business choice for the Malaysian palm oil industry.

In 2001, the Philippine Coconut Bureau announced a domestic plan to use coconut oil biodiesel as an alternative fuel. Coconut oil contains 45–53% of lauric acid, a saturated short-chain fatty acid (12:0) with more than 14.9% of oxygen content. On the one hand, high oxygen standards result in lower energy content and lower engine efficiency. On the other hand, they have good combustion performance and lower emission standards.

Biodiesel has been recognized as an alternative fuel. It is estimated that the tax relief program will be launched within 2–3 years. In South Korea, two small biodiesel production plants with a total capacity of 8,000 tons per year. NEOENERGY is building a large biodiesel plant with 100,000 tons per year, intending to use soybean oil biodiesel as a non-trademark B20 blended fuel is on the market. Some local governments used biodiesel for testing operations in the fleet, which was completed in July 2004.

In the past few years, Thailand has blended a variety of unesterified vegetable oils (coconut and palm oil) with diesel or kerosene, called "biodiesol". Many of these oils do not meet official standards for commercial use.

References

1. Byrne J, Shen B, Li X (1996) The challenge of sustainability: balancing China's energy, economic and environmental goals. Energy Policy 24(5):455–462
2. Sahir MH, Qureshi AH (2008) Assessment of new and renewable energy resources potential and identification of barriers to their significant utilization in Pakistan. Renew Sustain Energy Rev 12(1):290–298
3. Bryce R (2011) Power hungry: the myths of "Green" energy and the real fuels of the future. PublicAffairs
4. Ahmad AL, Yasin NHM, Derek CJC et al (2011) Microalgae as a sustainable energy source for biodiesel production: a review. Renew Sustain Energy Rev 15(1):584–593
5. Rong F, Victor DG (2011) Coal liquefaction policy in China: explaining the policy reversal since 2006. Energy Policy 39(12):8175–8184
6. Mutlu H, Meier MAR (2010) Castor oil as a renewable resource for the chemical industry. Eur J Lipid Sci Technol 112(1):10–30
7. Bailey AE (1951) Industrial oil and fat products. Industrial oil and fat products, 2nd edn
8. Jaworski J, Cahoon EB (2003) Industrial oils from transgenic plants. Curr Opin Plant Biol 6(2):178–184
9. Lu C, Napier JA, Clemente TE et al (2011) New frontiers in oilseed biotechnology: meeting the global demand for vegetable oils for food, feed, biofuel, and industrial applications. Curr Opin Biotechnol 22(2):252–259
10. Santosa SJ (2008) Palm oil boom in Indonesia: from plantation to downstream products and biodiesel. CLEAN Soil Air Water 36(5–6):453–465
11. Abdurakhmonov IY, Abdukarimov A (2008) Application of association mapping to understanding the genetic diversity of plant germplasm resources. Int J Plant Genomics
12. Beuselinck PR, Steiner JJ (1992) A proposed framework for identifying, quantifying, and utilizing plant germplasm resources. Field Crops Res 29(3):261–272
13. Zichao L, Hongliang Z, Chuanqing S et al (1999) Status and prospects of core collection in plant germplasm resource. J China Agric. Univ. 4(5):51–62
14. Rao VR, Hodgkin T (2002) Genetic diversity and conservation and utilization of plant genetic resources. Plant Cell Tissue Organ Cult 68(1):1–19
15. Demirbaş A (2001) Biomass resource facilities and biomass conversion processing for fuels and chemicals. Energy Convers Manag 42(11):1357–1378
16. O'brien RD (2008) Fats and oils: formulating and processing for applications. CRC Press
17. Schöb C, Kerle S, Karley AJ et al (2015) Intraspecific genetic diversity and composition modify species-level diversity–productivity relationships. New Phytol 205(2):720–730
18. Huang X, Zhao Y, Li C et al (2012) Genome-wide association study of flowering time and grain yield traits in a worldwide collection of rice germplasm. Nat Genet 44(1):32
19. Ragauskas AJ, Williams CK, Davison BH et al (2006) The path forward for biofuels and biomaterials. Science 311(5760):484–489
20. Chen Z, Klessig DF (1991) Identification of a soluble salicylic acid-binding protein that may function in signal transduction in the plant disease-resistance response. Proc Natl Acad Sci 88(18):8179–8183
21. Chinoy JJ (1962) Indian J Plant Physiol 5:172 (F.G. Gregory Memorial Volume)
22. Mitchison JM (1970) Physiological and cytological methods for Schizosaccharomyces pombe. Methods in cell biology. Academic Press, vol 4, pp 131–165
23. Hubbert MK (1949) Energy from fossil fuels. Science 109(2823):103–109
24. Navarro RM, Pena MA, Fierro JLG (2007) Hydrogen production reactions from carbon feedstocks: fossil fuels and biomass. Chem Rev 107(10):3952–3991
25. Bimbo AP (2007) Current and future sources of raw materials for the long-chain omega-3 fatty acid market. Lipid Technol 19(8):176–179
26. Paschke RF, Peterson LE, Wheeler DH (1964) Dimer acid structures. The thermal dimer of methyl 10-trans, 12-trans, linoleate. J. Am. Oil Chem. Soc. 41(11):723–727

27. Liu S, Zhu Q, Guan Q et al (2015) Bio-aviation fuel production from hydroprocessing castor oil promoted by the nickel-based bifunctional catalysts. Biores Technol 183:93–100
28. Hari TK, Yaakob Z, Binitha NN (2015) Aviation biofuel from renewable resources: routes, opportunities and challenges. Renew Sustain Energy Rev 42:1234–1244
29. Metzger JO (2009) Fats and oils as renewable feedstock for chemistry. Eur J Lipid Sci Technol 111(9):865–876
30. Kirakosyan A, Kaufman PB (2009) Recent advances in plant biotechnology. Springer, Dordrecht
31. Shay EG (1993) Diesel fuel from vegetable oils: status and opportunities. Biomass Bioenerg 4(4):227–242
32. Wakker E, Watch S, Rozario J (2004) Greasy palms: the social and ecological impacts of large-scale oil palm plantation development in Southeast Asia
33. Amiruddin MN (2003) Palm oil products exports, prices and export duties: Malaysia and Indonesia compared. Oil Palm Ind Econ J 3(2):15–20

Chapter 2
Industrial Plant Oil Extraction

Zhihong Xiao, Rukuan Liu, Hong Wu, Qiang Liu, and Changzhu Li

Abstract This chapter is concerned with industrial plant oil extraction from oilseeds. It includes a brief account of the principle of oil extraction, storage and preparation of oilseeds, mechanical extraction, solvent extraction, and oil refining. Different purposes of use between edible plant oil and industrial plant oil have led to the different technology, crafts and equipments during each production process. The industrial plant oil extraction has a history, but it did not form a systematic theory and mechanism. How to get a suitable industrial plant oil is the first step. So far, the industrial plant oil extraction is similar to edible plant oil. For example, the solvent extraction process consists of the unit operations of solvent extraction, meal desolventizing, meal drying and cooling, miscella distillation, and solvent recovery. Finally, some novel technology and equipment of oil extraction about especial oilseeds are provided and discussed.

Keyword Industrial plant oil · Pretreatment · Storage · Extraction

The oil extraction from oilseeds involves some steps, depending on the particular commodity to be processed, the scale of operation, and so on. In the early days, industrial plant oils were extracted by pressing, and the functions were often little consideration on the impacts made by subsequent processes. At present solvent extracting method has been widely used in the oil industry. Extracting process has

Z. Xiao (✉) · R. Liu · H. Wu · Q. Liu · C. Li
Hunan Academy of Forestry, Changsha 410004, Hunan, China
e-mail: xzhh1015@163.com

Z. Xiao · R. Liu · H. Wu · C. Li
State Key Laboratory of Utilization of Woody Oil Resource, Changsha 410004, Hunan, China

C. Li
Key Laboratory of National Forestry and Grassland Administration on Utilization Science for Southern Woody Oilseed and Hunan Provincial Key Laboratory of Oils & Fats Molecular Structure and Function, Changsha 410004, Hunan, China

Z. Xiao
Nankai University, Tianjin 300071, China

Hunan Grain Group Co., Ltd, Changsha 410004, Hunan, China

C. Li et al. (eds.), *Industrial Oil Plant*,
https://doi.org/10.1007/978-981-15-4920-5_2

extensive adaptability for raw material, and oil yield efficiency is high; the meal is of good quality and low processing cost. But the extracting process has the poor quality of raw oil, solvent is generally flammable and explosive, and there is certain toxicity. In order to make sure of the quality of the oil, refining the oil is a necessary process. So this chapter will refer to the production technology of industrial plant oil mainly such as the principle of oil extraction mechanical crushing method, solvent extracting method, and oil refining.

2.1 Principle of Oil Extraction

2.1.1 Principle of Mechanical Extraction

Mechanical extraction of the oilseeds is the process of extruding the oil from the blank with mechanical force. In the process of pressing, the extraction of oil can be regarded as the movement of incompressible liquid in the deformed porous medium, following the movement of mucilage fluid. The average velocity of the oil movement is mainly determined by the friction and driving force in the liquid layer. At the same time, the thickness of the liquid layer and the length of the oil pipeline are also important factors affecting the oil discharge rate. Generally speaking, the smaller the oil viscosity, the greater the pressure, the faster the oil flows from the pores. At the same time, the longer the flow path, the smaller the pore which will reduce the flow rate and make the press more slowly.

Under strong pressure, the pressed particles will leave a single molecular oil layer or a multi-molecular oil layer similar to a single molecule on the surface of the pressed particles. Because of the huge molecular force field on the surface, this oil layer is completely combined between the surfaces, and no longer follows the general law of fluid dynamics to flow, nor can it be squeezed out from the gaps between the surfaces. At this point, the oil molecules are in a very thin adsorption film state. In fact, the amount of oil that remains in the cake is much higher than the amount of monomolecular oil that remains on the surface of the particles. This is mainly because the inner and outer surfaces of the particles are not all tightly packed. At the same time, the surface of individual particles contacts directly, so that a part of residual oil remains in the sealed oil circuit.

In addition, under the action of pressure, the particles will squeeze more tighter with the oil discharged. The plastic deformation will be made between the particles contacting each other. Especially the press particles will be integrated when the oil film breaks. In this way, at the end of the press, the press particles are no longer loose and began to form a complete plastic body, called cake. The cake is not bound to all the particles, but is an incompletely bound gel porous body with a large number of pores. That is to say, apart from the continuous gel skeleton of the cake formed by the partial binding action of the particles, there are still many pores between the particles or between the groups of combined particles. Some of these pores are likely to be disconnected, while the other part is connected to each other to form a channel; it is still possible to continue to press oil. Therefore, the residual oil in the pressed cake is composed of the oil enclosed in the pores by the closed oil circuit, the oil combined on the inner and outer surfaces of the particles, and the residual oil in the undamaged oil plant cells. The uneven distribution of residual oil in the cake will be caused by the uneven distribution of pressure and the inconsistent flow rate in the actual pressing process.

2.1.2 *Principle of Solvent Extraction*

Solvent extraction is according to the principle of extraction of the organic chemical separation process. And it is the use of organic solvents that can dissolve oil, through the role of wetting, infiltration, molecular diffusion; the oil in the blank will be extracted. The oil extraction process is a mass transfer process from the solid phase to the liquid phase. This mass transfer process is accomplished by molecular diffusion and convective diffusion. The miscibility between oil and organic solvents can easily happen, with good mutual diffusion, using n-hexane as extracting agent. Oil extraction happened between the solvent and the material under the condition of relative motion, so in addition to the oil molecular diffusion process, there are solvent flow and convection–diffusion processes. General technological process of the extraction method is shown in Fig. 2.1.

Oil embryo or bread by squeeze are dipped in the selected solvent, to make the oil dissolved in a solvent (composed of mixed oil), and then we can separate the mixed oil and solid residue (wet meal). Mixed oil is according to the different boiling point of solvent and steam gasification to separation, then get extraction raw oil. After condensation, cooling recycling solvent steam continues to recycle. Wet meal also contains a certain amount of solvent, after desolventizing drying treatment for a finished product. The solvent in the process of desolventizing drying treatment was condensed cooling recycling.

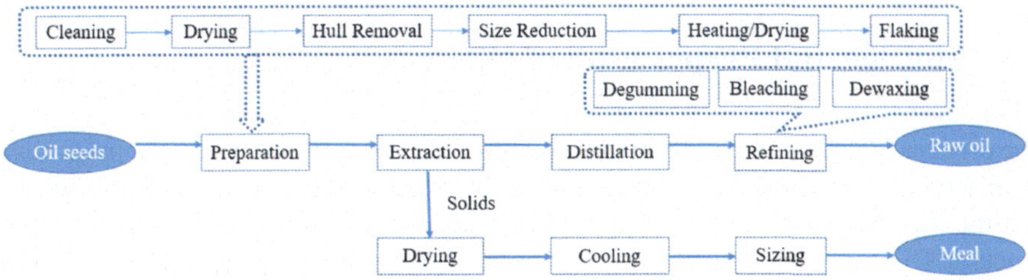

Fig. 2.1 Solvent extraction oil process flow diagram

2.1.2.1 Molecular Diffusion

Molecular diffusion refers to the transfer of substances in the form of a single molecule, which is caused by the random thermal motion of molecules. When the oil is in contact with the solvent, the oil molecule in the oil seeps out of the oil and diffuses to the solvent by means of its thermal movement, thus forming a mixed oil. At the same time, solvent molecules also permeate and diffuse into the oil. So the two different concentrations of mixed oil are formed on both sides of the interface between oil and solvent. Because of the heat movement of the molecules and the difference of the oil concentration between the two sides, the oil molecules will continuously transfer from the higher concentration region to the lower concentration region, until the two sides of the molecular concentration reach equilibrium.

2.1.2.2 Convection–Diffusion

Convection–diffusion refers to the transfer of a substance solution in the form of a smaller volume. Like molecular diffusion, the quantity of diffusions is related to diffusion area, concentration difference, diffusion time, and diffusion coefficient. In the convection–diffusion process, the larger the volume of convection, the more the volume per unit area per unit time; and the greater the convection–diffusion coefficient, the greater the amount of mass transfer.

The essence of the oil leaching process is the mass transfer process. The mass transfer process is completed by both molecular diffusion and convective diffusion. In molecular diffusion, matter is transferred by the kinetic energy of molecular thermal motion. Increasing the extracting temperature properly is helpful to improve the molecular diffusion coefficient and accelerate molecular diffusion. In convection–diffusion, matter is mainly dependent on the energy provided by the outside world for transfer. Generally, the liquid level difference or the pressure produced by the pump is used to make the solvent or mixed oil and oil stock in a state of relative movement to promote convection and diffusion.

2.2 Storage and Preparation of Oilseeds

The process of producing oil from plant oil has certain requirements for the properties of oilseeds. Therefore, the oilseeds should be treated before extracting the oil, so that the oilseeds have the best adaptability to meet the requirements of the relevant process. The function of the preparation process is to properly prepare the oilseeds for extraction of the oil, either by solvent or mechanical methods and, if applicable, remove the hulls and other materials from the seed kernel or meal. While a particular seed may contain from 20 to 50% oil, the oil is tightly bound within the cell and mechanical action must be taken to either forcefully remove the oil or to make the oil more accessible to subsequent solvent extraction. So the common functions of storage operations refer to receiving, sampling, drying, storage, and cleaning. Although there are some variants, for example, some processors may clean the oilseed both before and after the drying operations, the basic operations are designed to accomplish the same task, which is to provide a safe haven for the oilseed and deliver it at the proper time and condition at the processing facility.

As the fresh oilseeds are from the field, they will contain some contaminants, consisting of sticks and pods, metal and rock accumulated during handling, and contamination from weed seeds and other seeds. People will sample the oilseeds and make adjustments in the price paid based on the moisture, heat damage, and other factors. Typically the oilseeds receipts are segregated based on these quality factors, with the moisture content being one of the prime factors for separation. For proper storage and subsequent processing, the contaminants must be removed and the oilseeds must be dried prior to storage. As oilseeds freshly harvested may have a moisture content of up to 40%, they must generally be dried to around 13% moisture for safe extended storage. High moisture damage typically results in reduced oil content, decreased protein, and increased color, and refining loss of the extracted oil. A precursor to this damage is often indicated by a rising oilseed temperature, and many storage facilities are equipped with a series of temperature cables embedded in the oilseeds with indicating and recording equipment located in the manager's office. If left unchecked, the oilseeds will spontaneously heat and become damaged and under extreme conditions, a fire may develop. It is routine practice to monitor the temperature of the oilseeds daily and, if heating is occurring, immediately process the oilseeds. If this is not possible, or the degree of heating is not severe, the manager may simply move the oilseeds from one place to another.

The oilseed is usually scaled when arriving at the preparation facility. After weighing the seed, it then undergoes a thorough cleaning process to remove sand, stalks, plant debris, and any other foreign matter. Cleaning is by rotary or table sieves, usually with air aspiration by fans, and cyclones for dust removal from the air. After cleaning, some seeds with hull should be decorticated by impacting the seed, breaking and separating the hulls. The traditional process continues with the cracking rolls, which are a set of two or three high corrugated rolls turning at relatively high speeds that break the oilseeds into several pieces. After decortication, the oilseeds are delivered to the conditioner where heat is gently applied to make the cracks soft and pliable for the subsequent flaking operation. Conditioning is carried out in a cooker equipped to heat the seeds by indirect steam, adding water and direct steam as required. Flaking of the seed particles is done in roller mills with smooth roller surfaces. The flake thickness in a typical case is between 0.2 and 0.3 mm. To achieve this reduction in thickness, the rollers of the mill are pressed together, usually by hydraulic pressure, and rupture of the seed cells occurs, which makes oil recovery possible.

2.3 Mechanical Crushing Method and Process

Crushing method is an effective way to get oil from the fruit, with the aid of the role of external mechanical forces. Now crushing method is the main method of oil extraction plant in China. Crushing method is adaptable, simple operation process, production equipment maintenance convenient, flexible production scale, product quality good, light color, good flavor, suitable for all kinds of vegetable oil extraction, safe production at the same time, but crushing method is residual oil high in press cake, low efficiency, large power consumption, parts easy to wear and tear.

Now the crushing equipment mainly includes two categories: intermittent production of hydraulic press and continuous production of spiral press. Hydraulic press is according to the theory of fluid transmitting pressure, the fruit in the cake circle is squeezed, and the oil is out of the fruit. The crushing machine has a simple structure, convenient operation, low power consumption, and oil cake with good quality. It can process a variety of fruits, but is only suitable for mall scale processing. But hydraulic press process is the labor strength, strict process conditions, has been gradually replaced by continuous squeezing device. Screw press was widely used in the world's more advanced continuous oil equipments. Its working principle is with continuously forward, at the same time due to the squeeze of spiral lead to shorten or root diameter increases, the pressing chamber space is shrinking volume and pressure, and the billet material is pressed into bread pieces, oil flows out from a crack in the press cage.

In the pressing process, the seed is subjected to extreme heat and pressure with oil mechanically forced from the oil cell. A typical pressing operation involves cooking, pressing, cake cooling and finishing, and oil filtration. The operation and configuration of extraction in different plants is different. In a generic operation, pressed meats from the flakes enter a cooker where heat is applied and protein is denatured. After the cooker, the hot meats enter a mechanical press, where about 60% of the available oil is removed by the application of intensive mechanical pressure. Using a full press operation, the oil exits through drainage bars during the process of pressing, there is about 90% of the available oil. Because the raw material is cooked, compressed, passed through the end plate, where through the application of high friction, so that the cake is often quite hard. After the pressing operation, the cake is normally broken and cooled, with the prepressed cake usually sent to the solvent extraction unit for final oil removal. If the economics do not permit solvent extraction, a higher degree of oil removal is performed. This is also known as a full press operation. And the cake can be used directly as animal feed after removing the harmful substances.

2.4 Solvent Extraction

After leaving the preparation process, the flakes are delivered to the solvent extraction operation. Extraction of the oil from the seed by solvent that leaves behind the non-oil constituents, such as protein and fibre, is practically the only method to achieve an almost complete recovery of oil. Hexane is widely accepted as the most effective solvent used today; there are concerns about its flammability, exposure, and environmental impacts. As this process typically uses a flammable solvent, the operation is usually somewhat removed from other facilities, and access to the controlled area is restricted. Research has focused on various alternative solvents in the hopes of finding one with acceptable performance while providing greater safety. Alternative solvents that have received some attention include isopropyl alcohol, supercritical carbon dioxide, and other fluids. However, no economical alternative to n-hexane has been accepted at this point.

The typical unit operations are associated with solvent extraction, which include extraction, solvent distillation, and liquid-phase recovery. Upon discharge from the extractor, solid-phase extracted material is desolventized, toasted, dried, and cooled prior to meal finishing. And so, the solvent extraction plant is usually consisted of three parts. The main equipment is the extractor, in which the oilseed, after pretreatment and, if applicable after pre-expelling, makes contact with the solvent. The other two basic sections involve the removal of the solvent from the extracted meal and from the miscella, or solvent–oil mixture.

In the extractor, which is a countercurrent flow device, the solid oil-bearing material is conveyed in an opposite direction to the solvent. The extracted meal leaves at one end, and the miscella at the other end. As the oil-bearing material to be extracted enters the unit, it is contacted with the miscella at nearly full oil concentration. After this first wash, the miscella, containing around 25%–30% oil, leaves the extractor for solvent distillation and recovery. After passing through the various washing stages, finally being contacted with fresh solvent and allowed to drain for a brief period, the extracted material, commonly known as white flakes, is removed from the extractor and is conveyed to the desolventizing process.

The vast majority of extractors in operation are based on a continuous conveyor on which the solid prepared seed is carried from the feed end to the discharge end. The conveying element may be an endless belt or chain carrying buckets. The belt, or the bottom of the buckets, consists of sieves, or perforated material, that allow free drainage of the solvent from the oilseeds. Fresh solvent is sprayed over the oil-meal from near the meal discharge end of the extractor; it is enriched with oil as it percolates through the oil-meal and is collected in a series of troughs.

A countercurrent movement of the solvent is achieved in stages by a series of pumps that spray the miscella over the part of the conveyor near the entry of the oilseed. At the final stage, the miscella percolates through the oil-rich feed and leaves the extractor containing the maximum concentration of oil.

There are several types of extractors today. The deep-bed extractor and the shallow-bed extractor are popular. The deep-bed extractor operates largely in a semi-continuous fashion with some individual baskets. The baskets are initially filled with the flakes about depths of 2–3 m. The flakes are supported on a drainage screen, which allows the miscella to pass but retaining the solid-phase material. As the rotary extractor is constantly moving at a slow speed, the extract time in the extractor is about 30–45 min. Shallow-bed extractors, whose bed depths are generally less than 1 m, provide similar drainage screens in the form of slotted bars. The oilseeds pretreated are continuously conveyed through zones where the countercurrent washing of the miscella occurs. After extracting, all the solid-phase material is dropped into a hopper and is conveyed to the desolventizing system.

Temperature control is extremely important during the extraction process. As extractability is enhanced with high temperatures, the operator generally desires to keep temperatures as high as possible. However, phospholipase enzyme activity affects the oil quality at these elevated temperature conditions, partially causing an increase in non-hydratable phosphatides. While this trade-off is generally not a problem, when processing field-damaged seed it may be necessary to reduce extraction temperature, sacrificing residual fats somewhat, in order to produce oil that can be acceptably processed without undue refining losses. As mentioned earlier, several processes have been developed to help inactive enzyme activity prior to extraction.

The extracted meal is drained in the extractor for some minutes, but, after discharge, it is still wet with residual solvent around 25–35%. The wet meal is conveyed from the extractor to the Desolventizer Toaster (DT). In the traditional system, steam is used to counter-currently flash the volatile solvent from the solid phase. The vapor phase then passes through the distillation system on its way to condensing and collection.

The solvent from the DT is condensed, but it is usual to utilize the heat in the solvent vapors to assist in distilling the solvent from the miscella. The air and vapors from the solvent condensers and other parts of the extraction plant are finally freed from the solvent before being discharged into the atmosphere.

2.5 Oil Refining

Preparing with crushing or extracting, without refined oils, is known as raw oil. Raw oil's main ingredient is triglyceride oil, commonly known as neutral oil. In addition, the oil also exists in the other ingredients that are collectively referred to as impurities. So vegetable oil products need raw oil to be refined. The general process is as shown in Fig. 2.2.

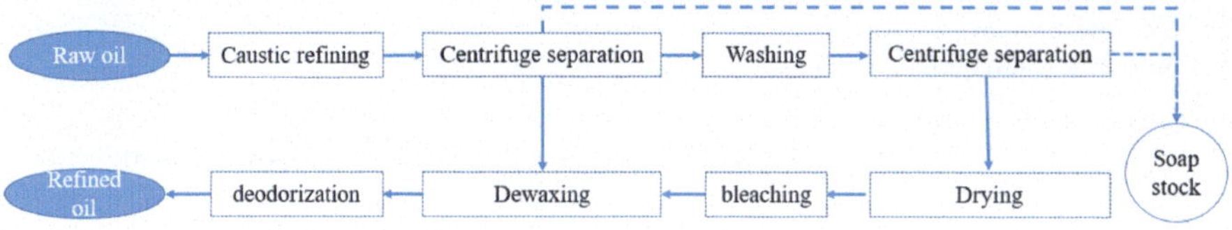

Fig. 2.2 Vegetable oil refining process flow diagram

2.5.1 *Raw Oil Suspension Impurities Removal*

Raw oil usually contains a certain amount of suspended solids. Screw press machine of raw oil contain suspended matter mostly, some even as high as 15% or above. The suspended matter may accelerate oil hydrolysis and rancidity, which caused excessive floating in the alkali refining. The result of centrifuge will frequently down for cleaning, if using it to separated soap. Thus, it is essential to remove suspended impurities whether it is storage, processing, or use.

Sedimentation is often called precipitation, which is the simplest method in oil refining. It is a method that separates suspended solids and oil by using the different density of suspended impurities and oil. Therefore, relatively clear oil floating on it and heavier impurities sink to the device.

Filtration is a solid–liquid separation operation that the suspension pass through the filter media (within refineries is generally filter cloth, screen) and the solid particles are retained. Grease factory essentially used intermittent filter. Raw oil has slow filtration in the low temperature and high viscosity. Lifting temperature is a simplest and effective measure to reduce the viscosity, but it is more easier to make oil oxidation due to higher temperature. Thus the filtration temperature of raw oil generally does not exceed 70 °C.

Centrifugalization is a separation method that separates suspended impurities by using centrifugal force. Among many forms of centrifuge equipment, horizontal spiral decanter centrifuge has long mechanical products in light chemical applications. Part of the oil factory obtained better process results with raw oil in removing suspended impurities using centrifuge recently.

2.5.2 *Degumming*

Degumming is the process of removing rubber impurities in raw oil. The gum of raw oil mainly is phospholipids, thus it also called dephosphorization. It will reduce the use value and storage stability of oils because of the presence of glial, such as phospholipids, and it also has a series of negative effects in oil refining and processing that eventually decline the quality of the finished oil. Thus it becomes imperative ahead of the removal of glial. Degumming has many methods, among them hydration is the universal application of degumming and acid degumming is used in oil process.

Fig. 2.3 Intermittent hydration process diagram

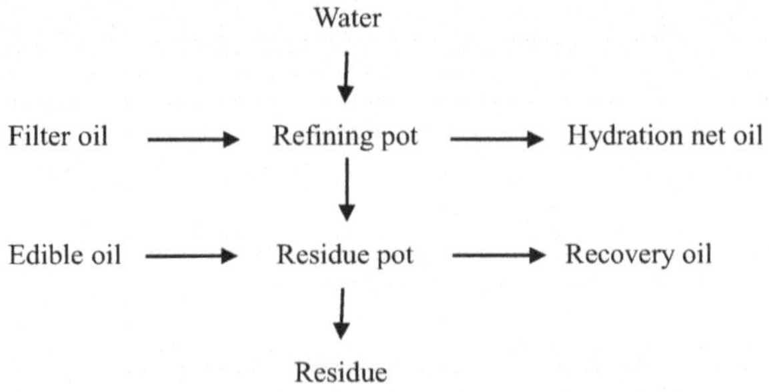

In hydration degumming, there is a certain amount of water or dilute solution added in raw oil under stirring, then it makes gum impurities condense and separate from raw oil by swelling because of the hydrophilic of phospholipids and other impurities. This refining method is called hydration degumming. Those condensed substances mainly are phospholipids during the hydration degumming process. Besides, there is phospholipid binding with protein, mucus, and micro metal ions. The raw oil is preheated, hydration adds water, settles, and the separation of oil residue in refining pot is done. The intermittent hydration process is illustrated in Fig. 2.3.

According to the difference of operating temperature, Intermittent hydration can be divided into high-temperature hydration (about 85–95 °C), middle-temperature hydration (about 60–65 °C) and low-temperature hydration (about 20–30 °C). High-temperature hydration is hydration preheated in higher temperature for raw oil, which is benefit for improve refining rates. It just as the name implies that low-temperature hydration is often called room temperature hydration in china.

2.5.3 Dewaxing

Most vegetable oils contain wax, which mainly comes from fruit and seed hull. The wax content of vegetable oil generally is 0.06–5.00%. Wax will dissolve in oil at 40 °C. Wax is a kind of advanced monocarboxylic acid and has higher alcohols ester formation, which makes biodiesel coke easier in the combustion process. The process of dewaxing has the conventional method and solvent method. The conventional dewaxing method is throught frozen crystals alone, and then separated by mechanical methods. Solvent dewaxing choose solvent added to the crystallization of oil, then separate wax and oil, and evaporated solvent.

2.6 Novel Technology of Oil Extraction

2.6.1 Simultaneous Extraction of Oil and Tea Saponin from Camellia Oleosa Seeds Using Subcritical Water

Camellia oleifera is an oilseed with high oil content and it is an excellent source of polyunsaturated fatty acids. The world's total plant area of Camellia is around 3.3 million ha, of which approximately 3.0 million ha in China stands for about 90% of the global total. Camellia oil is considered to be nutritious and healthy, owing to its fatty acid composition and the presence of bioactive compounds, such as tea polyphenol, tea saponin, and squalene. The content of these bioactive minor components varies with geographic origin, seasonal variation, stress experienced by the plant due to soil and environmental factors, and oil processing methods.

Camellia oleifera seed is a good source of oil and tea saponin. Generally, Camellia seed contains 13.7–42.84% of edible oil and 7.28–16.24% tea saponin. During the production of camellia oil in the past, the tea saponin were often discarded with the oil cake or used for low-value-added fertilizer in the traditional oil processing method. Actually, the tea saponin has good characters such as a strong foaming, emulsifying, dispersing, and wetting performances and anti-cancer, anti-inflammatory, antibacterial, and other biological activities; it could be widely used in food, medicine, pesticides, and other fields.

In recent years, there is steady progress in extraction technology with the development of new and simpler sample preparation methods such as supercritical fluid extraction, microwave-assisted extraction, and subcritical water extraction which have been widely used in extracting bioactive compounds such as flavonoids, polyphenol, and essential oil. And more and more studies of subcritical water extraction (SWE) focus on oil and oil-soluble compounds because SWE provided significant advantages in terms of crude oil yields and simultaneously maximization of valuable bioactive compounds.

Subcritical water, which is defined as liquid water at 100–374 °C under pressurized conditions, had attracted worldwide attention due to its own merits such as shorter extraction time with higher quality and productivities, low energy consumption as well as being environmentally friendly.

Since the increase of temperature has multiple positive effects on mass transfer during SWE, extractions should be performed at the highest temperature not causing significant degradation of target compounds. There have been some reports on SCW extraction of vegetable oils such as soybean oils, palm oils, and sunflower oils on a lab scale. However, Camellia oleosa seeds oil and tea saponin were simultaneously extracted under SWE conditions by varying the temperature, time, pressure, and the liquid-to-solid ratios. Three levels, three variables, Box–Behnken experimental design (BBD) with the response surface methodology (RSM) was used for optimization of the subcritical water extraction process. Temperature (X_1: 110–150 °C), extraction time (X_2: 20–40 min), and solvent to material ratios (X_3: 5:1–15:1) were investigated as independent variables. The response variables were fitted to a second-order polynomial model. The highest extraction yield of oil is 94.07% at the condition of 133.59 °C, 32.03 min, 10.79 mL/g with the tea saponin yield of 71.38%, and the highest extraction of tea saponin is 74.21% with the oil yield of 91.27% of the total content at the condition of 121.11 °C, 32.07 min, 8.33 mL/g. It is concluded that Camellia oil and tea saponin could be co-extracted with a good yield at optimal conditions for camellia oil and tea saponin. The chemical profile of the oils was determined by GC–MS. The fatty acid compositions were compared to those obtained by the Soxhlet extraction (SE) method. The fatty acid profiles were of no differences with those of the cold SE oils. FT-IR analysis indicates the similarity of SCW oils and SE oils is 96.54%. This similarity indicates that subcritical water also extracts the same FAs as those extracted by the conventional SE methodology. Furthermore, it was determined that the oil extracted by SCW was even more resistant to lipid oxidation and more abundant valuable bioactive compounds than the cold-pressed oils. It is proved that the SWE method is an alternative and greener processing method for the simultaneous extract of oil and saponin from the Camellia oleosa seeds.

2.6.1.1 Single-Factor Experimental Analysis

To study the effect of extraction conditions on the Camellia oil and the saponin yield, the single-factor was first adopted. Effects of extraction temperature (°C), extraction time (min), extraction pressure (MPa), and solvent/solid ratio (mL/g) on the extraction yield of oil and saponin are depicted in Fig. 2.4. A series of experiments were performed under the conditions of extraction time of 30 min, pressure of 3 MPa, solvent to material ratio of 10:1 (mL/g) in the presence of a nitrogen flow to evaluate the effect of temperature on the yield of the oil and the saponin. The optimal temperature for the oil extraction and tea saponin is 125 °C, while the extraction ratio at 100 °C (81.53%, 66.42%) and 140 °C (87.64%, 64.18%) were a little lower than that of 120 °C (92.42%, 71.67%, Fig. 2.4a). As the temperature increased, the yield of the saponin and oil were reduced. The increase of temperature could lead to the degradation of the saponin and the oil, or the denaturation of starch and protein. The deformed starch and protein would wrap and adsorb the oil in the cake, and hence reduce the oil yield. Therefore, 125 °C was considered to be the optimum. The effect of reaction time on the yield of oil and saponin was investigated under the condition of temperature 125 °C, pressure 3 MPa, solvent to material ratio 10:1 (mL/g), and the results are shown in Fig. 2.4b. The oil and saponin yield reached 93.24%, 73.28% at 30 min, respectively. Furthermore, the yield of the oil slightly decreased when the extraction time was increased from 30 to 60 min, the corresponding decrease in saponin yield was from 73.28 to 59.31%, indicating that under the above experimental conditions, the saponin had the tendency to be hydrolyzed. Therefore, 30 min was selected as the optimal extraction time. The influence of pressure and solvent to material ratio on the yield of oils and saponin were investigated by varying the pressure and solvent to materials ratio respectively and the results are displayed in Fig. 2.4c, d, respectively. It was first observed the yield of both oils and saponin

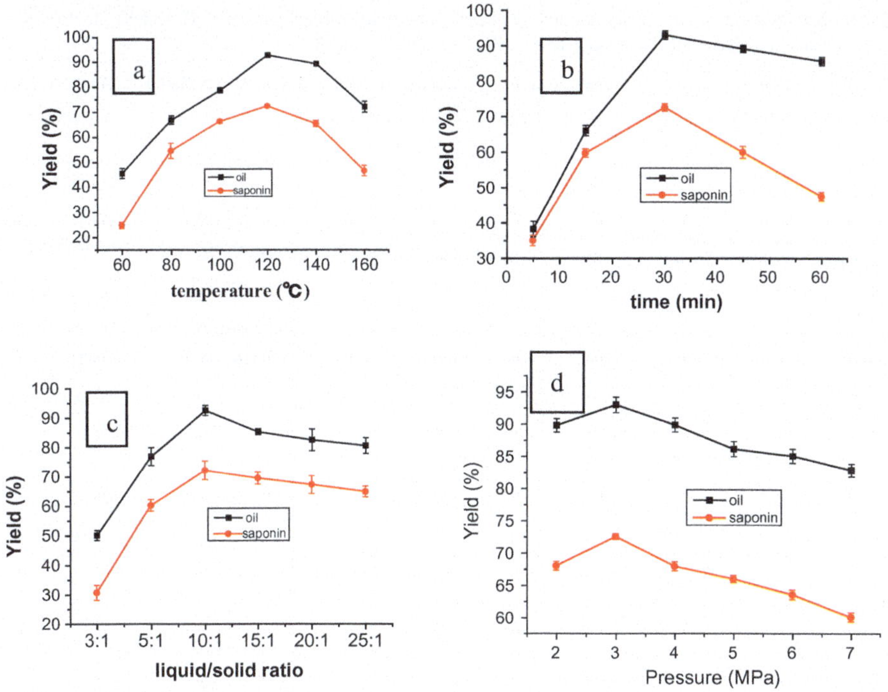

Fig. 2.4 Effect of extraction conditions on the yield of oils and tea saponin

was significantly increased with the increase of the pressure and solvent to material ratios with a similar trend. Similarly, at the pressure above 3 MPa and the solvent to material ratio above 10:1 mL/g, a minimal degree of yield increase was observed, indicating that 3 MPa and 10:1 mL/g were enough to make the oil and saponin fully dissolved.

2.6.1.2 SWE Optimization

Experimental results of oil yield and tea saponin yield obtained under different SWE conditions are presented in Table 2.1. Single-factor experimental design used for SWE contained temperature, extraction time, pressure, and solvent to material ratios as independent variables. However, the pressure effect on oil yield and tea saponin yield was rather insignificant. Moreover, it has been previously reported that the influence of pressure in SWE could be neglected in extracting other compounds. Therefore, temperature, extraction time, and solvent to material ratios were chosen as the independent variables in order to potentially enhance hydrolysis and co-extraction of oil and tea saponin, which has been previously performed in subcritical water or hot water.

Table 2.1 Natural and coded values of independent variables for Box–Behnken design and experimentally observed responses (oil yield, tea saponin yield)

Run	Factor 1	Factor 2	Factor 3	Response 1	Response 2
	A:temperature (°C)	B:time (min)	C:solvent/material (mL/g)	Y_1 (%)	Y_2 (%)
1	110	30	15	82.16	69.59
2	130	40	5	86.89	66.69
3	130	30	10	92.96	73.68
4	110	20	10	77.74	69.85
5	150	30	5	86.18	62.56
6	150	40	10	86.45	50.19
7	150	20	10	82.86	60.96
8	150	30	15	87.16	59.61
9	130	20	15	85.93	66.61
10	130	40	15	89.08	68.06
11	130	30	10	93.86	72.66
12	110	40	10	82.18	73.18
13	130	30	10	94.12	73.56
14	130	20	5	84.89	70.96
15	130	30	10	93.28	74.02
16	110	30	5	78.92	70.12
17	130	30	10	93.69	72.88

Experimentally obtained oil yield and tea saponin yield of Camellia oleosa seeds obtained by SWE varied from 77.74 to 94.12%, 50.19 to 74.02, respectively. The lowest yield was observed at a lower level of temperature (60 °C), lower level of extraction time (5 min), and lower level of solvent to material ratio (5:1 mL/g), while higher yields were obtained at the middle level of temperature and middleextraction time (130 °C, 30 min). This suggests that temperature and extraction time together were the very important factors influencing this response. Influence of investigated SWE parameters toward individual responses was expressed as significant ($p < 0.05$) and insignificant ($p > 0.05$) according to the p-values for the regression coefficients in the second-order polynomial model (Tables 2.2, 2.3). Therefore, linear and quadratic terms of temperature and extraction time had a significant influence ($p < 0.05$) on oil yield, while solvent to material ratios were insignificant. For the tea saponin yield, only the temperature was significant (Fig. 2.5).

Table 2.2 shows the results obtained by ANOVA for the responses of oil yields and tea saponin yields extracted by the SCW method. The ANOVA for the sum of oils yields (Y_1) of Camellia oil extracted by the SCW process showed that temperature and extraction time factors were significant for the model. The ANOVA for the sum of tea saponin yields (Y_3) of camellia oil extracted by the SCW process showed that temperature was significant for the model. The analysis of the effects indicated

Table 2.2 Analysis of variance (ANOVA) of the fitted models for investigated responses

Source	Sum of squares	*DF*	Mean square	*F* value	*p* value
Model	440.17	9	48.91	112.42	<0.0001
X_1-temperature	58.59	1	58.59	134.67	<0.0001
X_2-time	21.71	1	21.71	49.91	0.0002
X_3-solvent/material	6.94	1	6.94	15.95	0.0052
X_1X_2	0.18	1	0.18	0.42	0.5399
X_1X_3	1.28	1	1.28	2.94	0.1304
X_2X_3	0.33	1	0.33	0.76	0.4122
X_1^2	217.27	1	217.27	499.41	<0.0001
X_2^2	70.47	1	70.47	161.98	<0.0001
X_3^2	32.86	1	32.86	75.52	<0.0001
Residual	3.05	7	0.44		
Lack of fit	2.19	3	0.73	3.41	0.1336
Pure error	0.86	4	0.21		
Cor total	443.21	16			
Model	631.64	9	70.18	16.54	0.0006
X_1-temperature	305.29	1	305.29	71.95	<0.0001
X_2-time	13.16	1	13.16	3.1	0.1216
X_3-solvent/material	5.22	1	5.22	1.23	0.3042
X_1X_2	49.7	1	49.7	11.71	0.0111
X_1X_3	1.46	1	1.46	0.35	0.5754
X_2X_3	8.18	1	8.18	1.93	0.2076
X_1^2	162.51	1	162.51	38.3	0.0005
X_2^2	54.64	1	54.64	12.88	0.0089
X_3^2	11.85	1	11.85	2.79	0.1386
Residual	29.7	7	4.24		
Lack of fit	28.4	3	9.47	29.17	0.0035
Pure error	1.3	4	0.32		
Cor total	661.34	16			

Table 2.3 Similarity analysis of camellia oils obtained using SWE and other extraction methods

Extraction condition	Similarity (%)
Soxhlet	100
Cold pressed	98.18
SWE 80 °C	94.67
SWE 100 °C	95.16
SWE 120 °C	96.18
SWE 140 °C	95.94
SWE 160 °C	95.31

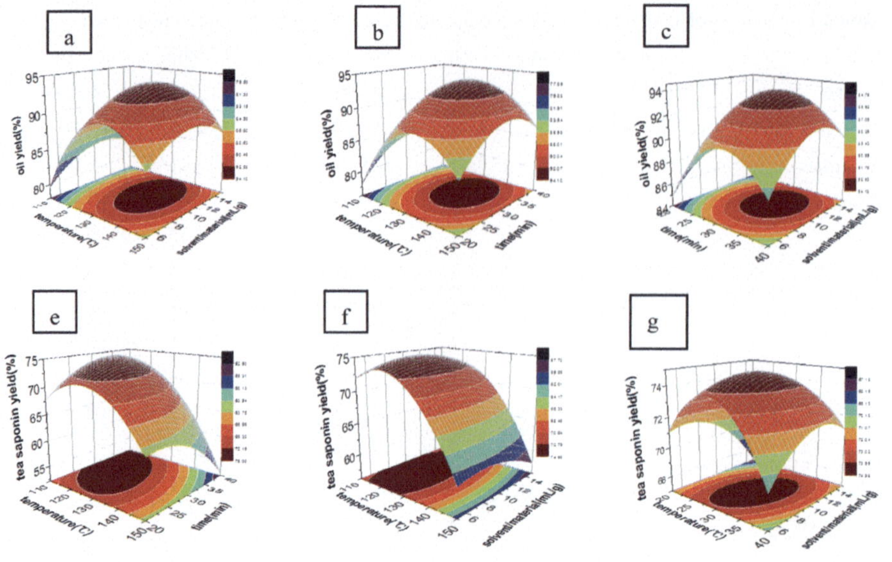

Fig. 2.5 Response surface plots showing combined effects of temperature, time, and solvent to material ratio on oil yield (**a**)–(**c**), tea saponin yield (**d**)–(**f**)

that the temperature and the temperature × time interaction effect were the most significant factors of the extraction.

Via Eqs. (2.1), (2.2), it was possible to conclude that all studied factors contribute to the increase of the extraction of oil yields and tea saponin yields, and the greatest yield of oils was obtained at 133.59 °C, 32.03 min, 10.79 mL/g, while the highest yield of tea saponin was obtained at 121.11 °C, 32.07 min, 8.33 mL/g. The ANOVA also indicates that the model was significant and confirms that factors such as temperature and time in the SCW may influence.

$$Y_1 = 93.58 + 2.71X_1 + 1.65X_2 + 0.93X_3 - 0.21X_1X_2 - 0.57X_1X_3 + 0.29X_2X_3 - 7.18X_1^2 - 4.09X_2^2 - 2.79X_3^2 \quad (2.1)$$

$$Y_2 = 73.36 - 6.18X_1 - 1.28X_2 - 0.81X_3 - 3.53X_1X_2 - 0.61X_1X_3 + 1.43X_2X_3 - 6.21X_1^2 - 3.60X_2^2 - 1.68X_3^2 \quad (2.2)$$

2.6.1.3 Evaluation of Changes in Specific Functional Groups and Chemical Bonds

In addition to the oil extraction experiments under different temperature, characterization of the oil was performed to understand the functional group and the composition differences of the oils. Fourier transform infrared spectroscopy (FT-IR) analysis

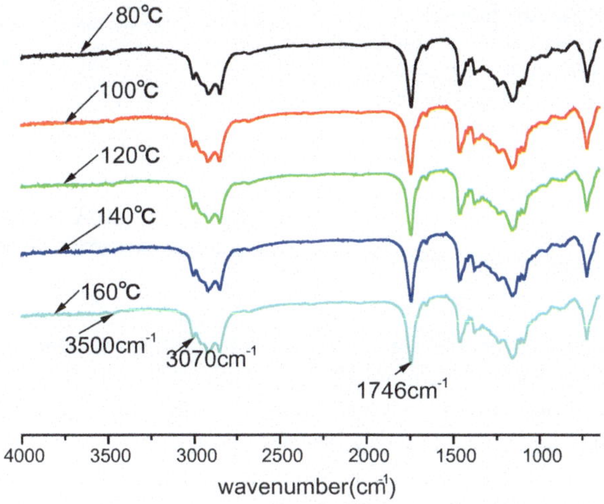

Fig. 2.6 FT-IR analysis of SCW oils extracted at different temperatures

was performed using a Nicolet iS10 FT-IR spectrometer by an attenuated total reflection method. The IR spectrum for Camellia oil is shown in Fig. 2.6. The intense peak at 1746 cm^{-1} corresponds to the C = O stretch of the ester. The C–H stretches are mainly below 3000 cm^{-1}, indicating the dominance of alkanoic protons. An obvious signal that appeared at 3070 cm^{-1} reveals the presence of a great amount of C = C double bond. A very weak peak appeared at 3500 cm^{-1}, which should come from O–H stretches of free fatty acids, showing a small amount of lipid had hydrolyzed. Almost the same trend was observed for both the SCW oils obtained at (80, 100, 120, 140, 160 °C) and the Soxhlet extraction oils. The similarity was calculated by similarity analysis using the OMNIC software; the results are shown in Table 2.3. Above 120 °C, a negative correlation was observed between temperature and similarity. But for all obtained oils minimal degree of difference could be identified. The SCW oil at 120 °C is very similar to that obtained by SE oil (similarity 96.18%). This similarity indicates that subcritical water at a different temperature extracts nearly the same compositions.

2.6.1.4 Fatty Acid Profile

The performed conditions of the components were as follows: HP-77 (100 m × 0.25 mm × 0.25 μm), carrier gas, helium; flow rate, 1.0 mL/min; injector temperature, 220 °C; transmission line temperature, 200 °C; ionization source temperature, 250 °C; the oven temperature program, initial temperature, 50 °C, held for 1 min, increase at 20 °C/min to 170 °C, held for 2 min, then increased at 5 °C/min to 220 °C, held for 15 min. The quantification of single FFA was done from calibration curves obtained from standards (Table 2.4). Gas chromatography combined with mass spectrometry (GC-MS) allows the visualization of the composition of Camellia oil according to the retention time and the peak area. The fatty acid content was

Table 2.4 Internal calibration curve virus $C_{17:0}$, correlation coefficient and recovery of FAMEs in *Camellia oleiferas*

Analyte	Regression equation	R^2	Recovery (%)
$C_{16:0}$	$Y = 0.99X + 0.001534$	0.9997	98.0 ± 0.3
$C_{16:1}$	$Y = 0.87X + 0.002673$	0.9996	96.5 ± 0.5
$C_{18:0}$	$Y = 1.01X + 0.001648$	0.9998	97.6 ± 1.2
$C_{18:1}$	$Y = 0.97X + 0.003345$	0.9997	98.9 ± 0.2
$C_{18:2}$	$Y = 1.04X + 0.002147$	0.9996	97.2 ± 0.8
α-$C_{18:3}$	$Y = 3.42X + 0.001527$	0.9997	96.2 ± 1.1

obtained according to the internal standard shown in Table 2.5. The difference of retention time among FFA(s) obtained by CP, Soxhlet, and the SCW extraction is insignificant ($p > 0.005$), indicating that the Camellia oil would not isomerize or degrade using subcritical water extraction.

Oils obtained by CP, Soxhlet, and the SCW extraction have the same fatty acids profile, which shows that these extraction methods provide similar FA compositions. As shown in Table 2.6, the main composition of Camellia oil is palmitic acid ($C_{16:0}$), stearate ($C_{18:0}$), oleic acid ($C_{18:1}$), linoleic acid ($C_{18:2}$), and a-linolenic acid ($C_{18:3}$). The predominant FA in the lipid composition of Camellia oil (728.73 mg/g) was found to be oleic acid ($C_{18:1}$). A relatively higher concentration of α-linolenic acid was obtained at the optimal condition of the SWE method, showing a preservation against α-linolenic acid for the SWE method.

Table 2.5 Camellia fatty acids composition using different extraction ways

Extraction way	$C_{16:0}$ (mg/g)	$C_{18:0}$ (mg/g)	$C_{18:1}$ (mg/g)	$C_{18:2}$ (mg/g)	$C_{18:3}$ α (mg/g)
Soxhlet	115.17	30.1	674.18	171.16	4.72
Cold press	122.8	26.01	720.62	115.39	6.99
SCW-120 °C	115.26	24.74	728.73	118.06	9.12
SCW-optimal	112.78	26.36	727.14	114.54	10.16

Table 2.6 Fatty acids composition of camellia oil at different extraction conditions

Extraction condition	$C_{16:0}$ (mg/g)	$C_{18:0}$ (mg/g)	$C_{18:1}$ (mg/g)	$C_{18:2}$ (mg/g)	$C_{18:3}$ α(mg/g)
Soxhlet	115.17	30.1	674.18	171.16	4.72
Cold press	122.8	26.01	720.62	115.39	6.99
SCW-120 °C	115.26	24.74	728.73	118.06	9.12
SCW-optimal	112.78	26.36	727.14	114.54	10.16

Table 2.7 Physicochemical properties of SCW oil and cold-pressed oil

Physicochemical properties	SCW oil	Cold-pressed oil	Oil-tea Camellia seed oil
Transparency	Clarify, transparent	Clarify, transparent	Clarify, transparent
Color	Pale yellow	Yellow	Yellow
Odor	*Camellia* oil fragrance	*Camellia* oil fragrance	*Camellia* oil fragrance
Specific gravity (g/mL)	0.916 ± 0.002	0.921 ± 0.007	0.912–0.922
Refractive index (25 °C)	1.462 ± 0.003	1.467 ± 0.007	1.460–1.464
Acid value (mg/g)	2.96 ± 0.007	3.39 ± 0.009	≤4.0
Peroxide value (mmol/kg)	2.86 ± 0.026	4.09 ± 0.042	≤6.0
Unsaponifiable matter (g/kg)	25.35 ± 0.16	12.36 ± 0.24	≤15
Trolex eq. (mmol/L)	112.17 ± 0.24	45.26 ± 0.19	–

2.6.1.5 Physicochemical Properties

The values of some physicochemical properties measured for SCW oil and CP oil are summarized in Table 2.7. Acid value and peroxide value of SCW oil and CP oil were (2.96 ± 0.007 mg/g, 3.39 ± 0.009 mg/g), (2.86 ± 0.026 mmol/kg, 4.09 ± 0.042 mmol/kg), respectively. Indicating a minimal degree of triglyceride hydrolysis to free fatty acid and glycerol, and minimal triglyceride was oxidized. These values of SCW oil is lower than those of CP extraction oil, and also lower than the Chinese national camellia oil standard (GB 11765-2003). Unsaponifiable matters including high-level fatty alcohols, sterol, and hydrocarbon are the most important nutrient composition in Camellia oil. The unsaponifiable matter of SCW oil was measured to be 25.35 ± 0.16 mg/kg, while the CP oil was 12.36 ± 0.24 mg/kg. The DPPH scavenging activity was made for the reference standard Trolox. The Trolex equation value obtained by SCW was, however, higher than that of CP oil. This could be due to the SCW extraction method that could collect more fat-soluble micronutrients such as fatty alcohols, sterol, and hydrocarbon, this is in accordance with the unsaponifiable matter obtained. Compared with CP, it proved that the SCW oil obtained has even better quality.

2.6.1.6 Conclusions

The study demonstrated the utilization of subcritical water as an effective method for the simultaneous extraction of camellia oil and tea saponin from camellia oleosa seeds. Compared with CP, the yield of SCW was improved, functional nutrition such as a-linolenic acid and unsaponifiable matter were better preserved. The SCW oil is similar to the CP oil in composition and the similarity reached up to nearly 96%. For all SCW oils, there is a minimal degree of hydrolysis of the triglycerides. Subcritical water extraction proved to be an alternative and greener method for simultaneous separation and production of oil and tea saponin.

*2.6.2 Aqueous Enzymatic Extraction of Castor (***Ricinus communis***) Seeds Oil*

Ricinus communis (Castor), a member of *Euphorbiaceae*, is an important non-food oilseed crop predominantly grown in subtropical and tropical regions worldwide. The high oil yield (contains 37.2–60.6% oil) and unique fatty acid composition (the ricinoleic acid constitutes around 90% of the total fatty acids) of the castor seeds make it widely be used for lubricating oil and bioenergy production. Castor oil is mainly extracted by pressing or solvent extraction (SE), sometimes seeds are mechanically pre-pressed followed by SE. Unfortunately, severe heat treatment during conventional processing not only affects the oil quality but also denatures the protein in the meal which is ultimately used as fertilizer. On the other hand, the flammability, possible contamination of the oil, and air pollution caused by solvent residues made it urgent to find a nonhazardous and environmental-friendly way to solve the problem.

Alternatively, aqueous enzymatic extraction (AEE) has been regarded as a promoting method for simultaneous extraction of oil and protein, which based on enzymatic hydrolysis of cellulose and hemicellulose to weaken oilseeds tissue defenses by digesting the cell walls promote the efficiency of algae lipid extraction, and help to release the desired compounds into the extracting solution. It is thought to be environmentally-friendly, cheap, and safe, which avoids serious damage to the oil and proteins of the seed and has been extensively applied for the extraction of oils from seed crops such as sesame, sunflower, pumpkin, soybean, camellia, and peanut.

And the effects of solvent extraction (SE) and aqueous enzymatic extraction (AEE) on the oil's physicochemical properties were compared. The enzyme of nutrase was found to be the best one in extracting free oil compared to other enzymes (cellulase, hemicellase, pectinase, and amylase), and the highest oil recovering rate of 77.53% was achieved under optimal experimental conditions of hydrolysis temperature (47 °C), enzyme concentration (3%), ratios of material to water was (1:3.7) (w/v), and hydrolysis time (4.5 h). Compared with SE oil, AEE oil was more acidic and had a higher hydroxyl value but did not affect the fatty acids composition. SEM results illustrated that the nutrase mainly breaks down the seed cells and oil bodies' membranes structure of castor, which resulted in more free oil releasing from seeds cell. The results demonstrated that AEE is a promising environmentally friendly technology for oil extraction of castor seeds in the oil industry.

2.6.2.1 Aqueous Enzymatic Extraction (AEE)

The oilseed material was weighed (w_0, g) and mixed with distilled water at a ratio of 1:3(w/v) using a flask. The mixture was subjected to boiling (100 °C) for 10 min and allowed to cool down to room temperature (25 °C). The pH was then set to the suggested point for each enzyme or enzymatic combination using 0.5 N aqueous NaOH or 0.5 N aqueous HCl solutions. Then, an amount (2.0% by seed kernels material weight) of each of the five enzymes or enzymatic combinations (cellulase, hemicellase, pectinase, amylase, and nutrase) was added. The conditions of temperature, pH, and reaction time used for these enzymes and enzymatic combinations are given as ranges in Tables 2.8 and 2.9. The mixture was incubated at the suggested temperature for corresponding pH and time with regular shaking at 120 rpm (e.g., in Table 2.9, nutrase + cellulase, temperature 55 + 45 °C, pH 7.0 + 4.5, time 2 + 2 h,

Table 2.8 Effect of different enzymes on the oil recovering rate of castor seeds

Enzyme	EC number	Enzymatic activity (U/mg)	Initial pH	Temperature (°C)	Oil recovering rate (%)
Cellulase	EC 3.2.1.1	≥400	4.5	45	45.35 ± 2bc
Hemicellase	EC 3.1.1.73	0.3–3.0	4.5	45	53.01 ± 2b
Pectinase	EC 3.2.1.15	≥500	8.0	60	55.12 ± 3b
Amylase	EC 3.2.1.1	380	7.0	45	37.31 ± 2c
Nutrase	EC 3.4.24.39	50	7.0	55	63.23 ± 3a
No enzyme	–	–	7.0	45	9.16 ± 3d

Table 2.9 Effect of different enzymatic combinations on the oil recovering rate of castor seeds

Enzyme	Temperature (°C)	Initial pH	Hydrolysis time (h)	Oil recovering rate (%)
Nutrase + cellulase	55 + 45	7.0 + 4.5	2 + 2	60.62 ± 1a
Nutrase + hemicellase	55 + 45	7.0 + 4.5	2 + 2	55.23 ± 1b
Nutrase + pectinase	55 + 60	7.0 + 8.0	2 + 2	56.67 ± 2b
Cellulase + pectinase	45 + 60	4.5 + 8.0	2 + 2	50.55 ± 2bc
Cellulase + hemicellase	45	4.5	4	48.02 ± 1bc
Pectinase + hemicellase	60 + 45	8.0 + 4.5	2 + 2	56.22 ± 2b

that means add Nutrase into seed mixture first, the mixture was incubated at 55 °C, pH 7.0 for 2 h, and then add Cellulase into the same mixture, temperature and pH were adjusted to 45 °C, and 4.5, respectively, and the mixture was incubated for 2 h more).

Followed by centrifugation for 30 min at 10,000 rpm (Sigma, 3 K30, Osterode am Harz, Germany). After centrifugation, the upper oily phase was withdrawn using a pipette and weighted (w_1, g). Emulsion was demulsified by vacuum dewatering at 90 kPa and at 50 °C for 4 h, followed by centrifugation for 30 min at 10,000 rpm, upper oily phase was obtained and weighted (w_2, g). The control samples were prepared and treated identically without addition of enzymes. The free oil content (%) and total oil recovering rate (ORR) of AEE compared to SE were calculated by Eqs. (2.3) and (2.4):

$$\text{Free oil content}(\%) = \frac{w_1 + w_2}{w_0} \times 100\% \tag{2.3}$$

$$\text{ORR}(\%) = \frac{\text{Free oil content }(\%)}{\text{Total oil content}(\%)} \times 100\% \tag{2.4}$$

As shown as Table 2.8, a higher free oil recovery from castor seeds, in the range of 37.31–63.23% for the aqueous extraction process involving enzymes was observed compared to only 9.16% for the control (oil obtained without enzyme treatment). Among the enzymes studied, nutrase was found to be the best enzyme offering the highest oil yield (63.23%), which was significantly ($p < 0.05$) higher than other enzymes, whereas the oil recovering rate was minimum (37.31%) for the enzyme of amylase treated seed samples. Overall, the tested enzyme mixtures exhibited efficacy in the order: nutrase > pectinase > hemicellase > cellulose > amylase. The high efficacy of proteolytic enzymes for free oil extraction is in agreement with the findings of soybean and rapeseed. The enzyme action of nutrase during aqueous extraction can be attributed to the degradation of cytomembrane, breakdown of the protein networks of cotyledon cells and oleosin-based membranes that surround lipid bodies, which results in liberating more oils than other enzymes.

The treatments of enzymatic combinations might improve the oil recovering rate, nutrase, cellulose, and pectinase were chosen to combine with other enzymes to evaluate their cooperative effects on oil yield in this study. As shown in Table 2.9, the enzymatic combinations obtained were in the range of 48.02–60.62% oil recovering rate. Among the enzymatic combinations, nutrase coupled with cellulase had the highest oil recovering rate of 60.62%, whereas the oil recovering rate was minimum (48.02%) for cellulase combined with hemicellase. Overall, all enzymatic combinations did not increase the oil recovering rate compared to 63.23% for using the enzyme of nutrase only. Similar results were also observed on peanut seeds oil enzymatic extraction, which reported that the compounding alcalase with other enzymes did not improve the peanut seeds oil yield significantly. Therefore, nutrase was chosen for subsequent enzymatic hydrolysis process optimization in the present study.

2.6.2.2 Optimization of Enzymatic Hydrolysis

Response surface methodology was applied to identify optimum levels of four key independent variables including hydrolysis temperature (°C), enzyme concentration (%), ratios of material to water (w/v), and hydrolysis time (h) for the selected enzyme. Mono-factor tests were done prior to determining the independent variables range for BBD. Three variables were kept constant at their respective central test range values and the other variable varied within its experimental ranges.

After a series of preliminary mono-factor tests, a Box–Behnken design (BBD) was used to survey the effects of independent variables at three levels on the dependent variable (oil recovering rate). A total of 29 randomized experiments including 24 factorial and 5 zero-point tests were designed. The regression analysis was carried out to evaluate the response function as a quadratic polynomial:

$$Y = \alpha_0 + \sum_{j=1}^{k} \alpha_j X_j + \sum_{j=1}^{k} \alpha_{jj} X_j^2 + \sum\sum_{i<j} \alpha_{ij} X_i X_j (k = 4) \quad (2.5)$$

where Y is the predicted response; α_0, α_j, α_{jj}, and α_{ij} are the regression coefficients for intercept, linearity, square, and interaction, respectively. X_i and X_j represent the different independent variables, and k represents the number of variables. The actual and coded levels of the independent variables used in the experimental design are summarized in Table 2.10. The experiment data were analyzed statistically with Design-Expert 8.0 (Stat-Ease, Inc., Minneapolis, MN). And the interaction response surfaces and planar contour plots were obtained from OriginPro 8.0 (OriginLab Corporation, Massachusetts, USA).

Enzymatic hydrolysis parameters including hydrolysis temperature (°C), enzyme concentration (%), ratios of material to water (w/v), and hydrolysis time (h) for nutrase on free oil recovery were optimized by mono-factor tests prior to determining the independent variables range for BBD.

The effect of ratios of material to water (w/v) (range from 1:2 to 1:6) on the oil recovering rate was studied, the reaction condition of hydrolysis temperature (°C), enzyme concentration (%), and hydrolysis time (h) were maintained at 50 °C, 2%, and 5 h, respectively. The results are presented in Fig. 2.7a; the ratio of 1:4 is the best one offering the highest oil recovering rate of 69.46% ($p < 0.05$), whereas lower

Table 2.10 Independent variables and their levels used in Box–Behnken design (BBD)

Independent variables		Factor levels		
		−1	0	1
X_1	Enzyme concentration (%)	2.5	3	3.5
X_2	Ratios of material to water (w/v)	1:3	1:4	1:5
X_3	Hydrolysis time (h)	4	5	6
X_4	Hydrolysis temperature (°C)	45	50	55

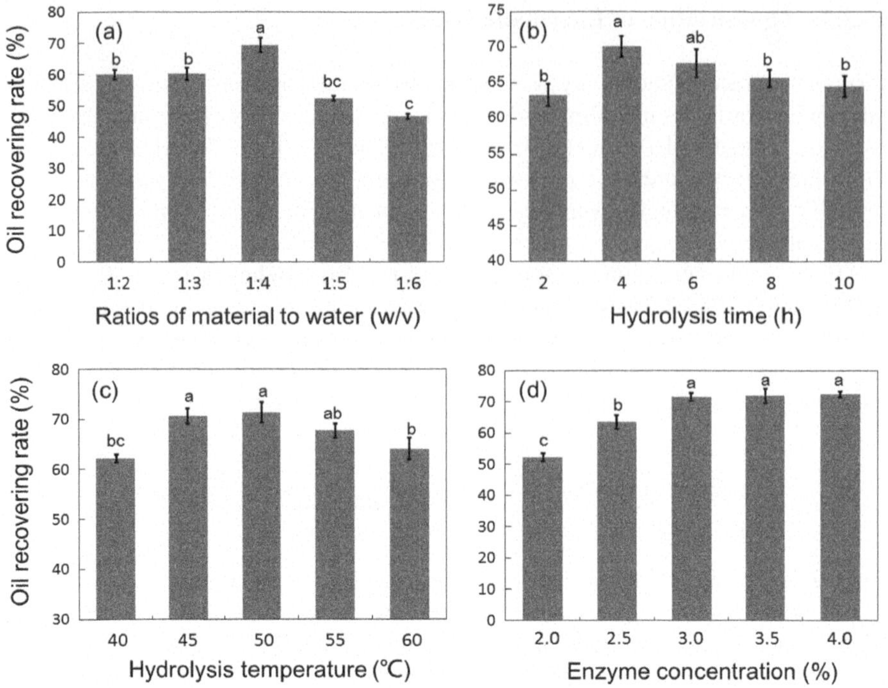

Fig. 2.7 Effect of variables (hydrolysis temperature, enzyme concentration, ratios of material to water and hydrolysis time) on oil recovering rate of castor seeds by mono-factor tests. Results were expressed as the mean value ± standard deviation ($n = 3$), and the same lowercase letters labeled indicate oil recovering rate which are not significantly different among treatments by Tukey's honestly significant difference ($P < 0.05$)

castor seed material to water ratios (1:2 and 1:3) obtained the lower oil yield, because the mixture was too viscous and difficult to homogenize. And an excessive amount of water in the mixture from higher ratios (1:5 and 1:6) would inevitably lower the concentration and hydrolysis efficacy of the enzyme; other studies reported similar results that the liquid to material ratio being too low or too high may not favor the movement of the plant cells to the active site of the enzyme and the polysaccharides to the medium.

The effect of enzymatic hydrolysis time (range from 2 to 10 h) on the oil recovering rate was investigated (Fig. 2.7b), the reaction condition of hydrolysis temperature (°C), enzyme concentration (%), and ratios of material to water (w/v) were maintained at 50 °C, 2%, and 1:4, respectively. The oil recovering rate improved significantly when the hydrolysis time was increased from 2 to 4 h (from 63.26 to 70.05%) ($p < 0.05$). However, the oil recovering rate was not found continuing to increase in longer reaction time from 6 to 10 h, which can be explained due to the enzymatic dosage limit and the reverse inhibition of the reaction product.

As shown in Fig. 2.7c, when the reaction condition of hydrolysis time (h), enzyme concentration (%), and ratios of material to water (w/v) were maintained at 4 h, 2%, and 1:4, respectively. The hydrolysis temperature (range from 40 to 60 °C) has a big influence on the oil recovering rate which dramatically effects the activity of enzymes. In this study, the temperature ranges from 45 to 55 °C had the best performance on the oil recovering rate ($p < 0.05$). When the hydrolysis temperature was below 45 or exceeded 55 °C, it led to the lower oil yield which might be attributed to the deactivating of the enzymes.

The reaction condition of hydrolysis time (h), hydrolysis temperature (°C), and ratios of material to water (w/v) were maintained at 4 h, 50 °C, and 1:4, respectively. The concentration of enzyme is a vital factor in aqueous enzymatic extraction. The oil recovering rate increased with enzyme concentrations, whereas, there was no significant increase in the oil recovering rate when the enzyme concentration increased from 3 to 4% ($p > 0.05$) (Fig. 2.7d). In the perspective of industrial applications, enzyme dosage reduction would decrease the cost under the stable oil yield achievement, therefore, 3% would be more suitable for aqueous enzymatic extraction.

2.6.2.3 Physicochemical Properties Determination

Physicochemical properties of castor seed oil like acid, iodine saponification, and hydroxyl values are the most important characteristics of oil. The acid, saponification, iodine, and hydroxyl value and of castor seeds oil samples were determined according to AOCS standard methods (1998). The acid was determined by Sodium hydroxide titration and calculated in terms of milligrams of potassium hydroxide necessary to neutralize the free acids in 1 g of oil sample. The iodine value is expressed in grams of iodine which react with 100 g of the oil sample to saturates, saponification value is represented the number of milligrams of potassium hydroxide required to saponify 1 g of oil sample, and is defined as the number of milligrams of potassium hydroxide required to neutralize the acetic acid taken up on acetylation of one gram of a chemical substance that contains free hydroxyl groups. The refractive index (R.I.) was determined using an RM40 automatic refractometer (Mettler Toledo Co., Ltd., Switzerland), and the results were standardized at 25 °C.

Fatty acid components of the oil were analyzed using Clarus 600 gas chromatograph-mass spectrometer (GC-MS, Perkin Elmer Instrument Co., Ltd., Shanghai, China). Oil sample was methyl esterified in 0.5 mol/L blending solution of potassium hydroxide (KOH) and methyl alcohol, and then stratified in petroleum ether with deionized water, and centrifuged under 3000 r/min for 5 min, the saponified fruit oil injected into a free fatty acid polyester (FFAP) column (0.3 mm × 25 m). Oven temperature was programmed as follows: held at 40 °C for 1 min, increased to 100 °C at 20 °C per min and held at 100 °C for 2 min, increased to 220 °C at 20 °C per min and held at 220 °C for 2 min, increased to 240 °C at 20 °C per min and held at 240 °C for 5 min. Carrier gas, helium (He), was provided at a flow rate of 1.1 per min in the column. The injector temperature was 210 °C, and the injection volume was 1 μL. GC and MS interface temperature and ion source temperature was 230

Table 2.11 Physicochemical properties of castor seed oils obtained by different exaction methods

Physicochemical properties	SE	AEE
Acid value (mg KOH/g)	4.9b	10.8a
Iodine value (g/100 g)	93a	92a
Saponification value (mg KOH/g)	197a	196a
Hydroxyl value (mg KOH/g)	155b	166a
Refractive index (25 °C)	1.4841a	1.4824a

and 200 °C, respectively. Electron impact energy of mass spectrometer was 70 eV, and scanning quality ranged from 15 to 500 amu. The fatty acid components and relative content were identified on the basis of mass spectrometer results.

As shown in Table 2.11, differences of the five physicochemical characteristics (acid, iodine, saponification, hydroxyl values, and refractive index) of castor seed oil obtained by AEE and SE were found to be statistically insignificant ($p > 0.05$) except for acid and hydroxyl values ($p < 0.05$). The relatively higher acid value (10.8 mg KOH/g) was observed in oil obtained by AEE as against SE oils (4.9 mg KOH/g). A similar trend was observed in hydroxyl value as a significantly higher value (166 mg mg KOH/g) was observed in oil obtained by AEE than SE (155 mg KOH/g). It was previously reported that higher acid value was found in oil from AEE, which suggests a higher amount of free fatty acid in the oil.

Same lowercase letters within a row and the data are not significantly different among treatments by Student's t test at $p < 0.05$.

The fatty acid compositions were analyzed between AEE and SE of castor seed oils. As shown in Table 2.12, fatty acid of the castor seed oil was mainly composed of saturated fatty acid including palmitic acid $C_{16:0}$ (0.94–1.12%), stearic acid $C_{18:0}$ (0.38–0.64%), and unsaturated fatty acid including ricinoleic acid 12-OH $C_{18:1}$ (86.84–87.23%), oleic acid $C_{18:1}$ (3.88–4.19%), linoleic acid $C_{18:2}$ (4.96–5.74%), and linolenic acid $C_{18:3}$ (0.24–0.31%).

Table 2.12 Fatty acids composition of castor seed oils obtained by different exaction methods

No.	Fatty acids	Fatty acids composition (%)	
		SE	AEE
1	Palmitic acid ($C_{16:0}$)	0.94	1.12
2	Stearic acid ($C_{18:0}$)	0.64	0.38
3	Oleic acid ($C_{18:1}$)	3.88	4.19
4	Linoleic acid ($C_{18:2}$)	5.74	4.96
5	Linolenic acid ($C_{18:3}$)	0.24	0.31
6	Ricinoleic acid (12-OH $C_{18:1}$)	86.84	87.23
Total	Saturated fatty acids (SFA)	1.58	1.5
Total	Monounsaturated fatty acids (MUFA)	90.72	91.42
Total	Polyunsaturated fatty acids (PUFA)	5.98	5.27

These fatty acid compositions revealed in this study are similar to those reported by Salimon et al. and Ogunniyi. Among them, ricinoleic acid ($C_{18}H_{34}O_3$) was the most predominant fatty acid; it is a structurally *cis*-12-hydroxyoctadeca-9-enoic acid, 18-carbon hydroxylated fatty acid having one double bond, which can be easily oxidated, hydrogenated, and dehydrated, making it widely be used for industrial applications like bio-lubricating oil and biodiesel with excellent viscosity and fluidity. There was no significant difference ($p > 0.05$) in the amount of major fatty acids of the AEE oil compared with SE. The same results were observed in *Camellia*, *Pumpkin*, and *Balanites aegyptiaca* kernel oil. Suggesting AEE would not affect the fatty acid composition of castor seed oil.

2.6.2.4 Scanning Electron Micrographs (SEM)

The castor seed kernels samples before and after extraction were fixed in the glutaraldehyde solution for 2 h and stored in −4 °C for 5 min; after ethanol dehydration and vacuum desiccation, the samples were sputtered with a thin layer of carbon and gold. The morphological alterations of these samples were observed with a JEM-1200 scanning electron microscope system (JEOL Company, Japan).

To gain a better understanding of the AEE mechanism of castor seeds oil, the cell surface structure of castor seed kernel samples were investigated by SEM. As shown in Fig. 2.8a, a smooth and intact cell surface was observed before extraction, some protein bodies were released out due to the physical grounding pretreatment. After aqueous extraction process (without addition of enzymes), more protein and oil were diffused, but the external surface of seed tissues was still intact and smooth (Fig. 2.8b). After enzymes were applied, the cell walls of seed tissue begin to rupture (Fig. 2.8c). A significant destruction of the morphological structure of seed tissues was observed as the extraction time increasing from 1 to 4 h, and most inner cells were disorganized and broken (Fig. 2.8d).

2.6.2.5 Statistical Analysis

All the experiments were performed in triplicate whereas the statistical analysis of the data was done by analysis of variance (ANOVA). A probability value at $p < 0.05$ was considered statistically significant. Mean separation among treatments was conducted using Tukey's honest significant difference (HSD) multiple comparison. All statistical analyses were performed using JMP (version 12, SAS Institute Inc., Cary, NC).

Based on the results of mono-factor tests, the enzyme concentration from 2.5 to 3.5%, hydrolysis temperature from 45 to 55 °C, hydrolysis time from 1 to 6 h, and ratios of material to water from 1:3 to 1:5 were selected for BBD (Table 2.13). Moreover, the significance of each coefficient of the model was also determined using the *t* test and *p*-value.

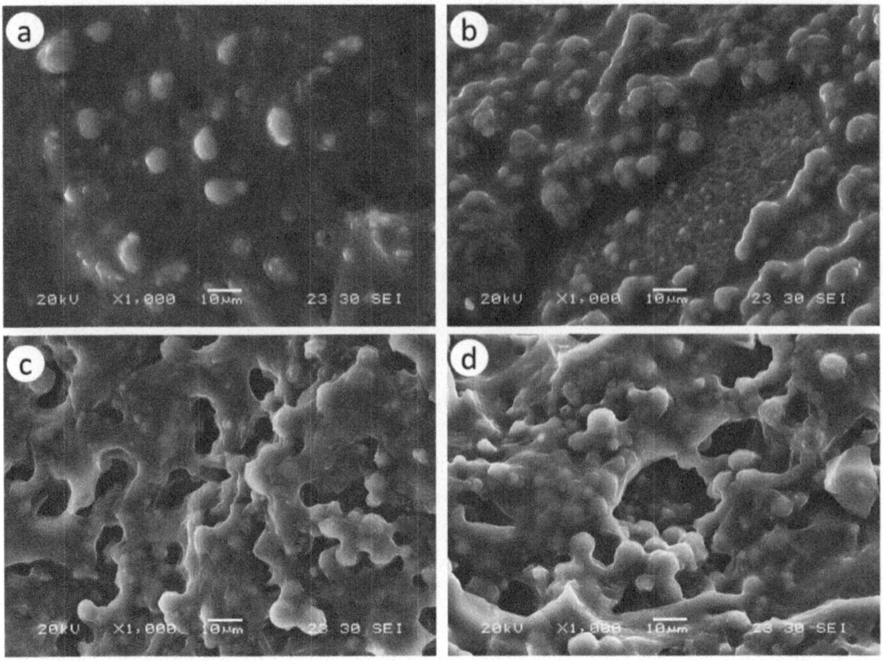

Fig. 2.8 Cell surface morphological alterations of castor seed samples by scanning electron microscopy (SEM). **a** Non-extraction, **b** aqueous extraction, **c** enzymatic hydrolysis extraction for 1 h, **d** enzymatic hydrolysis extraction for 4 h

Table 2.13 Independent variables and their levels used in Box–Behnken design (BBD)

Independent variables		Factor levels		
		−1	0	1
X_1	Enzyme concentration (%)	2.5	3	3.5
X_2	Ratios of material to water (w/v)	1:3	1:4	1:5
X_3	Hydrolysis time (h)	4	5	6
X_4	Hydrolysis temperature (°C)	45	50	55

The ANOVA results for response surface quadratic model and multiple regression analysis are evaluated using the corresponding *F*- and *p*- values, and results were presented in Table 2.14.

The regression model *F*-value of 14.13 and the *p*-value of the model ($p < 0.0001$) imply the model is significant, and the *p*-value of lack of fit model ($p = 0.4860 > 0.05$) implies the lack of fit model is not significant relative to the pure error, and the determination coefficient of the model (R_2) was 0.9339, which suggests the model is applicable and supporting a good accuracy and ability of the established model. In addition, Pre-R_2 (0.6911) is in reasonable agreement with the Adj-R^2 of

Table 2.14 Analysis of variance for the extraction regression model of oil recovering rate

Source	Sum of squares	Degree of freedom	Mean of square	F-value	p value	Significance
Model	364.91	14	26.07	14.13	<0.0001	**
X_1	0.25	1	0.25	0.14	0.7186	NS
X_2	2.57	1	2.57	1.39	0.2578	NS
X_3	1.98	1	1.98	1.08	0.3172	NS
X_4	4.54	1	4.54	2.46	0.1390	NS
X_1X_2	0.04	1	0.04	0.022	0.8850	NS
X_1X_3	0.14	1	0.14	0.074	0.7892	NS
X_1X_4	39.25	1	39.25	21.28	0.0004	**
X_2X_3	1.99	1	1.99	1.08	0.3167	NS
X_2X_4	41.54	1	41.54	22.52	0.0003	**
X_3X_4	42.64	1	42.64	23.12	0.0003	**
X_1^2	73.35	1	73.35	39.77	<0.0001	**
X_2^2	176.09	1	176.09	95.48	<0.0001	**
X_3^2	48.93	1	48.93	26.53	0.0001	**
X_4^2	30.59	1	30.59	16.59	0.0011	**
Residual	25.82	14	1.84			
Lack of fit	19.15	10	1.91	1.15	0.486	NS
Pure error	6.67	4	1.67			
Cor total	390.73	28				
R^2	0.9339		Adequate precision	11.014		
Adj. R^2	0.8678		Std. dev.	1.36		
Pred. R^2	0.6911		C.V. %	2.2		

* $0.01 \leq p < 0.05$
** $p < 0.01$

0.8678 (Adj-R^2-Pre-R^2 = 0.1767 < 0.2), and reveals a high degree of correlation between the observed and predicted data from the regression model. The coefficient of variation (C.V.) expresses the standard deviation as a percentage of the mean, and was found to be 2.2% (<5.00%) for the oil recovering rate, which indicates that the model is reproducible. Factors of the regression model with significant effects on oil recovering rate ($p < 0.05$) were interaction terms of X_1X_4, X_2X_4, and X_3X_4, and all quadratic terms of X_1^2, X_2^2, X_3^2, and X_4^2, whereas, all linear terms (X_1, X_2, X_3, and X_4) and other three interaction terms (X_1X_2, X_1X_3, and X_2X_3) were not significant ($p > 0.05$). The insignificant model terms reduction can improve the productivity of the model. Therefore, the modified regression equation was applied to express the free oil yield of castor seeds in terms of actual factors which is given below:

$$Y = 77.28 - 3.13X_1X_4 + 3.22X_2X_4 + 3.27X_3X_4 - 3.36X_1^2 - 5.21X_2^2 - 2.75X_3^2 - 2.17X_4^2 \quad (2.6)$$

where Y is the free oil yield (%); X_1, X_2, X_3, and X_4 are enzyme concentration (%), ratios of material to water (w/v), hydrolysis time (h), and hydrolysis temperature (°C), respectively.

2.6.2.6 Conclusions

Enzymatic hydrolysis extraction was applied to obtaining the free oil from castor seeds in this study. The enzyme of nutrase was found to be the best one in extracting free oil compared to other enzymes (cellulase, hemicellase, pectinase, and amylase), and the highest oil recovering rate of 77.53% was achieved under optimal experimental conditions of hydrolysis temperature (45 °C), enzyme concentration (3.5%), ratios of material to water was (1:3.7) (w/v) and hydrolysis time (4.5 h). Compared with hexane extracted oil, aqueous enzymatic-extracted oil was more acidic and had a higher hydroxyl value but did not influence the fatty acids composition. SEM results illustrated nutrase mainly breaks down cells and oil bodies' membranes structure of castor seeds resulting in high free oil releasing. Suggesting the technology of AEE is a promising environmental friendly technology for oil extraction of castor seeds in the oil industry.

2.6.3 Cold Pressing Process of Swida wilsoniana *Fruits*

Swida wilsoniana is a kind of woody oil plants with ecological adaptability, in which the natural dried fruits contain more than 32% oil, and the average production is greater than 100 kg/(667 m^2).

Containing more than 70% unsaturated acids, fruits oil has a high nutritional value in application. The cold processing processes with core equipment of screw or hydraulic press can guarantee the original nutrients and flavor of oils. There are a few equipments suitable for special oilseeds. It is necessary to design new type of cold pressing equipment to get high-quality oil.

In order to guarantee a low temperature during the whole pressing process of fruits, this research adopted an experimental device simulating diameter equidistant screw extrusion in which only pure shear extrusion. Pressures of oil squeezed out, fruits crushed, and critical compression were extracted and performances of fruits in pressing process were studied by one-dimensional pressing device, which can provide data for the hydraulic or spiral press in cold pressing process.

Swida wilsoniana is an important woody oil plant of which the whole fruits contain oils. Main contents of fruits from *Swida wilsoniana* were analyzed through chemical means using 46 samples from different areas. These fruits have an average moisture content of about 8.19% after they are naturally dried. The oil content is about 28.02% (dry basis) or more. The protein content is 7.25% (dry basis) and the cellulose content is about 14.18%. Through theoretical calculation and the one-dimensional pressing device, the conditions of cold pressing process and performances of fruits from *Swida wilsoniana* had been studied. The results show that *Swida wilsoniana* fruits dried through natural weather are more suitable for direct pressing, which guarantees a higher quality of oils. The volume of fruits fed into pressing device had little effect on pressures of oil squeezed out, fruits crushed, and critical compression, and the value of those pressures were 55.7, 71.5, and 80 MPa individually. At the critical pressure of 80 MPa and the compression ratio of 3.14, residual oil of meal is 6.5%. The fruits oil has about 72% unsaturated fatty acids, of which approximately oleic acid is 29.83%, linoleic acid is 39.15%, and linolenic acid is 2.59%.

2.6.3.1 Main Contents of *Swida wilsoniana* Fruits and the Theoretical Basis of Direct Pressing

Oilseeds mainly consist of fat, protein, cellulose and water, and these substances affect the physical strength of oilseeds. *Swida wilsoniana* fruits from different habitats and species are collected and measured by conventional chemical analysis methods for contents of crude oil, moisture, crude protein, and cellulose. The main contents of fruits from *Swida wilsoniana* were analyzed through chemical means using 46 samples from different areas. The fruits have an average moisture content of about 8.19% after they are naturally dried. The oil content is about 28.02% (dry basis) or more. The protein content is 7.25% (dry basis) and the cellulose content is about 14.18%.

For oilseeds, moisture content is a convenient adjustment factor, whose change will directly lead to changes of contents in oil, protein, cellulose, and other indicators. Besides, physical strength will be changed. Therefore, control of moisture content of *Swida wilsoniana* fruits is an important means to regulate conditions of the cold pressing process. Changes of moisture content generally follow the rule that high temperature is suitable for low moisture content while low temperature is suitable for high moisture, and the relation between moisture content and temperature can be calculated according to the N.B. empirical formula, which is as follows:

$$c = (14 - 0.1 * T) * k \tag{2.7}$$

In the above formula, C (%) stands for oil content; T (°C) stands for pressing temperature; k stands for correction factor, which can be expressed by the formula $k = (100\text{-M})/55$, in which M (%) stands for the oil content of oilseeds.

Taking a low temperature of 70–80 °C, we can get the best conditioned fruits whose moisture content ranges from 7.85 to 9.16%, which is just similar to the natural moisture content of fruits. So, it is suitable for *Swida wilsoniana* fruits to be directly fed into the press, and it provides theoretical support for us to focus on the direct pressing process.

2.6.3.2 Parameters of Pressing Process for Oilseeds

Figure 2.9 illustrates the influence of the volume of oilseeds on different pressing pressure. As shown in Fig. 2.9, pressure of oil squeezed out (P_1) and pressure of fruits crushed (P_2) are changing slowly with the increase of oilseeds. It means the amount of oilseeds has a low influence on the two kinds of pressures. P_2 is about 16 MPa higher than P_1. The average values of the pressure from 30 to 50 mm are 55.7 MPa for P_1 and 71.5 MPa for P_2.

Figure 2.9 illustrates the influence of pressing time on compression ratio in different volumes of oilseeds. The curves showed obvious changes in compression ratio with the increase of pressing time. The final compression ratios are too close which is about 3.1. With the increasing volume of oilseeds, pressing time for achieving a final uncompressed stage increases gradually. All in all, it is easy for oilseeds to achieve a relative uncompressed stage when pressed within 30 s. Through Matlab fitting, equation of relation between pressing time and volume of oilseeds ranging from 9.424 to 15.707 cm^3 could be got as follows

$$t = 2.7370\,V_1 - 0.0328\,V_2 - 0.001\,V_3 + 0.0011 \tag{2.8}$$

The actual compression ratio, closely related to the pressing pressure and pressing time, can directly reflect the actual stage of both the oilseeds and the meal when pressed. Figure 2.10 illustrates the influence of pressing pressure on compression ratio in different volumes of oilseeds. As shown in Fig. 2.10, curves are similar. The actual compression ratio increases with the increase of pressing pressure. But when the pressing pressure increases to a certain value, the increase of compression ratio is smaller, and when it reaches 80 MPa, the slope of the curve is 0.0056 which means the pressing pressure has so little influence on compression ratio that can be neglected, and the oilseeds entered a non-compressible phase. Taking 80 MPa as

Fig. 2.9 Influence of volume of oilseeds on different pressing pressure

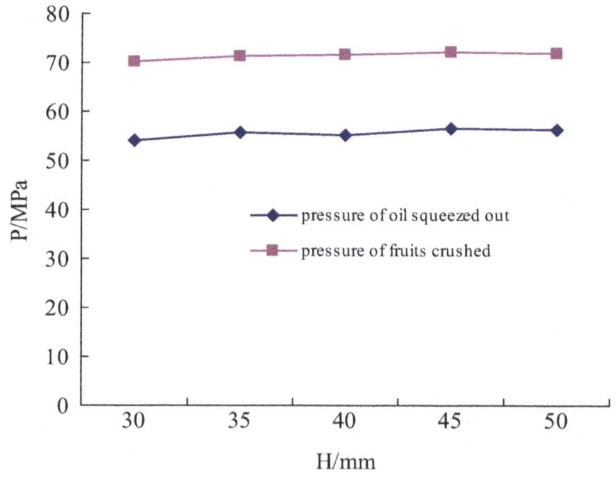

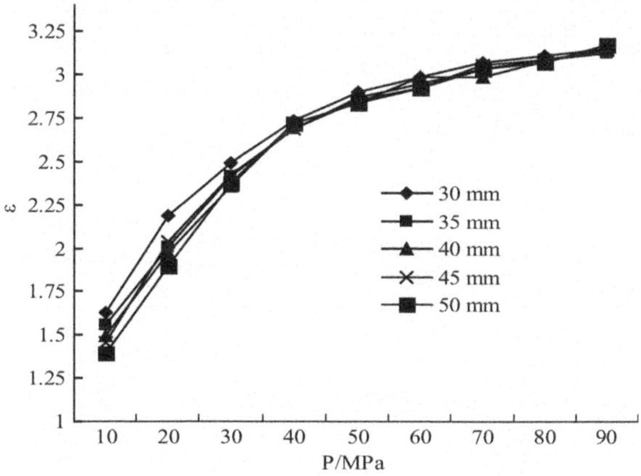

Fig. 2.10 Influence of pressing pressure on compression ratio in different volumes of oilseeds

the critical pressing pressure (P_3), the compression ratio is 3.14 and the residual oil content of meal is about 6.5% (dry basis).

2.6.3.3 Analysis of Properties of Oilseeds by Cold Press and Properties of Fruits Oil

Figure 2.11 shows changes of main contents between *Swida wilsoniana* fruits and its meal. Oil content decreases greatly from about 30% in *Swida wilsoniana* fruits to about 6% in the meal after being pressed. And other contents change differently. The shapes of raw oilseeds and the meal are shown in Fig. 2.12.

For fresh fruits, timely drying is necessary, which is conducive to both the preservation of fruits and high quality control of crude oil. Table 2.15 illustrates the physicochemical property of *Swida wilsoniana* oil and analysis of the major fatty acids by gas chromatography. Newly pressed *Swida wilsoniana* oil has a deep red color, which needs to be further refined before being used as edible oil. Gas chromatography analysis showed that *Swida wilsoniana* oil is a kind of high-quality woody oil, which is rich in unsaturated fatty acid, of nearly 72%, altogether 29.83% oleic acid, 39.15% linoleic acid, and 2.59% linolenic acid.

2.6.3.4 Conclusions

Swida wilsoniana is an important woody oil plant in which the whole fruits contain oils. The main contents of fruits from *Swida wilsoniana* were analyzed through chemical means using 46 samples from different areas. These fruits have an average moisture content of about 8.19% after they are naturally dried. The oil content is about 28.02% (dry basis) or more. The protein content is 7.25% (dry basis) and the cellulose content is about 14.18%.

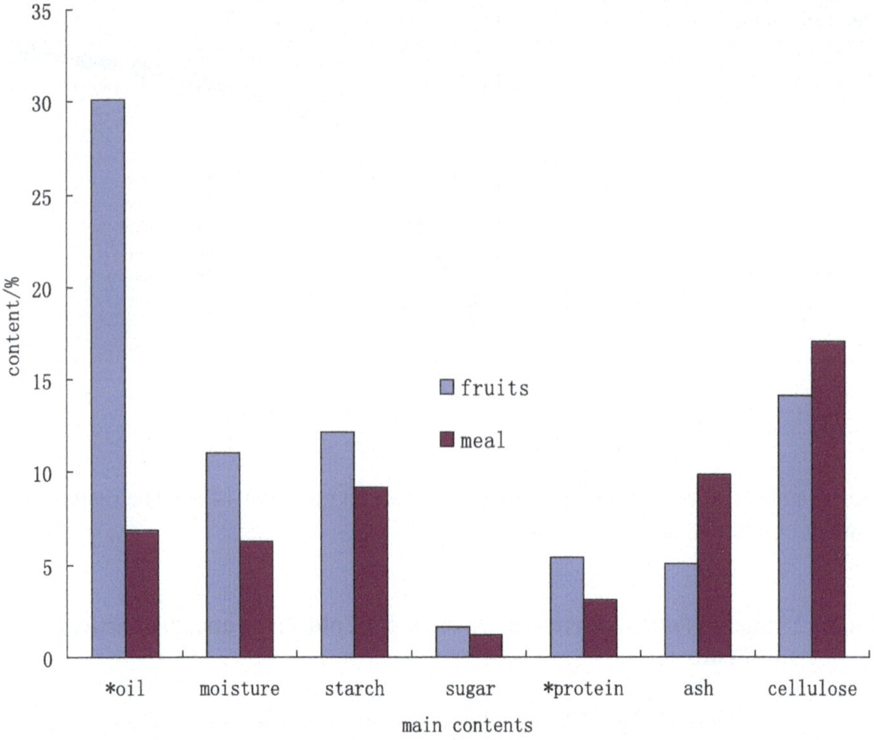

Fig. 2.11 Comparison of main contents between *Swida wilsoniana* fruits and its meal

Fig. 2.12 *Swida wilsoniana* fruits (left) and its meal (right)

Table 2.15 Physicochemical property of *Swida wilsoniana* oil and analysis of the major fatty acids

Program	Index	Unit	Result
Property	Acid value	mg KOH/g	1.27
	Peroxide value	mmol/Kg	7.6
	Color (25.4)	–	$R = 40.0$, $Y = 15.0$, $B = 15.0$
	Iodine	g I_2/100 g	104.48
	Saponification value	Mg KOH/g	199.04
Composition of fatty acids	$C_{18}H_{32}O_2$	%	39.15
	$C_{18}H_{34}O_2$	%	29.83
	$C_{16}H_{32}O_2$	%	21.79
	$C_{18}H_{30}O_2$	%	2.59
	$C_{18}H_{36}O_2$	%	1.61
	Others	%	5.03

Through theoretical calculation and the one-dimensional pressing device, the conditions of the cold pressing process and performances of fruits from *Swida wilsoniana* had been studied. *Swida wilsoniana* fruits dried through natural weather are more suitable for direct pressing, for that guarantees a higher oil quality. The volume of fruits fed into pressing device had little effect on pressures of oil squeezed out, fruits crushed and critical compression, and the value of those pressures were 55.7, 71.5, and 80 MPa individually. At the critical pressure of 80 MPa and the compression ratio of 3.14, residual oil of meal is 6.5%.

Swida wilsoniana oil is a kind of high-quality woody oil, which is rich in unsaturated fatty acid, of nearly 72%, altogether 29.83% oleic acid, 39.15% linoleic acid, and 2.59% linolenic acid.

2.7 Summary

The oil industry has evolved from small operations to larger processing facilities over the past decades. And the individual operation must include the impact of the other operations. This chapter is concerned with industrial plant oil extraction from oilseeds. It includes a brief account of the principle of oil extraction, storage and preparation of oilseeds, mechanical extraction, solvent extraction, and oil refining. Different purposes of use between edible plant oil and industrial plant oil have led to the different technology, crafts, and equipments during each production process. The industrial plant oil extraction has a history, but it did not form a systematic theory and mechanism. So far, the industrial plant oil extraction is similar to edible plant oil. For example, the solvent extraction process consists of the unit operations of solvent extraction, meal desolventizing, meal drying and cooling, miscella distillation, and solvent recovery. How to get a suitable industrial plant oil is the first step. Finally, some novel technology and equipment of oil extraction especially about oilseeds are provided and discussed.

Chapter 3
Bio-liquid Fuels in Industrial Plant Oil

Aihua Zhang and Changzhu Li

Abstract With the depletion of fossil energy in the world and the continuous expansion of energy demand, people are actively seeking new alternatives to new energy fuels. Among these new energy sources, bio-based fuels with high biodegradability and non-toxicity are more attractive as alternative fuels. The use of non-edible woody oil as the main raw material for the preparation of bio-based fuels can alleviate the pressure of edible oil resources, so the development of non-edible woody oil plants using marginal land has become an inevitable trend. This section describes the main types, potentials, and developments of non-edible woody vegetable oils as bio-based fuels.

Keywords Non-edible woody oil · Microalgae oil · Biodiesel · Hydrocarbon-rich fuel · Bio-jet fuel

Industrial plant oil is an important raw material for bioenergy, chemicals, oil-based materials, and pharmaceuticals. With the continuous improvement of people's living standards, energy and chemical products are urgently required to develop in the direction of renewable, low toxicity, environment-friendly to the environment. It is one of the main developing directions of industrial oil to replace the chemical products with petroleum, natural gas, and coal as the main raw materials [1–5]. Because industrial plant oil have the advantages of low toxicity, easy biodegradation, and well adaptability to the environment. Nowadays, the world is facing the challenge of resource shortage. The industrial plant oil is a renewable resource with great potential. As an indispensable industrial raw material, its important value is highly valued [6, 7].

A. Zhang (✉) · C. Li
Hunan Academy of Forestry, Changsha, 410004, China
e-mail: zhangaihua909@163.com

C. Li et al. (eds.), *Industrial Oil Plant*,
https://doi.org/10.1007/978-981-15-4920-5_3

3.1 Preparation of Biodiesel from Industrial Vegetable Oil

The use of plant oils in diesel engines is nearly as old as the diesel engine itself. The inventor of the diesel engine, Rudolf Diesel, reportedly used groundnut (peanut) oil as a fuel for demonstration purposes in 1900 [8]. Some other work was carried out on the use of plant oils in diesel engines in the 1930s and 1940s. The fuel and energy crises of the late 1970s and early 1980s, as well as the accompanying concerns about the depletion of the world's nonrenewable resources, provided the incentives to seek alternatives to conventional, petroleum-based fuels. In this context, plant oils as fuel for diesel engines were remembered. They now occupy a prominent position in the development of alternative fuels. Hundreds of scientific articles and various other reports from around the world dealing with plant oil-based alternative diesel fuels (biodiesel) have appeared in print. They have advanced from being purely experimental fuels to initial stages of commercialization [9, 10]. Nevertheless, various technical and economic aspects require further improvement of these fuels. Numerous different plant oils have been tested as biodiesel. Often the plant oils investigated for their suitability as biodiesel are those which occur abundantly in the country of testing. Therefore, soybean oil is of primary interest as biodiesel source in the United States while many European countries are concerned with rapeseed oil, and countries with tropical climate prefer to utilize coconut oil or palm oil. Other plant oils, including sunflower, and safflower, have also been investigated [11, 12]. Furthermore, other sources of biodiesel studied include animal fats and used or waste cooking oils. Sources of biodiesel with some emphasis on developing countries have been discussed [13]. However, the direct use of plant oils in fuel engines is problematic. Due to their high viscosity (about 11–17 times higher than diesel fuel) and low volatility, they do not burn completely and form deposits in the fuel injector of diesel engines [14].

3.1.1 Research Progress on the Conversion of Plant Oil to Biodiesel at Home and Abroad

3.1.1.1 Development of Biodiesel Abroad

At present, some countries and regions, such as the United States, Europe, and Asia have begun to set up commercial biodiesel production bases and widely use biodiesel as alternative fuels [15].

Biodiesel is widely used in Europe, and the main raw materials are rapeseed oil, soybean oil, and waste oil. The European Parliament relieved 90% of the tax on biodiesel and enacted laws to support alternative fuels. There are differential taxes and subsidies for rapeseed production that have helped boost the rapid growth of the biodiesel industry. In 2003, European countries produced more than 1.76 million tons of biodiesel, and by 2010, biodiesel production in Europe will reach 8.3 million thousand tons. Germany is the most widely biodiesel used country, producing and consuming 1.1 million tons of biodiesel till now, account for 3.5 million tons of the world's total production of 1/3.

The United States was the first country to study biodiesel. At present, there are many biodiesel production plants and suppliers in the United States. The main raw materials are soybean oil, the annual production of biodiesel is more than 300 thousand tons, and the tax rate of biodiesel is zero.

While producing soybean oil biodiesel, the United States also actively explored other ways to produce biodiesel. The National Laboratory of Renewable Resources of the United States made "engineering microalgae" through modern biotechnology. Under laboratory conditions, the lipid content can reach 40% and 60%. It is estimated that "engineering microalgae" can produce 6400–16,000 L biodiesel per acre, which opens a new way for the production of biodiesel [16].

The pure form of biodiesel, also known as net biodiesel, has been officially listed as a vehicle alternative fuel by the United States Energy Policy Act. Depending on the difference of raw materials and producers, the price of U.S. net biodiesel is less than 0.515–0.793 $/L; 80% of the market price of biodiesel, which costs 7.93–10.57 cents per liter more than conventional diesel [17, 18].

Japan mainly uses fried oil as raw material to produce biodiesel, the current annual production capacity of biodiesel reached 400 thousand tons. The production of biodiesel from castor oil in Brazil is being popularized.

3.1.1.2 Development of Domestic Biodiesel

The research and development of biodiesel in China is a little later than that in foreign countries. In 1981, the experimental research on the production of biodiesel from vegetable oil, such as rapeseed oil and fruit oil of light bark tree, was started [19]. In recent years, some scientific research institutions and universities have carried out the research of biodiesel, and successfully developed the technology of using rapeseed oil, soybean oil, and waste fried oil as raw materials to produce biodiesel. The government has also formulated a series of preferential policies and measures to support the research and development of biodiesel, which is conducive to the rapid development of China's biodiesel industry.

The raw material of biodiesel in China will mainly come from two aspects: waste oil and woody plant oil. The waste oil mainly includes animal oil, food waste oil, and gutter oil, and the woody oil plants mainly include *Camellia oleifera Abel, Olea europaea Linn., Elaeis guineensis Jacq*. In recent years, the seed oil of *Jatropha Curcas, Xanthoceras sorbifolia Bunge, Pistacia chinensis*, and *Acer truncatum Bunge* has been studied as biodiesel feedstock. China is rich in oil-bearing plants with 1553 species belonging to 151 families, there are 154 species of which with more than 40% oil content, approximate 30 species tree and shrub species can be used to establish large-scale biomass fuel oil raw material base and there are about 10 kinds of biomass fuel oil plants that set up a large-scale supply base for improved varieties.

In the Tenth Five-Year Plan, the Ministry of Science and Technology, proceeding from the integrated system of planting resources and energy, selected fine fuel oil plants suitable for different regions, and development base of fuel oil plant planting resources and the demonstration project of biofuel oil production were established. Genetic engineering and other modern techniques were used to improve species characters and to cultivate new plant varieties with high fuel oil production. The experiment of introducing, domesticating, and small area cultivation of foreign oil plant species and fine clones with an important economic value has been carried out through the demonstration of excellent provenance and rapid high yield cultivation of clones [20–22].

3.1.2 Challenges in the Development of Biodiesel

(1) Domestic biodiesel has some shortcomings, such as small production scale, backward technology, bad follow-up development, and so on.

The cost of biodiesel production is high because there is no tax exemption on biodiesel in China (75% of which is the cost of feedstock), about three times as much as mineral diesel, making it difficult to achieve mass production. At present, scientific research institutes and enterprises mainly focus on the production technology of biodiesel.

(2) The production of biodiesel is relatively small, and the condition of using biodiesel as pure fuel is not yet met.

At present, the average oil content and yield per unit yield of soybean and rape in China are lower than the world average, which makes the production of vegetable oil in China have a certain potential, but its development is still limited, even though the oil content and yield level of oil plant is completely up to the international high level. As estimated, this will still be difficult to make up for the current huge gap in food consumption. It is difficult to achieve the development of biodiesel by expanding the planting area of oil crops.

(3) The cost of feedstock

The main problem in large-scale production of biodiesel is the cost of raw materials. About 75% of the cost of biodiesel production is the raw materials, and the choice of feedstock route is very important to the competitiveness of biodiesel. Although the raw materials for the production of biodiesel in China come from rapeseed oil, cottonseed oil, Chinese tallow oil, wood oil, tea oil and gutter oil, and so on, but the supplement of raw materials in the future will be mainly non-grain.

3.1.3 Solutions

(1) Biodiesel is produced from a variety of waste oils and oily residues. These resources include waste oil from cooking and frying foods, animal fats from roasting foods, and animal grease treated from waste material produced during normal temperature. Cooking waste oil also known as swill oil mainly refers to the oil from the remaining meal separated by oil, gutter oil, kitchen hood cold condensed oil, etc.
(2) The seed of woody oil plants has the potential to produce biodiesel. There are abundant woody oil plant resources in our country, including *Jatropha Curcas, green jade trees, Pistacia chinensis, Cornus wilsoniana, oil trees, Chinese tallow,* and so on. They have the characteristics of wild, cold tolerant, and barren tolerant. Although it takes a lot of labor to collect these raw materials, considering the relatively cheap labor resources in China, the development of bio-refining and chemical plants using woody plant oil as raw material is in line with the national conditions of our country.

3.2 Microalgae Biofuel

Assessing the sustainability of algal liquid biofuels requires an understanding of the individual components that make up potential supply chains. This chapter focuses on the basic processes of algal biofuel production from the biology and traits of the organisms, to methods for cultivation, and to processing into liquid fuels. It discusses algal strains and the attributes of those strains critical for biofuel production, the photo-autotrophic methods for algae cultivation through open-pond and closed photobioreactor systems, the processes for collection and dewatering if necessary, and the processing of algal lipid, biomass, or secreted products into fuels. It provides the basic descriptions of the supply chain components used in later chapters and summarizes some critical process improvements that could enhance the overall sustainability of algal biofuels [23].

3.2.1 Algal Feedstocks

The organisms considered as a potential feedstock for algal biofuel production belong to a vast and diverse assemblage of aquatic organisms that carry out oxygen-evolving photosynthesis and lack the stems, roots, leaves, and embryos of plants [24]. The category includes eukaryotic species that are related to the plant lineage and can be further categorized as macroalgae that are large structured species (for example, kelps) or microalgae that are microscopic species (for example, *Nannochloropsis spp.*). In the context of biofuel, the term “microalgae” also includes cyanobacteria, a diverse prokaryotic lineage whose ancestor gave rise to the plant chloroplast [25]. More than 40,000 species of microalgae have been described, and they collectively cover a comprehensive spectrum of habitats and tolerances of ranges of pH, salinity, and temperature [26–29] estimated that prokaryotic and eukaryotic microalgae are responsible for more than 40% of net primary productivity on Earth [26–29]. Algae can be a more appealing biofuel feedstock than land plants because of their faster biomass doubling cycle, their more accessible forms of stored carbon than the lignocelluloses used for cellulosic biofuels, and their ability to thrive on water sources and on land sites that are unsuitable for terrestrial farming.

Microalgae contain diverse pigments and metabolites that are desirable as nutritional supplements and colorants. Examples of such products include astaxanthin, an antioxidant derived from the alga Haematococcus, and a high-protein powder derived from cyanobacterial species of *Spirulina* (Arthrospira) [30, 31]. Commercial scale algal ponds that grow these and other microalgae have operated for more than a decade [32]. However, the scale of deployment for algae cultivation for fuel is expected to be much larger than the scale of algae cultivation for nutraceuticals or other specialty products currently available in the market.

Generating biofuels from algae requires exploiting and expanding the demonstrated commercial scale growth of algal biomass, and harvesting the relatively accessible carbon stored therein. Carbon is stored within algal cells in various forms, and these molecules can be accessed by different technologies. Both eukaryotic and prokaryotic algal cells are rich sources of polar lipids that are associated with membranes; in some cases, the photosynthetic thylakoid membranes are extensive. Carbon is such a crucial element for algae that it is typical for them to store surplus carbon when cellular division is restricted by some factor other than carbon availability—this situation is termed unbalanced growth. In many eukaryotic microalgae, photosynthetic carbon fixation continues under unbalanced conditions. Under extended periods of environmental stress, the excess fixed carbon is stored in the form of neutral lipids called triacylglycerols (TAGs). TAGs are hydrocarbon chains terminated in a carboxylic acid group. The three carboxyl groups are bound to glycerol through an ester linkage. Biofuels containing hydrocarbon chains longer than six carbons are particularly valued because of their high heats of combustion, volatility, and compatibility with existing engines. As discussed later in this chapter, extracted TAGs can be converted to biodiesel using a number of technologies, including transesterification and hydrotreating. Even algal species that do not store large amounts of TAGs can be converted to biofuels through various chemical conversion technologies. For example, species that store polysaccharides can be fermented to yield ethanol, and other biomass processing technologies, such as gasification, pyrolysis, and hydrothermal liquefaction, have shown great utility for the conversion of whole biomass into biofuels.

The incipient algal biofuel industry is emerging and evolving from its early foundations in algae cultivation for fish feedstuff and for human nutraceuticals. Early technology development of processing algae to fuels emphasized the conversion of neutral lipids (TAGs) to biodiesel. Choices of algal feedstock have been expanding to address the goals of fuel production rather than nutritional content and to exploit new technologies for processing biomass that extend beyond those that focus on TAGs. Ideal attributes for algal feedstock for fuels include rapid and dense growth; efficient use of nutrients, light, and carbon dioxide (CO_2) under a range of temperatures; resistance to pests and predators; accumulation of desirable macromolecules that can be processed into fuels; ease of harvest; and the absence of undesirable by-products.

Commercial and research interest in the United States has focused on microalgae, and these species are emphasized in this report. Microalgae have been reported to reach short-term maximum productivities of 50–60 g dry weight per square meter per day in CO_2 enriched open ponds in Hawaii and California [33]. These and other data on productivity from laboratory scale experiments have promoted the reputation of microalgae as prime candidates for providing cheap biomass feedstocks for food, feedstuff, or energy. Some authors have extrapolated values of maximal biomass productivity and combined them with maximal oil content to predict oil yields of 100 tons per hectare (ha) per year. Such reports have spurred investment in intensive research on algal biofuel production. However, such high productivity projections have yet to be obtained in large-scale, long-term experiments. Serious barriers remain for reproducing optimal growth and productivity conditions at a commercial scale. They include maintaining the stability of the culture and delivering the required nutrients and other resources in an efficient manner at such scales. Current yields from large-scale operations range from 40–60 tons dry weight of algal biomass production per year, and conservative projections anticipate up to 100 tons dry weight of biomass, or 30 tons of biodiesel per year in subtropical or tropical, sunny climates [34].

3.2.2 Processing Algal Biomass into Biofuels

Fuel production from algal biomass is most commonly assumed to involve the cultivation of microalgal species that can process the lipid into biodiesel and have high lipid productivity. In this case, production of biofuel requires the algae to be concentrated and subsequently treated to cause the release of the intracellular lipids. The concentration or harvest step involves the separation and typically drying of the algal cells to prepare them for lipid collection. Lipid collection usually is accomplished by rupturing the algal cells. Subsequent extraction of the biomass might be required for economical oil recovery. Thus, biodiesel production from algae requires two distinct separation steps—harvest and product collection—regardless of whether growth occurs in open or closed photobioreactors.

The important feature in harvest and extraction is that the algae and the lipids are insoluble in water. The technical problem in the production of biodiesel is simply producing a pure, dry triacylglycerol stream for subsequent processing to biofuels. Because the algal biomass and the algal oils are immiscible in water, harvest can be completely spontaneous, and there is no key thermodynamic separation energy to be overcome. The constraints on the system are purely engineering-related, and better engineering can reduce the energy expenditure required for separating the algal biomass from the culture water and drying it for subsequent oil collection. Relatively low algal biomass concentrations and the small size of microalgae make separation challenging and energy intensive. A meta-analysis of published studies shows that more than 40% of the total energy required for biodiesel production can be attributed to harvest and product collection [35].

Purity of the algal lipid is an important parameter for processing into liquid transportation fuel. Inorganic materials that stay with the oil are a concern, and the method of harvest and collection can influence the impurity levels. Inorganic salts and phospholipids are two known impurities that could affect processing. Inorganic salts are in the culture medium and occur naturally in algae, but they also can be introduced as flocculants.

Evaluating the sustainability of algal cultivation systems for biofuel production requires examining the various material and energy inputs needed for the cultivation systems to maintain scalable productivity, maximize system robustness, and minimize costs. Scalable productivity refers to a cultivation system's ability to maintain productivities with respect to algal biomass and algal product (mass/area-time or mass/volume-time) from the laboratory scale to the commercial scale. System robustness refers to a cultivation system's ability to reliably and dependably deliver consistent productivity and avoid system crashes or failures as a result of either biological or physicochemical causes. Costs pertain to capital and operating costs for a cultivation system.

3.3 Hydrocarbon-Rich Fuel Produced by Pyrolysis of Vegetable Oil

3.3.1 Reaction Mechanism of Catalytic Pyrolysis

Pyrolysis of oil refers to the process of breaking the chemical bond in triglycerides to produce small molecule substances under almost oxygen-free and high-temperature conditions. The molecular weight of animal and vegetable oil, which is generally larger, can be decreased by pyrolysis that helps to reduce the viscosity of raw oil. The pyrolysis of oil, known as a complicated process, includes molecular bond rupture, molecular isomerization, and polymerization, and we can obtain the products in different distributions of molecular weight by altering pyrolytic temperature, pyrolysis time, reaction atmosphere, and heating rate [36].

Catalytic cracking of oil refers to the high-temperature cracking of animal and vegetable oils and waste oils to produce low-carbon alkanes, olefins, etc., after catalyst addition. The addition of the catalyst not only reduces the activation energy of the cracking reaction, increases the reaction rate, increases the yield of the low-carbon alkyl alkene, but also increases the flexibility of the distribution of the cracked product, changes the distribution of the reactants, and improves the selectivity of the target product.

Vegetable oil pyrolysis has been studied for many years. Due to the existence of multiple secondary reactions and different reaction channels, the exact reaction mechanism of the pyrolysis mechanism of triglycerides remains unclear. The products obtained by catalytic cracking of oils are relatively complex organic hydrocarbons and other organic mixtures. Under high-temperature catalytic cracking, the C–C bond cleavage of triglycerides is the most basic first reaction, and then various monomers are undergoing many interlacing reactions, i.e., the second reaction. In two pyrolysis reaction, increasing the pyrolysis temperature will increase the yield of gas products and form a more complex reaction system.

Although the pyrolysis mechanism is somewhat complicated, there are still some studies suggesting that there are two different steps in the cleavage reaction. The first step is that the triglyceride breaks first at high-temperatures, producing the corresponding acidic substances (mainly carboxylic acids) and others. This step is called the main cracking. The second step, also known as secondary cracking, is the decomposition of the material obtained in the first step, thus forming hydrocarbons with shorter carbon chains, including saturated and unsaturated.

Due to the complex reaction of catalytic cracking, there are few studies at home and abroad. Here, only the main thermal cracking mechanism of vegetable oil is expounded. The main processes are as follows.

Primary cleavage can be illustrated by the mechanism of γ-hydrogen transfer and β-elimination. Figure 3.1 shows the mechanism of gamma-hydrogen transfer during the thermal decomposition of triglycerides. According to this mechanism, terminal olefins and glycerol triacetate are used as their by-products. Figure 3.2 shows the mechanism of β-elimination in the thermal decomposition of triglycerides. The β-hydrogen interacts with a bis-acyl group with a free electron on the oxygen

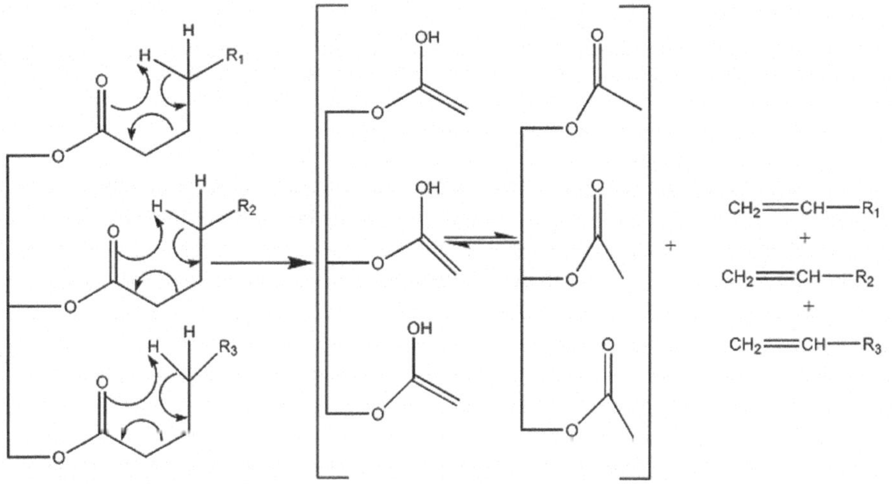

Fig. 3.1 The mechanisms of γ—hydrogen transfer (R_1, R_2, R_3 indicate saturated or unsaturated carbon chains)

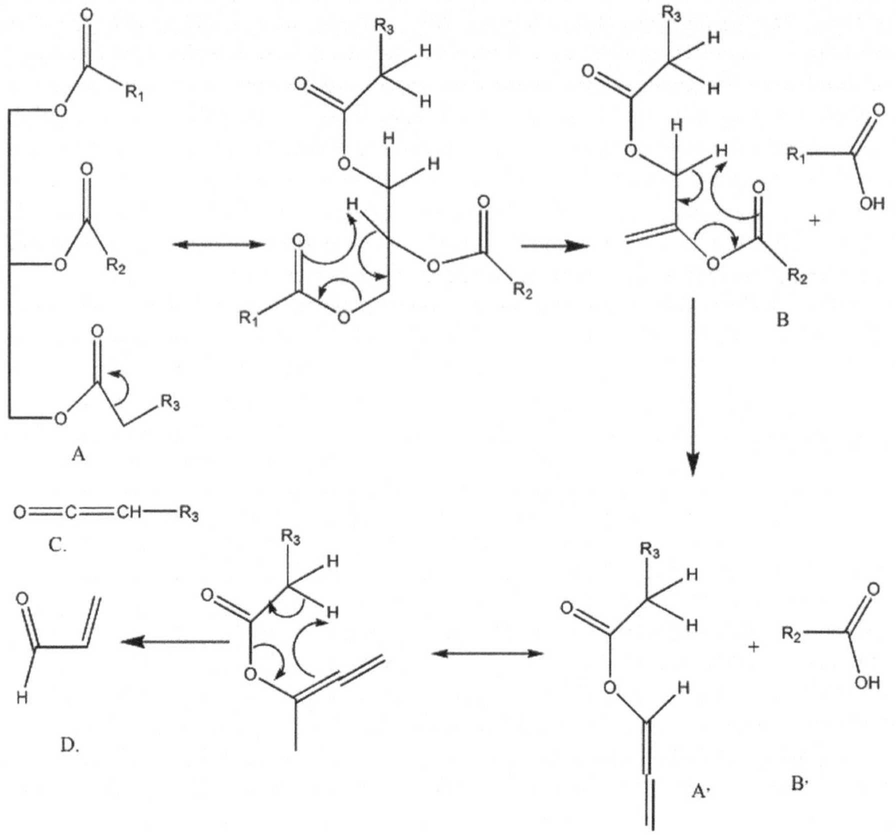

Fig. 3.2 β-elimination mechanism
(R1, R2, and R3 represent saturated or unsaturated carbon chains, respectively; the decomposition process forms extremely unstable intermediates including A1, fatty acid B, ketene C, and acrolein D.)

atom, resulting in a C–O cleavage between the glyceride and the carbon chain, thus also forming an intermediate structure (A1) and two fatty acid molecules (B). The intermediate structure is rearranged to form an enone molecule (C) and an acrolein molecule (D). The length of the fatty acid chain obtained by pyrolysis and the molecular weight of the ketene is determined by the composition of the triglyceride R group.

Secondary cracking is characterized by deoxidation (by thermal decomposition). The carboxylic acid produced during the primary cracking process can form a linear saturated or unsaturated hydrocarbon by deoxygenation. There are two different possible reaction routes for the deoxygenation reaction: decarboxylation and decarbonylation. The acid organic compound may undergo hydroxy O–H bond cleavage (a) or acyloxy bond C–O cleavage (b) at a high-temperature. The acid molecule undergoes cleavage of the hydroxyl O–H bond to obtain a carboxyl radical, which

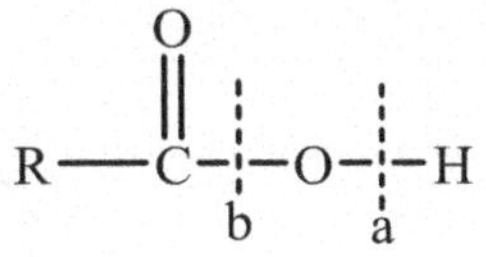

Fig. 3.3 Hydroxyl bond and acyloxy bond cleavage

releases a molecule of carbon dioxide, and R and H combine to form a hydrocarbon compound. This reaction process is called decarboxylation. The acid organic compound is cleavable by an acyloxy bond to obtain an acyl radical, and a molecule of carbon monoxide is released, and R and H combine to form a hydrocarbon compound, and the reaction process is called a decarbonylation reaction. Since decarboxylation or decarbonylation of an acid compound is required to be carried out at a high-temperature, the decarboxylation or decarbonylation reaction is often accompanied by one of the reactions and the other reaction is accompanied by the other reaction (Fig. 3.3).

Figure 3.4 shows the specific process of the secondary cracking reaction mechanism. In the decarboxylation reaction process (a) a linear alkane and carbon dioxide is produced; in the decarburization reaction process (b), an olefin, carbon monoxide, and water have emerged. However, there is evidence that under high-temperature conditions, acid species tend to form free radicals for other reactions (c), while gamma-hydrogen transfer carboxylic acids condense to form symmetric ketones.

Taking stearic acid and linoleic acid as examples, the possible pathways for the cleavage of acid substances are explained. Since linoleic acid is an unsaturated fatty acid, it is easily cleaved, and stearic acid is a saturated fatty acid, so it is hard. As shown in the figure, Figs. 3.5 and 3.6 show the cleavage reaction process of linoleic acid and stearic acid, and infer the possible cleavage reaction mechanism.

Under the condition of no hydrogen, the microporous Pd/C catalyst cracks the unsaturated raw material to catalyze, and the cleavage of linoleic acid will undergo

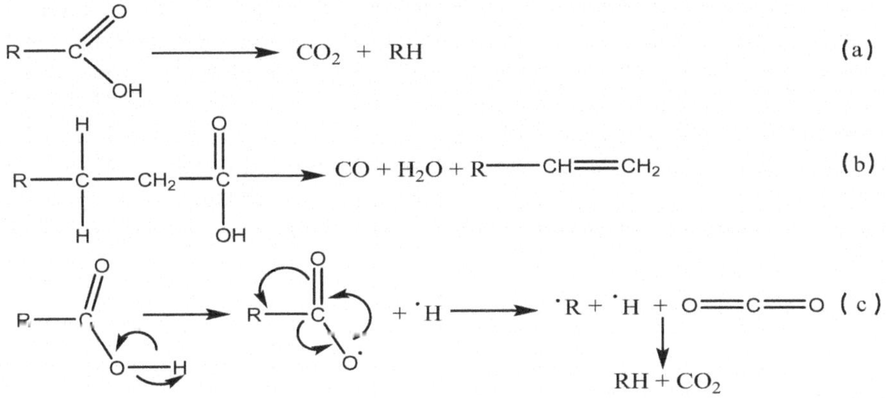

Fig. 3.4 Three ways of secondary cracking

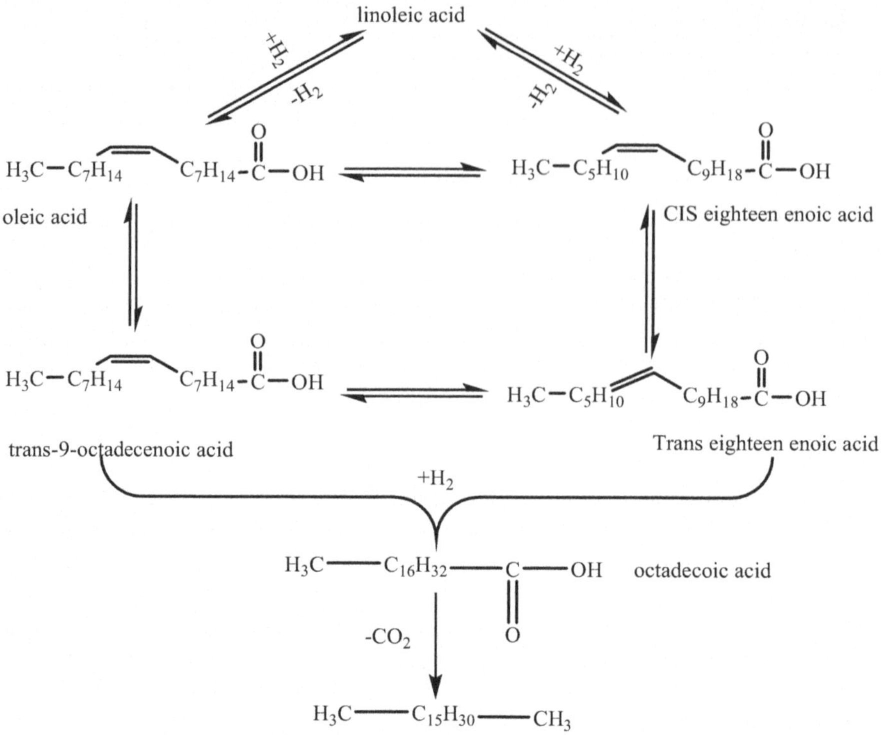

Fig. 3.5 Linoleic acid deoxidation

various deoxygenation reactions, namely isomerization, hydrogenation, and deoxidation (Fig. 3.5). The reaction conditions are from 300 to 360 °C under a pressure of 15–17 Pa; isomerization and hydrogenation first occur, followed by decarboxylation. Under inert conditions, hydrogenation generally occurs on hydrogen treated catalysts, however, since there is no supply of hydrogen, the rate of hydrogenation will decrease. For the cleavage of unsaturated esters, the products are primarily the corresponding saturated esters, and some anaerobic organic products. Figure 3.6 shows the catalytic deoxygenation of stearic acid in Pd/C or Pt/C with high selectivity. The main gas products obtained by online and off-line gas analysis are CO and CO_2. These reactions may undergo decarbonylation, decarboxylation, hydrogenation, dehydrogenation, isomerization, cyclization, and dimerization during fatty acid conversion.

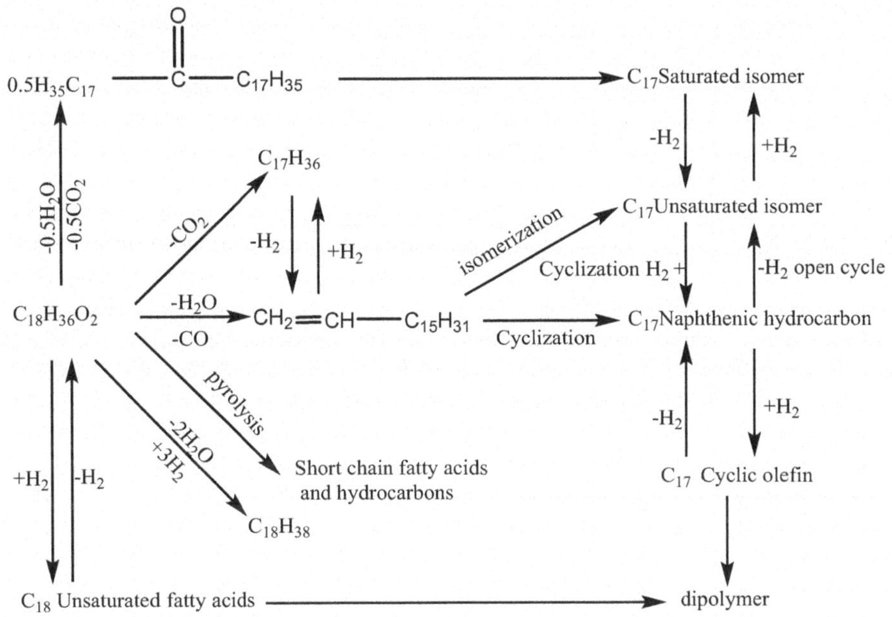

Fig. 3.6 Stearic acid deoxygenation

3.3.2 Research Progress on Cracking of Industrial Plant Oil

Silicon oxide, magnesium oxide, and alumina were used in the early catalytic cracking of vegetable oils to produce hydrocarbon fuels. Among them, alumina had high catalytic activity and showed good catalytic activity for decarboxylation of fatty acids. With the deepening of research, people began to use zeolite catalysts with molecular selectivity, the first choice is in the petroleum industry has been widely used in the ZSM-5 zeolite. It was found that HZSM-5 zeolite with a higher specific surface area (300–500 m^2/g) had higher activity in the pyrolysis of triglycerides. However, due to its strong acidity and small pore size, a large number of small gaseous molecules were found in the pyrolysis products. Therefore, it is inappropriate to use hydrocarbons above C5 (as liquid fuels) as target products for cracking. When these catalysts were used to catalyze the pyrolysis of triglycerides, a mixture of alkanes, olefins, and carboxylic acids with 33% C5 or more was obtained, and the other main products were gaseous products below C5. Alena and other people used rape and soybean oil as raw materials to explore the pathway of triglyceride cracking. Alumina was used as the catalyst in this experiment. It contains 15–25% C2–C10 saturated monocarboxylic acid and about 30% alkane. In the early stage of the project group, the batch pyrolysis reactor was used to study the catalytic cracking of the bark oil, with KF/CaO catalyst in the range of 350–50 temperature. The results showed that the recovery rate was 81.5% and the acid value was 46 mg KOH/g under the conditions of 1.0% catalyst, 420 temperature, and 50 h.

Mesoporous molecular sieve catalysts with regular pore structure and shape-selective catalytic effect have been developed. Mesoporous molecular sieves as catalysts can significantly reduce the formation of non-condensable gasses and improve the yield of liquid products. It has been proved in petrochemical, waste plastic cracking, and other research fields, and similar results have been obtained in oil cracking. However, there are many weaknesses, such as the acid content of pyrolysis products is high, and the acid value is about 120 mg KOH/g. According to the literature, the development of catalysts is limited to the synthesis of catalysts used in petroleum refining, and ignoring the negative impact of high acid value of pyrolysis products on fuel performance. Twaiq and Sang et al. attempted to carry out triglyceride catalytic cracking using a small pore-mesoporous composite catalytic material. It was found that the composite material with this structure has better gasoline fraction selectivity. The reaction conditions for obtaining the best gasoline fraction yield are: temperature 440 °C, the ratio of fatty acid to catalyst is 9.64, and the weight hourly space velocity (WHSV) is 3.66. H-1. Shinae Jun et al. reported the synthesis of a new material for mesoporous carbon molecular sieves and studied its structure. The results show that carbon molecular sieve is an inert material, which can ensure the stability of its own skeleton structure regardless of the acid-base environment. Mesoporous silica molecular sieves (such as MCM-41) will undergo structural instability under alkaline conditions and cause skeleton collapse. In addition, mesoporous carbon molecular sieves have similar mesoporous structure, high specific surface area and thermal stability to mesoporous silica molecular sieves (such as MCM-41), but their catalytic properties have not been studied. Zhang FQ and other studies have shown that carbon molecular sieve as a catalyst carrier has the advantages of high dispersion of active phase, reaction heat removal in time, reduction of polycondensation and condensation, etc. It is an ideal catalyst carrier. Therefore, if the basic metal compound is uniformly supported on the surface of the mesoporous carbon molecular sieve, the alkaline mesoporous carbon molecular sieve is created as a grease cracking catalyst, and the uniform mesoporous structure can be utilized to avoid the limitation of the molecular sieve having a smaller pore size in the cracking reaction. It can also produce >C5 hydrocarbons (for liquid fuels), ensure higher conversion of raw materials, reduce the occurrence of carbon formation, and improve the activity, selectivity, and service life of the catalyst. In addition, the uniform, highly dispersed surface alkali center can not only improve its catalytic activity, but also change the reaction process, convert the higher fatty acid produced by the cracking into carboxylate and decompose it to obtain low-oxygen bio-hydrocarbon fuel oil (alkane, olefin mixed fuel).), and the catalytic reaction kinetics, reaction mechanism, and surface physicochemical properties in this new catalytic system are different from the traditional catalytic system, many of which need to be studied and explored in-depth.

Recent studies have found that alkali metal elements can exhibit strong acid binding during catalytic cracking and can significantly reduce the carboxylic acid content of pyrolysis products. The mechanism is to form a carboxylate with the carboxylic acid produced by the cleavage, and then the carboxylate is decomposed at a set temperature to obtain an alkane or olefin mixture having better combustion properties. This reaction process effectively reduces the activation energy of the reaction, and the reaction rate is faster. However, the basic metal compound does not have a high catalytic specific surface and a developed pore structure, so that the amount of the catalyst is significantly increased and the reaction is incomplete, about 10–20% of the fatty acids are present in the cleavage product. Therefore, there are many shortcomings in the current cracking catalysts, but each has its own advantages. Aiming at the carboxylic acid components in the product which affect its fuel performance, how to effectively combine the above catalyst characteristics and exert synergy between the components is the key to solve the problems of the current cracking catalyst. In summary, domestic and international research lacks the basic work and preliminary experimental research on the preparation of bio-hydrocarbon fuels by industrial pyrotechnical directional pyrolysis, especially the basic catalytic reaction process mechanism, model data testing, and pyrolysis kinetics are subject to further research. In addition, most of the catalysts currently used are selected according to cracking catalysts in the petroleum industry, mostly strongly acidic microporous, mesoporous molecular sieves (ZSM-5, MCM-41, etc.), which are less stable in an alkaline environment, and the liquid product obtained in actual use has a low yield and a high acid value. Therefore, based on the synergistic action of the pore characteristics of the mesoporous material and the alkali active component, the oil cracking process (decomposition after forming a salt with the carboxylic acid produced by the cracking) can be converted to obtain a bio-hydrocarbon cracking fuel having a suitable molecular chain length. This helps to obtain a mixture of alkanes and olefins with better combustion properties, thereby achieving targeted cracking of the oil and fat on the molecular structure.

3.4 Aviation Liquid Fuels from Industrial Plant Oil

Biomass is the core component of the global renewable resources. It is one of the basic resources for the survival and development of human beings. Only second to coal, oil, and natural gas, it is also the most fundamental guarantee to maintain the sustainable development of the human economy and society. With the rapid development of economic globalization, the consumption of transportation fuel continues to increase. By 2013, the global energy consumption has increased by 2.3%, and oil consumption has increased to 1.4 million barrels per day, and the global crude oil reserves are now 1.6789 trillion barrels at the end of 2013, only 53.3 years' exploitation quantity can be satisfied. Under the double pressure of energy shortage and environmental deterioration, and with the multiple support of technology, policy and market, and so on, our government has attached great importance to the development and utilization of biomass resources, and has successively issued a series of strategic planning and guiding policies, which have put forward clear goals and requirements for aviation biofuels.

The essence of aviation biofuels is C8–C16 hydrocarbons, which, unlike all other transportation fuels, are mandatory, widely internationally versatile and highly safe, with suitable density (>0.775 g/cm^3) and high calorific value (>42886 kJ/Kg) and good combustion performance (stable, continuous, fast burning, carbon deposit), good low-temperature fluidity, excellent oxidation stability and thermal stability, no harmful substances such as mechanical impurities and moisture. At present, aircraft engines are mainly divided into the jet and internal combustion engines. Both civil aviation jet aircraft and military jet aircraft use jet fuel, usually Jet fuel No. 3; propeller aircraft, helicopters, etc., usually use aero piston engine fuel; the fighters use fighter 20 military aircraft as jet fuel.

Today, global climate change has become the biggest challenge. In 2012, the United Nations Framework Convention on climate change (UNFCCC) aims to keep global temperatures rising below 2 °C, while the United Nations Intergovernmental Panel on Climate Change(IPCC) predicts that global temperatures will rise to 1.4–5.8 °C, and controlling greenhouse gas emissions is one of the main goals to slow global temperature rises. According to IPCC, global aviation CO_2 emissions from 330 Mt CO_2/yr in 1990 to 480 Mt CO_2/yr in 2000, which have increased by about 1.5 times, and accounting for 2% of the total human emissions, and it is expected that by 2050 the emissions of air greenhouse gasses will be increased by 300–600%. As aviation CO_2 emissions continue to increase, the development and use of aviation biofuels will not only reduce the emissions of greenhouse gasses, but also alleviate the energy crisis, which have been supported by many government policies. In this context, the aviation biokerosene is concerned by the governments, oil companies, and airlines. As the world's second largest aviation power and the largest aircraft consumption market in the world, the aviation industry will face severe emission reduction and cost challenges. China has started early and made some progress in the research of aviation biokerosene.

Biological aviation kerosene is a kind of aviation kerosene that is produced from renewable resources. The renewable raw materials mainly include *comus wilsoniana* fruit oil, leprosy oil, linseed oil, algae oil, restaurant waste oil, halophytic vegetable oil, and so on. Compared with the traditional fossil, the carbon emission in the whole life cycle can be reduced by more than 35%, so aviation biokerosene is a kind of environment-friendly fuel and the most potential substitute for aviation kerosene based on the fossil in the future. At present, many countries have carried out no-load or commercial test flights of aviation biokerosene, which proves that bio-aviation kerosene can meet the requirements of the aviation industry for fuel performance.

Although the research of aviation biofuel has made relatively satisfactory progress, the aviation biokerosene is not competitive because of the factors such as high raw material price, unstable sources, and small market size. However, the adoption of biofuel is one of the important ways to reduce human dependence on fossil energy and achieve emission reduction. In the long run, aviation biofuels need to find stable raw material supply and more competitive processing technology in order to provide support for the healthy development of the industry. Among them, raw material supply is the most concerning problem in the development of aviation biofuel. It is necessary to realize "not competing with the people for grain, not competing with grain for land, not competing with farmland for a water source, not competing for feed with livestock.".

3.4.1 Second-Generation Biodiesel Synthesis Technology

Second-generation biodiesel does not contain oxygen and sulfur, It has a lower density and viscosity, a high cetane number and a lower cloud point, and a higher calorific value in the same mass unit.

The second-generation biodiesel uses a catalytic hydrogenation technology to hydrotreat animal and vegetable oils to obtain alkanes similar to diesel components. The preparation process comprises various chemical reactions, mainly including hydrogenation saturation, hydrodeoxygenation, hydrodecarboxylation, hydrodecarbonylation reaction, and hydrogen isomerization reaction of unsaturated fatty acids in animal and vegetable oils. The main component of animal and vegetable fats and oils is a fatty acid triglyceride, in which the fatty acid chain length is generally C12–24, Most of C16 and C18, the typical fatty acids in fats and oils include saturated acids, monounsaturated acids, and polyunsaturated acids, the degree of unsaturation varies greatly depending on the type of oil. Under catalytic hydrogenation conditions, the triglyceride will first undergo a hydrosaturation reaction of the unsaturated acid and further cracking to form an intermediate product including diglyceride, monoglyceride, and carboxylic acid. After hydrodecarboxylation, hydrodecarbonylation, and hydrodeoxygenation the final product of the normal paraffin reaction is mainly C12–24 normal paraffins, by-products including propane, water, and a small amount of CO, CO. The main reaction formula is as follows.

The biodiesel prepared by hydrogenation of oils can have a value of hexadecane of 90–100, no sulfur and oxygen, no aromatics, can be used as a high cetane number component in combination with petrochemical diesel in any ratio. However, due to the higher melting point of the normal paraffin, the prepared biodiesel has a higher cloud point and a lower low-temperature fluidity. We can convert some or all of the normal paraffins into isoparaffins by hydrogen isomerization to improve their performance at low-temperatures.

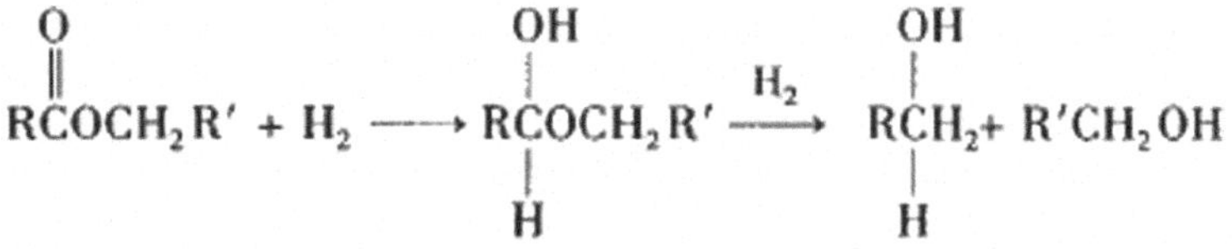

(3.1)

The second-generation biodiesel production process is based on catalytic hydrogenation. At present, the second-generation biodiesel production process mainly includes three processes of hydrogenation direct deoxidation, hydrodeoxygenation isomerization, and diesel blending. The reaction conditions and technical characteristics are shown in Chart 3.1.

It can be seen from Table 1 that the three processes of production, especially the hydrodeoxygenation and hydrogen isomerization process, are optimized in terms of technology and production cost. The biodiesel produced by this process has a high

Chart 3.1 The conditions of second-generation diesel oil process

Process	Temperature/°C	Pressure/MPa	Airspeed/h^{-1}	Catalyst	Technical characteristics
Hydrogenation direct deoxygenation	240–450	4–15	0.5–5.0	Co–Mo Ni–Mo	In the deep hydrogenation process of fats and oils under high-temperature and high pressure, the oxygen atoms in the carboxyl group combine with hydrogen to form water molecules, which are themselves reduced to hydrocarbons. The process is simple, and the product has a high cetane number, but the obtained diesel component is mainly a long-chain normal paraffin, which makes the product have a high cloud point, poor low-temperature fluidity, and is suppressed at high latitudes. Generally only as a high cetane diesel addition component
Hydrodeoxyisomerization	300–400	2–10	0.5–5.0	Co, Mo, Ni, Pt, Molecular sieve	The process comprises two stages, the first stage is similar to the conditions of direct hydrodeoxygenation in the hydrodeoxygenation stage, and the second stage is the isomerization of the normal alkane obtained in the first stage in the hydrogen heterogeneous stage. Isomerized products have lower density and viscosity, higher calorific value, and do not contain polycyclic aromatic hydrocarbons and sulfur. Moreover, the product has a high cetane number and good low-temperature fluidity, and can be blended with petrochemical diesel at any ratio in a low-temperature environment, so the range of use is further broadened
Diesel blending	340–380	5–8	0.5–2.0	Ni–Mo/Al_2O_3 Co–Mo/Al_2O_3	Blending animal and vegetable oils improves the cetane number of the product and saves investment in the oil-and-hydrogenation unit, which is simple and economical. However, since the hydrogenation of oils is a strong exothermic reaction and the hydrodeoxygenation reaction has a competitive factor with the hydrodesulfurization reaction of petrochemical diesel, these may affect the desulfurization and refining effect of the hydrogenation unit on petrochemical diesel, increase the operation difficulty and production cost of the processing unit

cetane number and is similar to petrochemical diesel. Viscosity and calorific value, low cloud point, can be used in high latitudes, and can greatly reduce engine fouling, significantly reduce noise, and significantly reduce the emission of nitrogen oxides and particulate matter, which is an ideal Petrochemical diesel alternative fuel.

As the world's demand for biodiesel has increased year by year, the second-generation of biodiesel has developed rapidly in recent years. According to the Global Biodiesel Market Report released by Global Data, global biodiesel production increased from 959 million L in 2001 to 15.76 billion L in 2009, with an average annual growth rate of 41.9%. The world's leading biodiesel producer, Finland Ness Neste Oil recently said that the annual global demand for renewable biodiesel will increase significantly from the current 10 million tons to 35 million tons by 2020. Currently, Nestlé Petroleum has two biodiesel units in Porvoo, Finland, with a combined capacity of 380,000 t/a. In addition, the company has invested 550 million euros to build a new 800,000 t/a renewable biodiesel plant in Singapore, and also invested 670 million euros to build a new 800,000 t/a renewable biodiesel plant in Rotterdam, the Netherlands.

References

1. Zhang Z, Jianbing JI (2014) Research progress of feedstocks and deep-rocepssing technologies for biodiesel. Chem Ind Eng Prog 33(11):2909–2999
2. Sang OY (2003) Biofuel production from catalytic cracking of palm oil. Energy Sources 25(9):859–869
3. Yang W, Liu W, Shi L et al (2011) Current progress and development trend of woody oil-plant derived biodiesel production in China. Forest Resour Manage 4:16–19
4. Wiggers VR, Meier HF, Wisniewski A, Barros AC, Maciel MR (2009) Biofuels from continuous fast pyrolysis of soybean oil: a pilot plant study. Bioresour Technol 100:6570–6577
5. Li CZ, Li PW, Xiao ZH et al (2012) Current progress in research and development of woody biodiesel oil feedstock and its industrialization prospect in China. J South China Agric Univ 17(06):165–170
6. Xu J, Jiang J, Lu Y, Chen J (2009) Liquid hydrocarbon fuels obtained by the pyrolysis of soybean oils. Bio Tech 100:4867–4870
7. Dias JM, Alvim-Ferraz MCM, Almeida MF (2009) Production of biodiesel from acid waste lard. Bioresour Technol 100(24):6355–6361
8. Kirakosyan A, Cseke LJ, Kaufman PB (2009) The use of plant cell biotechnology for the production of phytochemicals. In: Recent advances in plant biotechnology, pp 15–33. https://doi.org/10.1007/978-1-4419-0194-1_2
9. Wiggers VR, Wisniewski A Jr, Madureira LAS, Chivanga Barros AA, Meier HF (2009) Biofuels from waste fish oil pyrolysis: continuous production in a pilot plant. Fuel 88:2135–2141
10. Ngo TA, Kim J, Kim SK, Kim SS (2010) Pyrolysis of soybean oil with H-ZSM5 (Proton-exchange of Zeolite Socony Mobil) and MCM41 (Mobil Composition of Matter No. 41) catalysts in a fixed-bed reactor. Energy 35(6):2723–2728
11. Xu JM, Jiang JC, Lu YJ (2009) Liquid hydrocarbon fuels obtained by the pyrolysis of soybean oils. Bioresour Technol 100:4867–4870
12. Jie C (2010) Investigation of preparing liquid fuels from fats by catalytic cracking. Chin Acad Forest, Beijing
13. Xu J, Jiang J, Zhang T et al (2013) Biofuel production from catalytic cracking of triglyceride materials followed by an esterification reaction in a scale-up reactor. Energy Fuels 27(1):255–261

14. Jun S et al (2000) Synthesis of new, nanoporous carbon with hexagonally ordered mesostructure. J Am Chem Soc 122(43):10712–10713
15. Jin L, Wang J, Zheng Y, Hu H (2014) Research progress of hydrogen production by catalytic decomposition of methane over carbon catalysts. Chem Ind Eng Prog 33(12):3125–3132
16. Busson C., Billaud F., Dominguez V et al (1995) Production of hydrocarbons by pyrolysis of methyl esters from rapeseed oil 72(10)
17. Schwab AW, Bagby MO, Freedman B (1987) Preparation and properties of diesel fuels from vegetable oils. Fuel 66(10):1372–1378
18. Freedman B, Pryde EH, Mounts TL (1984) Variables affecting the yields of fatty esters from transesterified vegetable oils 61(10):1638–1643
19. Shen JJ, Chi XY, Yang QL et al (2006) Review on research progress of biodiesel. China Biotechnol 26(11):87–90
20. Liu A, Li Y, Feng T et al (2008) Vegetable oil supply and demand and the development of China'S biodiesel industry. Resour Sci 30(4):560–564
21. Ji X, Wang X, Zhang X et al (2007) Discussion on questions in the development of China's bio-diesel industry. Energy China 29(3):22–25
22. Wang CW, Cui FF, Song Y (2014) Research situation and development prospect of biodiesel. China Oils Fats (5):44–48
23. Zhao Z (2005) Toward cheaper microbial oil for biodiesel oil. China Biotechnol 25(2):8–11
24. Leliaert F, Verbruggen H, Zechman FW (2011) Into the deep: new discoveries at the base of the green plant phylogeny. BioEssays 33(9):683–692
25. Keeling PJ (2011) Endosymbiosis: bacteria sharing the load. Comments Comment Curr Biol Current Biol 21(16):1366–72
26. van den Hoek C, Mann DG, Jahns HM (1995) Algae an introduction to phycology. Cambridge University Press
27. Schofield O, Moline MA, Millie DF (1997) The physics and physiology of photosynthesis in aquatic ecosystems. J Phycol 33(6):1085–1086
28. Paerl HW, Pinckney JL, Steppe TF (2000) Cyanobacterial-bacterial mat consortia: examining the functional unit of microbial survival and growth in extreme environments. Environ Microbiol 2(1):11–26
29. Mckenzie M (2011) The nineteenth-century ecological and cultural transformation of cape cod. Clearing Coastline 4(2):361–364
30. Gershwin LA, Zeidler W (2008) Cladonema timmsii, a new species of hydromedusa (Cnidaria, Hydrozoa) from a salt lake in South Australia. ZOOTAXA 1826:59–68
31. Guedes AC, Amaro HM, Malcata FX (2011) Microalgae as sources of high added-value compounds-a brief review of recent work. Bio-technol Prog 27(3):597–613
32. Del Campo A, Greiner C, Álvarez I et al (2007) Patterned surfaces with pillars with controlled 3D tip geometry mimicking bioattachment devices. Adv Mater 19(15):1973–1977
33. Sheehan C (1998) How to profit by avoiding climate change. NZ Eng 53(2):18–19
34. Scott KM, Fox G, Girguis PR (2011) Measuring isotope fractionation by autotrophic microorganisms and enzymes. Methods Enzymol 494:281–299
35. Clarens AF, Resurreccion EP, White MA et al (2010) Environmental life cycle comparison of algae to other bioenergy feedstocks. Environ Sci Technol 44(5):1813–1819
36. Han JX (2011). Study on catalyst and reaction mechanism for preparing diesel oil hydrocarbon by deoxidation of vegetable oil. Doctoral thesis. Zhejiang University

Chapter 4
Current Status of the Biodiesel Industry and Its Assessment of Feedstocks

Changzhu Li, Aihua Zhang, and Peiyi Yu

Abstract Sustainability is a key principle in natural resource management, and it involves operational efficiency, minimization of environmental impact, and socioeconomic considerations. All of which are interdependent. It has become increasingly obvious that continued reliance on fossil fuel energy resources is unsustainable, owing to both depleting world reserves and the greenhouse gas emissions associated with their use. Therefore, there are vigorous research initiatives aimed at developing alternative renewable and potentially carbon neutral solid, liquid, and gaseous biofuels as alternative energy resources. Biodiesel is gaining more and more importance as an attractive fuel due to the easier access to feedstock with the Government support. This chapter given an overview of biodiesels, introduce the manufacturing status both China and abroad, focusing on detailed information of biodiesel and byproduct glycerol on both the preparation principle and commercial process. It was found that, whereas there are outstanding issues related to reproducible and environment-friendly, biofuels could progressively substitute a significant proportion of the fossil fuels required to meet the growing energy demand.

Keywords Biofuels · Biodiesel · Preparation principle · Commercial process

C. Li (✉) · A. Zhang · P. Yu
National Joint Laboratory of Oil Energy Plants High Efficient Conversion, Changsha 410004, China
e-mail: lichangzhu2013@aliyun.com

Laboratory of State Forestry Administration on Utilization Science for Southern Woody Oil Resource, Changsha, China

Hunan Academy of Forestry, Changsha 410004, China

Hunan Engineering & Technology Research Center of Lipids, Changsha 410004, China

C. Li et al. (eds.), *Industrial Oil Plant*,
https://doi.org/10.1007/978-981-15-4920-5_4

4.1 Introduction to Biodiesel

Global petroleum resources are facing depletion due to unrestrained exploitation and utilization. Waste gas pollution caused by petroleum fuel also poses a huge challenge to the human beings. With the improving living standard and an increasing awareness of environmental protection, people begin to realize the severe nature of air pollution caused by petroleum fuel. In particular, frequent occurrence of photochemical smog and acid rain endangers our survival, while the greenhouse effect brought about by an excessive amount of CO_2 produced destabilizes the ecological equilibrium of nature. Therefore, international petroleum organizations believe that it is urgent to develop a new energy to replace petroleum fuel. Researches have shown that biodiesel is a promising alternative for diesel fuel.

The National Biodiesel Board defines biodiesel as a clean alternative fuel made from renewable biological resources, including vegetable and animal oils and fats, which can be used in compression-ignition engines. Chemically, biodiesel includes a series of long-chain fatty acids and methyl esters. Most natural oils and fats consist of a linear fatty acid, fatty acid triglycerides. After transesterification, their molecular weight is reduced almost to the level of diesel. Meanwhile, they are an environment-friendly alternative fuel with diesel-like properties.

Biodiesel has significant advantages over petroleum diesel: (1) it is environmental. Biodiesel, with exceptional biodegradability, has a low sulfur content and does not contain aromatic hydrocarbons, so that its combustion exhaust causes less harm than that of diesel. (2) It has great lubricity. Even with only 0.4% biodiesel in petroleum diesel, biodiesel shows outstanding performance in abrasion resistance. It reduces abrasion caused by low sulfur content in the emerging clean fuels, and improves the overall performance of vehicle diesel. (3) It is safe. With a higher flash point than petroleum diesel, biodiesel is not regarded as a hazardous fuel. It has significant advantages in terms of transport, storage, and utilization. (4) It has good combustion performance. With a higher cetane number, it has better combustion performance than diesel. The combustion residues are slightly acidic, which lengthens the life cycle of catalyst and engine oil. As a renewable energy, it has an inexhaustible supply. (6) A biodiesel system requires only small investment. The original engine, refueling equipment, and maintenance equipment needs little adaptation. (7) Biodiesel can be mixed with petroleum diesel in a certain ratio, which reduces oil consumption, improves dynamics, and cuts exhaust pollution.

Biodiesel researches began in the 1970s, first with the aim to address the energy crisis. Back then, global petroleum price was soaring due to the oil crisis. Some developed countries initiated the development of biomass energy to solve the energy crisis of their own. Most researches focused on the direct use of vegetable oils or animal fats as fuel. However, due to their large molecular weight and long carbon chain, the consequent fuel has a high viscosity and poor low-temperature performance, and tends to carbonize coke to clog up the nozzle. Moreover, vegetable oils cannot be sufficiently oxidized, making it an unrealistic alternative to diesel.

With the pressing energy crisis and environmental problems in recent years, people are urged to search for clean, renewable, and environment-friendly energies. With similar combustion and dynamic characteristics compared to mineral diesel, biodiesel stands out for its environment-friendly nature. It is drawing attention from environmentalists and energy professionals, who are promoting the research and development of biodiesel to a new level.

(1) Status quo of biodiesel development in China

The R&D of biodiesel in China started later than many developed countries. It first began the biodiesel experiments with rapeseed oil and *Swida wilsoniana* fruit oil in 1981. In recent years, research institutes and universities also joined this effort. The biodiesel production processes using rapeseed oil, soybean oil, and waste frying oil have been developed, respectively. The Chinese government has formulated a series of preferential policies and measures to support the R&D of biodiesel, which boosts the rapid development of the biodiesel industry.

It has been reported that the raw material of biodiesel in China mainly comes from waste oils and fats, and oil plant seeds. More specifically, waste oils and fats include: (1) waste frying oil produced from cooking; (2) animal oils and fats produced from baking; (3) animal oils and fats made from waste and scrap of animal products produced during their process at normal temperature; (4) waste cooking oil consisting mainly of grease separated from leftovers; (5) "gutter oil", or waste oils discharged into the sewage system during dishwashing; (6) condensate oil of the kitchen ventilation system, including household range hood condensate grease or that of restaurants; (7) acidified oil residue.

As to the biodiesel industry in China, its large-scale production and application is just getting started. It is not mature yet to use biodiesel as the primary fuel. At present, manufacturers in China use their own industrial standards. China has placed biodiesel development in its national energy development plan with a focus on its long-term use. In order to strengthen the production and management of biodiesel, it is essential to formulate relevant national standards.

In China, biodiesel production is characterized by small-scale and insufficient follow-up development. Since China has not adopted a tax exemption policy on biodiesel, its production cost remains high (75% of which is the raw material cost)–about three times than that of mineral diesel, making large-scale production an unrealistic option. Therefore, most research institutes and enterprises are aiming at developing biodiesel production technologies that use inexpensive raw material. Have all developed relevant technologies with independent intellectual property rights. Beijing Sustainable Development Center is working with the China University of Petroleum to transform waste cooking oil in Beijing restaurants into biodiesel. Jiangxi Jubang Chemicals Co., Ltd. uses genetically modified soybean imported from the US and Chinese rapeseed oil to produce biodiesel. Its production facilities with a capacity of 10×10^4 tons/year. School of Life Science, Sichuan University is attempting to establish production facilities with a capacity of 2×10^4 tons/year using Jatropha Curcas oil as raw material.

(2) Status quo of international biodiesel development

The US and some European and Asian countries or regions have established commercial production bases of biodiesel, and have adopted biodiesel as a widely used alternative fuel [1, 2].

Europe takes the lead with the widest application of biodiesel. Its major raw material is rapeseed. European Parliament exempted 90% of the biodiesel tax. Many European countries enacted laws to support alternative fuels. Preferential tax and subsidy for rapeseed production also contribute to the fast-growing biodiesel industry. In 2003, the total biodiesel output of European countries exceeded 1.76 million tons. It is estimated that the number will reach 8.3 million tons in 2010. Biodiesel is the most widely used in Germany, with 1.1 million tons of biodiesel produced and consumed. This represents 1/3 of the world's total 3.5 million tons.

The US is the first country to have researched into biodiesel. There have been various biodiesel manufacturers and suppliers in the US. They mainly use soybean oil as raw material with an annual biodiesel output exceeding 300 thousand tons. Additionally, the US imposes zero tax on biodiesel.

The US is also exploring other ways to produce biodiesel. The National Renewable Energy Laboratory has engineered microalgae using modern biological technology. In lab conditions, its fat content can reach 40–60%. For every acre, the expected output of biodiesel produced from these microalgae is 6400–16,000 L. This paves a new path for the development of biodiesel. By April 2005, there were 60 biodiesel production factories (including those under construction). Its plan is to produce 1.15 million tons of biodiesel in 2011 and 3.3 million tons in 2016.

Pure biodiesel has been officially named in the Energy Policy Act as an alternative vehicle fuel [3]. The pure biodiesel prices in the US vary according to the raw materials and manufacturers, ranging from 0.515 to 0.793 USD/L; mixed biodiesel (80% biodiesel) is 7.93–10.57 cents more expensive than traditional diesel.

Japan mainly uses frying oil to produce biodiesel, with an annual output capacity of 400 thousand tons. Brazil uses castor oil as the main raw material to produce biodiesel, which is in the promotion and experimental stage.

Many countries have invested heavily in biodiesel development and biodiesel industry [2–5]. Significant progress has been made, as presented in Table 4.1.

4.2 Prospects of Biodiesel Application

Energy is the cornerstone of the national economy and foundation for human survival. Energy safety, including the safety of energy supply (petroleum, natural gas, and electricity) and treatment of environmental pollution caused by energy production and utilization, is an important aspect of national economic safety. It has a direct impact on national security, sustainable development, and social stability [6]. With the growing economy, China's energy demand relies more on the international market. Today, in the context of the global energy safety issue, China's energy safety problem

Table 4.1 Global biodiesel producers and their achievements

Country	Biodiesel yield (10 thousand tons)	Biodiesel percentage (%)	Raw material	Status quo
US	30	10–20	Soybean	Promoted and used
Germany	55	5–20,100	Rapeseed, soybean	Widely used
France	40	5–30	Various vegetable oils	Researched and promoted
Italy	33	20–100	Various vegetable oils	Widely used
Austria	5.5	100	Rapeseed, waste oils and fats	Widely used
Japan	40	5–20	Waste oils and fats	Widely used
Bulgaria	0.03	100	Sunflower, soybean	Promoted and used
Brazil			Castor oil	Experimented
Australia		100	Animal fats	Researched and promoted
Sweden		2–100	Various vegetable oils	Widely used
Belgium	24	5–20	Various vegetable oils	Widely used
Argentina		20	Soybean	Promoted and used
Canada		2–100	Tung oil, animal fats	Promoted and used
South Korea		5–20	Rice bran, recycled vegetable oil	Promoted and used
Malaysia			Palm oil	Researched and promoted

is becoming more challenging as well. Since China has a limited amount of crude oil resources [7], it will rely on petroleum imports in the long-term. Therefore, given China's own resources, it is of strategic importance to develop an alternative liquid fuel–biodiesel.

(1) Biodiesel produced from woody oil plants is a renewable oil feedstock

The manufacture, processing, and consumption of biodiesel produced from woody oil plants is an organic closed-circuit process. The woody raw materials of this biodiesel can transform solar energy into storable biological energy through photosynthesis [8]. The biological energy is made into biodiesel to be consumed by the humans.

The carbon in the biodiesel returns to the atmosphere in the form of carbon dioxide for future photosynthesis. Therefore, the production, manufacture, and consumption of biodiesel is a sustainable process. Biodiesel as a renewable fuel can address the energy crisis due to the depletion of some petrochemical energy resources. It ensures energy safety by providing an inexhaustible energy source.

(2) Biodiesel produced from woody oil plants is an important part of the energy strategy

As the world is threatened by the depleting petrochemical energy resources, the increasingly polluted environment, and global warming, people are shifting their attention again to the inexhaustible biological energy [9, 10]. The whole world is paying close attention to the development and utilization of renewable energy. In fact, many countries have made the development of renewable energy an integral part of their energy strategy. Biodiesel can be a strategic resource reserve. More specifically: ① to ensure fuel supply, which means to ensure the emergency supply of fuel in a certain period of time; ② to stabilize energy price, since a huge strategic fuel reserve is able to offset the power of the market. Biodiesel is a biomass, an alternative option to petroleum diesel and an emerging energy widely applicable in household life, production, and the military. In this sense, biodiesel represents well-structured energy strategies and makes a nation energy-independent. With biodiesel, countries can avoid intervention and manipulation of other parties, and reduce their reliance on the petroleum market. The development of biodiesel can help address a series of conflicts caused by petroleum, serving as a catalyst for international peace.

(3) Biodiesel produced from woody oil plants is a safe energy

Biodiesel produced from woody oil plants is a safe energy. Biodiesel has a high flash point. It is degradable. It is not radioactive. It is not susceptible to explosion, leakage or other accidents. It is safe in terms of production, transport, and utilization. Moreover, biodiesel is an environment-friendly energy that poses no threat to the human body or the environment.

China should take into consideration its own conditions in the development of biodiesel. It should take full advantage of peripheral, low-quality land to cultivate energy plants. It should utilize available animal and vegetable oil plants, and optimize the production process to prevent secondary pollution on the environment. It should also reduce energy consumption and streamline the production equipment to the fullest extent. The development of woody plants for biodiesel may follow the advice presented below.

① Formulate a general plan for the development of oil trees for energy, and for the development and utilization of biodiesel [11]. The cultivation of oil trees for energy is characterized by small investment, fast return, good reproducibility, and multiple harvests. It is a promising industry that benefits the country and its people and enterprises. As a saying goes, "small fruits make a big sustainable industry". Building a vast green oil forest is beneficial for resource industrialization, technology development, economic utilization, and market system building. A scientific approach should be adopted to develop and use resources. We propose that a supervising department should organize researches in different areas on the evaluation of oil trees to study their development potential, in order to provide a solid scientific basis for resource development and utilization. Based on resource distribution and development status, the government may determine the approach and goal of biodiesel development, and establish the layout and plan for biodiesel manufacturing and processing.

② Establish and perfect an industrialized production system and a market development system of biodiesel [12]. Researches should be conducted on the establishment of an industrialized system and a market-oriented operational mechanism that benefits biodiesel development. Such researches may involve selection and breeding of oil trees, cultivation management, operation and harvesting, transport and processing, as well as measures to standardize the production, commercialize the product, and marketize the operation. The goal is to incentivize enterprises to develop and produce biodiesel, mobilize local entities to cultivate oil trees on a large-scale, and attract wider investments in building cultivation bases [13, 14]. Ultimately, a mutually reinforcing and interdependent landscape will form to facilitate the sustainable development of biodiesel. China has implemented a series of preferential and subsidy policies. Combined with the Clean Development Mechanism (CDM) that aims to cut greenhouse gas emission during biodiesel production, they will become crucial factors for the development of the biodiesel industry and the reduction of processing cost.

③ Conduct further researches on the selection and breeding of oil trees and on the establishment of efficient energy tree bases. There are many types of oil trees for energy in China with contrasting qualities [15]. Additionally, the edible oil output in China is lower than the world average, while its price is higher than that of fuel oil. This discourages people from developing or cultivating trees for energy use. Therefore, it is not appropriate to produce biodiesel using plants for edible oil. A desirable trend is to utilize plants for nonedible oil. Therefore, it is imperative to further research the breeding of trees and the establishment of oil tree bases. In particular, it is necessary to develop and utilize potential plants for nonedible oil with high reproducibility, short growing seasons, large biomass, high productivity, and good environmental adaptation. All regular breeding approaches and biological technologies may be used to improve the breeding quality of available oil trees, so as to cultivate and select a number of high-quality species. Alternatively, plants for edible oil may be modified to become the main source for biodiesel production. The first major approach is to establish national breeding centers for oil trees in both southern and northern China, in order to accelerate the breeding and selection process of oil trees for energy. The second is to build bases for oil trees in suitable areas to ensure resource supply for industrialized development of biodiesel. To enhance the selection and breeding of oil trees for biodiesel production based on local conditions, afforestation should be combined with ecological development.

Biodiesel is an emerging hi-tech industry. China has stated its plan to develop various alternatives to petroleum, and recognized the development of liquid biofuel as a general direction of the industry. In this sense, biodiesel has a promising future (Table 4.2).

Table 4.2 General information on forest production capacity

Name of woody oil	Geographical distribution	Planting area	Average output ton/ha/year	Main area	Available land
Swida wilsoniana	Hunan, Hubei, Jiangxi, Guizhou, Sichuan, Guangdong, Guangxi	About 10,000 ha	4.5–9.0	In southern provinces only	At least 2,000,000 ha
Jatropha curcas	Guangdong, Guangxi, Yunnan, Sichuan, Guizhou, Fujian, Taiwan, Hainan	Over 16,000 ha in Sichuan	9.75	In tropical and subtropical zones mainly	At least 2,000,000 ha
Pistacia chinensis	Hebei, Henan, Anhui, Shaanxi	About 66,700 ha	7.5	In Taihang Mountains, central and southern Hebei and northern Henan mainly	At least 300,000 ha
Xanthoceras sorbifolia Bunge	Shaanxi, Shanxi, Hubei, Inner Mongolia, Ningxia, Gansu, Henan	About 25,000 ha	45 (biodiesel production only)	In northwestern and northern China	At least 4,000,000 ha

4.3 Feedstocks of Biodiesel Production

Biodiesel has become a hotspot of international new energy development because it is clean, safe, widely applicable, and degradable. Governments have introduced policies encouraging citizens and government departments to promote and use biodiesel. In reality, this process is pretty slow. A lack of support from the people and the weaknesses of biodiesel obstruct its promotion and utilization. First and foremost, the high cost of biodiesel restricts its utilization. Secondly, continuous large-scale production of biodiesel requires a stable raw material supply. The raw materials should meet the standards of both quantity and quality. Therefore, its raw material has become a crucial restrictive factor on its path to commercialization.

Since the outbreak of the Gulf War in 1990, the US began to focus on researches into composite raw material and process. In 1999, the US Department of Energy organized a research project on oil palm, algae, and some tropical plants with the participation of scientists from France, Netherlands, Germany, Austria, and Malaysia [16–18]. In 1994, the US Department of Energy commissioned research institutes, including the University of Kansas, to study the applicability of multi-feedstock biodiesel (Multi-Feedstock Biodiesel Project—Phase II, Mr. Steve A. Howell). The US has experimented with 40 vegetable oils on their short-term performance in an internal combustion engine. These include soybean oil, peanut oil, cottonseed oil, sunflower seed oil, rapeseed oil, and palm oil. Some researchers on energy plants have predicted that by 2050, 80% of the global liquid oil will originate from woody plants, herbal oil plants, and algae.

Today, the raw materials for large-scale biodiesel production include soybean (US), rapeseed (European Communities members), palm oil (Southeast Asian countries), and vegetable oil residues and recycled edible oil (Japan and Ireland).

China is home to a wide variety of oil plants for biodiesel production. A total of 138 families and 1159 species (including varieties, and with more than 10% available oils and fats content) can be used as biodiesel raw material. As the researches deepen and technologies improve, more plants will be placed on the biodiesel production list. China has abundant energy plant resources, which promises a bright future for biodiesel production. High-quality raw material of vegetable oil is the basis for high-quality biodiesel and high production yield. Therefore, a major challenge before us is to select appropriate species from all these types as raw material for cost-effective quality biodiesel production in China. With long-term experiment and research, combined with field study and market research, we have accumulated basic data for vegetable oil raw material selection. They can serve as evaluation indicators for relevant technologies and as reference for China's research, development, and utilization of biodiesel.

4.3.1 Research Methodology and Techniques

4.3.1.1 Data Source and Analysis

(1) Data source: research data or data collected by researchers during the eighth, ninth, and tenth five-year Plans; documents and data published by international economic organizations.

(2) Data processing and general analysis.

Based on the requirements for oil plant selection, the Delphi survey, cluster analysis, as well as quantitative and qualitative analysis can be used to determine the indicators and weight of raw material selection. Accordingly, resource classification and evaluation can be conducted to select the desired species.

(3) A comparison of oils and fats content of woody plants and herbal crops is presented in Table 4.1. By referring to the goal of developing and studying biodiesel in China, such data can be used to determine the appropriate raw materials for biodiesel production.

(4) Content indicators of the research object.

In this section, nine woody plants and three herbal plants produced or introduced in China are to be evaluated for researches on oil plants selection technology. The oils and fats composition and oil content of the research objects are presented in the Table 4.3.

Euphorbia tirucalli is an energy plant introduced in the US and South Africa. This is a species of genus Euphorbia, Euphorbiaceae, and its whole plant can be utilized. Its stalk and branch have 17.5% of latex, which contains 30–50% of olefin, terpenoid, alcohol, $C_{30}H_{50}O$, natural rubber, hentriacontane, and hentriacontanol. Its composition resembles that of petroleum (sulfur-free), which can be mixed with other substances to form crude oil. It is easy to reproduce. It can be propagated by cutting in subtropical and tropical areas, with a cutting cycle of about one to two years. In the fourth year, the average latex crude oil output per hectare is 1400 kg. One propagation promises harvest for 20 years (new branches grow after cutting, with a renewal cycle of 20 years).

4.3.1.2 Selection Principles and Evaluation Indicators

The selection of raw material is based on the following criteria: oil content, oil composition, oils and fats content and composition, oil-producing organs, propagation methods, propagation difficulty, people's planting habit, yield per unit area or fruit characteristics, suitable planting area, natural conditions in the suitable planting area, size of the plant, plant life form, plant ecotype, population structure of the plant, biomass ratio of canopy/plant or of aerial part/plant (herbal plants). By conducting the Delphi survey, expert consultation and fuzzy evaluation, five key indicators have been determined for evaluation.

(1) Oil content

Oil content is the primary indicator for the selection of biodiesel raw material. Oil content determines the value of the energy plant. Insufficient oil content obstructs the development and utilization of biodiesel raw material.

(2) Economic yield

Economic yield means the yield of fruits and seeds (latex for *Euphorbia tirucalli*). The economic yield of energy plant determines its development value and potential. The larger the yield, the better the value and potential.

(3) Suitable planting area

China with its vast territory features various landscapes from basin to the mountain range. From south to north, it spans across tropical, subtropical, warm temperate, temperate, and cold temperate zones. From east to west, it has moist, marine forest land, semiarid and semi-humid grassland, desert-steppe ecotone, continental semi-arid desert, and desert regions. The terrain stretches from the high-lying northwest to the low-lying southeast. If a given energy plant has a larger suitable planting area,

Table 4.3 Fatty acid composition and oil content of woody plants and herbal crops

		Fatty acid content (%)										Oil content (%)
		Lauric acid	Myristic acid	Palmitic acid	Stearic acid	Arachidic acid	Hexandcenoic acid	Oleic acid	Linoleic acid	Linolenic acid	Others	
Woody plants	*Camellia oleifera* Abel.I*	0.000	0.800	10.600	1.700	0.000	0.000	77.333	9.167	0.267		36.60
	*Pistacia chinensis**	0.000	0.013	20.867	1.500	0.567	1.200	46.400	29.367	0.007		34.38
	Jatropha curcas L.	0.000	0.000	18.2	1.8	0.000	0.000	47.3	32.7	0.000		52.25
	Cornus wilsoniana Wanger*	0.007	0.067	16.533	1.767	0.000	0.973	30.500	48.500	1.600		36.60
	Elaeis guineensis	19.500	18.600	27.000	12.300	0.000	0.000	22.600	0.000	0.000		67.95
	Eucalyptus	0.000	0.000	5.500	2.400	0.700	0.000	12.800	78.500	0.000		18.78
	Symplocos paniculata	0.000	0.000	20.733	0.873	0.000	0.000	48.233	30.167	0.000		32.41
	*Vernicid fordii**	0.167	0.007	5.733	2.567	0.000	0.007	16.400	22.067	0.300	Eleostearic acid 52	50.18
Herbal crops (control)	Rapeseed	0.000	0.040	3.567	1.133	0.007	0.140	14.500	15.467	13.600	Erucic acid 48	35.4
	Glycine max L.	0.000	0.000	13.000	2.900	0.000	0.000	19.350	58.050	6.700		22.1
	Euphorbia lathyris L.	0.010	0.000	5.800	1.900	0.300	1.100	70.250	16.200	2.000		47.05

it tends to grow in a wider region with potentially more utilization options and larger biomass for harvest.

(4) Propagation difficulty

To build large-scale biodiesel raw material bases, propagation difficulty is an important criterion. The application of an energy plant may be limited, if it has a low germination percentage despite its high oil content, if it does not survive easily by sexual or asexual reproduction, or if it is not feasible to establish large-scale biodiesel raw material bases for this plant.

(5) Oil composition

The various oil constituents of energy plants can be used to produce biodiesel through transesterification. The ideal vegetable oil composition should resemble that of petroleum fuel, or it can be easily esterified into monoesters or processed into hydrocarbons.

4.3.1.3 Expert Evaluation Method and Weight of Indicators

As the first step, an evaluation plan and an evaluation form is devised. Expertise and experience are utilized in combining quantitative and qualitative evaluations. Ten experts from the Hunan Academy of Forestry, Central South Forestry College, and Xiangtan University are invited to the evaluation process. These experts specialize in the economic forest, forestry, horticulture, chemical processing of forest products, food processing, and economic management. Given that the projects to be evaluated are multi-layered, they shall be divided into major projects and minor projects. As to the calculations, the scores in each minor project are added up, the results of which are then multiplied. These final scores serve as the basis for evaluation.

$$S = \prod_{i=1}^{m} \sum_{j=0}^{n} S_{\ddot{U}}$$

In the equation S_{ij}—Project j in Project i.
M—number of major projects n—number of minor projects.
I—$1 - m$; j—$1 - n$

4.3.1.4 Determination and Preselection of Indicators for Weight Calculation

Based on the data collected, preliminary indicators are designed to measure the effects of various factors on biodiesel production. These indicators are evaluated and scored by the ten experts. The weight of the five indicators, including plant oil content (A), yield per unit area or fruit characteristics (B), suitable planting area (C), level of propagation difficulty (D), and oil composition (E) is presented in Table 4.4.

Table 4.4 Weight of five indicators

Evaluation indicators	Weight of indicators
Raw material oil content (A)	0.237
Economic yield (B)	0.275
Suitable planting area (C)	0.163
Propagation difficulty (D)	0.187
Oil composition (E)	0.138

Note the weight factors above are determined by Delphi surveys and fuzzy evaluations

4.3.1.5 Results and Analysis

Overall Scores and Ranking of Preliminary Selection of Biodiesel Raw Material
After statistics processing, the scores and ranking of the ten biodiesel raw material plants are presented in Table 4.5.

Evaluation of General Functions and Secondary Selection
In terms of national development, biodiesel development is a massive systematic project. Its contributing factors are so complicated that they require strict, well-designed economic evaluation. To determine or optimize the technical projects, we cannot be satiated with the evaluation of a single technical indicator or economic indicator. Instead, multiple contributing factors should be analyzed and examined. That is, a scientific examination method and a quantity standard should be established to provide objective, accurate, and reliable evaluation.

Table 4.5 Scores and ranking of preliminary selection of biodiesel raw material

Raw material type	A	B	C	D	E	Overall score	Ranking
Euphorbia lathyris L.	1.067	1.375	1.304	1.496	0.828	6.070	1
Jatropha curcas L.	1.185	1.650	0.897	1.496	0.828	6.056	2
Cornus wilsoniana Wanger	0.830	1.788	1.275	1.029	0.828	5.750	3
Vernicid fordii	1.138	1.650	1.141	1.122	0.690	5.741	4
Elaeis guineensis	1.541	1.815	0.652	1.029	0.690	5.727	5
Euphorbia tirucalli	0.402	1.513	0.652	1.496	1.104	5.167	6
Camellia oleifera Abel.	0.830	1.238	0.978	1.122	0.828	4.996	7
Pistacia chinensis Bunge	0.782	1.100	1.141	0.935	0.690	4.678	8
Symplocos paniculata	0.735	1.045	0.978	1.122	0.690	4.570	9
Eucalyptus	0.426	1.100	0.652	1.029	0.966	4.173	10

As to the preselected biodiesel raw material plants (woody or herbal), their standing does not necessarily determine their value. Their production cost, adaptability to people's cultivation habits and ecological system, and social functions also require further analysis. Therefore, it is necessary to build on the preliminary selection to evaluate the raw materials from the perspectives of economic cost, ecological impact, and social benefits.

(1) Evaluation of economic cost (X_1)

The pretax production cost of biodiesel per liter by energy plants is used as the baseline for comparative analysis. The cost analysis of oil plants refers to the model in "Current Research Progress of Biodiesel Development and Its Commercial Application" (Li et al. 2002). The lower the cost, the higher X_1 scores.

$X_1 = 1$/pretax production cost of biodiesel per liter.

(2) Ecological functions (X_2)

The ecological functions of energy plants include soil and water conservation, environmental protection, and landscape making. Based on these categories, scores are obtained through expert consultations, and an average value (I) is calculated, which is multiplied by weight coefficient C ($C = 0.1$) to get the final X_2.

$$\mathrm{I} = \begin{cases} 1, \text{very weak;} \\ 2, \text{weak;} \\ 3, \text{medium;} \\ 4, \text{strong;} \\ 5, \text{very strong;} \end{cases}$$

$X_2 = C \times I.$

(3) Social functions (X_3)

The scoring of social functions is similar to that of ecological functions.

$$\mathrm{I}' = \begin{cases} 1, \text{very weak;} \\ 2, \text{weak;} \\ 3, \text{medium;} \\ 4, \text{strong;} \\ 5, \text{very strong;} \end{cases}$$

$X_3 = C' \times I', (C' = 0.1)$

(4) Overall score (Y)

$Y = X_1 + X_2 + X_3$

The overall scores are presented in Table 4.6.

Table 4.6 Overall evaluation and ranking of biodiesel raw material

Raw material type	X_1	X_2	X_3	Overall score	Ranking
Jatropha curcas L.	0.435	0.342	0.308	1.085	1
Euphorbia tirucalli	0.323	0.367	0.383	1.073	2
Elaeis guineensis	0.372	0.333	0.350	1.055	3
Euphorbia lathyris L.	0.485	0.308	0.258	1.051	4
Cornus wilsoniana Wanger	0.331	0.333	0.383	1.047	5
Brassica campestris L. (CK)	0.302	0.308	0.425	1.035	6
Vernicid fordii	0.249	0.342	0.442	1.033	7
Glycine max L. (CK)	0.214	0.379	0.438	1.031	8
Camellia oleifera Abel.	0.351	0.325	0.325	1.001	9
Pistacia chinensis Bunge	0.325	0.367	0.300	0.992	10
Eucalyptus	0.175	0.317	0.433	0.925	11
Symplocos paniculata	0.182	0.342	0.292	0.816	13

4.4 Final Selection

The results in Tables 4.1, 4.2, 4.3, and 4.4 indicate possible raw material options for China's biodiesel production. Considering China's biodiesel development goals, the 10 highest-scoring plants—*Jatropha curcas* L., *Euphorbia tirucalli*, *Elaeis guineensis*, *Euphorbia lathyris* L., *Swida wilsoniana* Wanger, *Brassica campestris* L., *Vernicid fordii*, *Glycine max* L., *Camellia oleifera* Abel., and *Pistacia chinensis* Bunge—can be regarded as suitable raw materials for vegetable oil.

Firstly, China has a limited per capita arable land area. If edible herbal crops, including rape plant and soybean, are utilized as the sole raw materials for biodiesel, it may result in insufficient edible vegetable oil production. On the other hand, China owns large hilly areas with abundant mountain resources. Statistics show that mountains, plateaus, and hills take up about 69% of China's total land area. Due to thin soil, low fertility, and inadequate water, herbal oil crops do not grow easily in mountainous areas. However, woody oil plants can grow in a dry, barren land, and one-time cultivation brings multiple harvests with low average raw material oil cost. Based on China's policy of "Returning Farmland to Forests", large-scale cultivation of biodiesel plants can turn the weakness of barren hills into strength. And in this sense, woody plants can make a big difference.

Secondly, compared to herbal crops, certain wooden plants have unique advantages. For example, the whole plant can be used (seeds, fruits, branches, and leaves can all be used for oil production), and bioactive substances can be extracted from their fruits, seeds, and even residues after deep processing for medical and chemical engineering use.

Thirdly, woody plants including *Jatropha curcas* L., *Euphorbia tirucalli*, *Swida wilsoniana* Wanger, *Vernicid fordii*, *Camellia oleifera* Abel, and *Pistacia chinensis*

Bunge has a high economic yield per unit area or large biomass. It is economically feasible to use such woody plants as biodiesel raw material.

Under the current economic and resource conditions, the development of biodiesel raw material forests should make full use of China's mountain resources by cultivating woody oil plants on a large-scale. This is the path China may have to choose in pursuit of biodiesel development.

4.5 Conclusion

Some experts from the industry have pointed out that as the amount of processed biodiesel continues to grow in China, it may have gasoline and kerosene surplus for export by 2005, while the demand for diesel will remain huge. By 2005, it is expected that the diesel production gap may reach 600 to 2400 kt. By 2010, the expected demand for diesel may exceed 100 Mt, a 24% increase compared to 2005; by 2015, the expected market demand may reach 130 Mt.

Therefore, research, development, and production of biodiesel are of strategic importance to energy supply and adjustment of energy structure in China.

The cost of biodiesel is an important factor in choosing appropriate raw materials. As to the selection and utilization of biodiesel raw material plants, comprehensive area-specific evaluations should be conducted, taking into consideration oil content, oils and fats composition, and conversion rate in the transesterification reaction. To construct a framework for the sustainable development of biodiesel in China, a two-step process should be taken based on the analysis below

(1) Short-term goal

Given that the oversupply of agricultural products and surplus labor force will continue to exist for some time in China, it may consider growing annual oil crops including rape plant and soybean in remote rural areas. The harvested crops can be used as biodiesel raw material. On the one hand, this will help sell excessive agricultural products. On the other hand, this will draw attention to promote the development of key technologies for biodiesel development, in order to establish an R&D and marketing system for biodiesel.

(2) Medium- and long-term goal

In the meantime, China should take the initiative to establish large-scale woody plants bases in different areas for biodiesel raw materials. This is because the use of annual plants will push up the cost of biodiesel raw material and affect negatively the industrial development. Additionally, it is unrealistic to scale down the use of edible oil plants. For subtropical areas in South Asia, plants that can be harvested in one to three years, such as *Euphorbia tirucalli* and *Jatropha curcas* L., are preferred. In the Yangtze River region, plants that can be harvested in five to seven years, including *Cornus wilsoniana* Wanger, *Vernicid fordii*, *Camellia oleifera* Abel, and *Pistacia chinensis* Bunge are a better choice.

References

1. Abdoli MA, Mohamadi F, Ghobadian B, Fayyazi E (2014) Effective parameters on biodiesel production from feather fat oil as a cost-effective feedstock. Int. J. Environ. Res. 8(1):139–148
2. Ackley B (2014) Research brief: applying green chemistry principles towards the sustainable synthesis of biodiesel from waste vegetable oil. Undergrad Rev
3. Alagu RM, Sundaram EG (2010) Notice of violation of IEEE publication principles nitrogen oxide emission in biodiesel fuelled ci engines: a review
4. Apostolakou AA, Kookos IK, Marazioti C, Angelopoulos KC (2009) Techno-economic analysis of a biodiesel production process from vegetable oils. Fuel Process Technol 90(7):1023–1031
5. Azam MM, Waris A, Nahar NM (2005) Prospects and potential of fatty acid methyl esters of some non-traditional seed oils for use as biodiesel in India. Biomass Bioenerg 29(4):293–302
6. Azócar L, Ciudad G, Heipieper HJ, Navia R (2010) Biotechnological processes for biodiesel production using alternative oils. Appl Microbiol Biotechnol 88(3):621–636
7. Barnwal BK, Sharma MP (2005) Prospects of biodiesel production from vegetable oils in India. Renew Sustain Energy Rev 9(4):363–378
8. Bendall S, Birdsallwilson M, Jenkins R, Chew YMJ, Chuck CJ (2015) Showcasing chemical engineering principles through the production of biodiesel from spent coffee grounds. J Chem Educ 92(4):683–687
9. Berchmans HJ, Hirata S (2008) Biodiesel production from crude *Jatropha curcas* L. seed oil with a high content of free fatty acids. Biores Technol 99(6):1716–1721
10. Bloomer S (2006) Biodiesel process. Int News Fats Oils
11. Boehman AL (2005) Biodiesel production and processing. Fuel Process Technol 86(10):1097–1107
12. Brásio ASR, Romanenko A, Fernandes NCP, Santos LO (2016) First principle modeling and predictive control of a continuous biodiesel plant. J Process Control 47:11–21
13. Cakmak T, Angun P, Demiray YE, Ozkan AD, Elibol Z, Tekinay T (2012) Differential effects of nitrogen and sulfur deprivation on growth and biodiesel feedstock production of *Chlamydomonas reinhardtii*. Biotechnol Bioeng 109(6):1947–1957
14. Canakci M, Monyem A, Gerpen JV (1999) Accelerated oxidation processes in biodiesel. Trans ASAE 42(6):687–706
15. Carmo FRD, Jr Sousa PM, Santiago-Aguiar RS, Sant'Ana HBD (2012) Development of a new model for biodiesel viscosity prediction based on the principle of corresponding state. Fuel 92(1): 250–257
16. Chew TL, Bhatia S (2008) Catalytic processes towards the production of biofuels in a palm oil and oil palm biomass-based biorefinery. Biores Technol 99(17):7911–7922
17. Damartzis T, Zabaniotou A (2011) Thermochemical conversion of biomass to second generation biofuels through integrated process design: a review. Renew Sustain Energy Rev 15(1):366–378
18. Damiani MC, Popovich CA, Constenla D, Leonardi PI (2010) Lipid analysis in *Haematococcus pluvialis* to assess its potential use as a biodiesel feedstock. Bioresour Technol 101(11):3801–3807

Chapter 5
Principle and Processing of Biodiesel Production

Changzhu Li, Aihua Zhang, and Rukuan Liu

Abstract At present, a major industrial technology for producing biodiesel is transesterification. The large-scale production of biodiesel mainly relies on soybean oil, rapeseed oil, palm oil, and waste cooking oil as raw materials. Transesterification takes place between these raw material oils and methanol with the help of catalytic agents, which produced biodiesel. This chapter elaborates on the principle and process of biodiesel production by transesterification. The principle and processes were emphasized in this chapter.

Keywords Biodiesel · Transesterification · Biodiesel production · Production process · Production principle

5.1 Introduction

Today, the production process of biodiesel is developing with three distinctive features—large production equipment; continuous, automatic, digitalized, and standardized production process; and "green" treatment of products and "Three Wastes".

The production process has a direct impact on the output and quality of biodiesel. Meanwhile, the output and quality of biodiesel produced from raw material oil per unit have a direct impact on the production cost of biodiesel. Researches have indicated that different raw materials have different production technologies. Therefore, countries all over the world are improving their biodiesel production process.

C. Li (✉) · A. Zhang · R. Liu
State Key Laboratory of Utilization of Woody Oil Resource, Changsha 410004, China
e-mail: lichangzhu2013@aliyun.com

Key Laboratory of State Forestry Administration on Utilization Science for Southern Woody Oil Resource, Changsha 410004, China

Hunan Academy of Forestry, Provincial Branch of Chinese Academy of Forestry, Changsha 410004, China

National Joint Laboratory of Oil Energy Plants High Efficient Conversion, Changsha 410004, China

C. Li et al. (eds.), *Industrial Oil Plant*,
https://doi.org/10.1007/978-981-15-4920-5_5

The researchers developed the VN process to produce high quality biodiesel and purify glycerol, the by-product of the reactions, for medical use [1–3]. This process enhanced the practical value of raw materials and reduced cost. Other studies discovered that branched fatty acid esters have better low-temperature performance than straight fatty acid esters [4–7]. Therefore, products using biodiesel produced from branched alcohols have significantly better low-temperature performance. The scientists managed to use saponite as raw material to produce high-performance biodiesel through enzymatic reaction [8–11]. A two-stage reaction device to achieve a high output, as well as low energy consumption in producing biodiesel by transesterification at a low-temperature [12]. In addition, Denoo (Germany) developed a two-stage catalytic technology that transforms waste oils and fats emitted from the catering industry, food industry, and paper industry into biodiesel [13].

At present, a major industrial technology for producing biodiesel is transesterification. The large-scale production of biodiesel mainly relies on soybean oil [14], rapeseed oil [15], palm oil [16], and waste cooking oil [17, 18] as raw materials. Transesterification takes place between these raw material oils and methanol with the help of catalytic agents, which produced biodiesel. This chapter elaborates on the principle and process of biodiesel production by transesterification.

5.1.1 Principle and Reaction Mechanism of Transesterification

5.1.1.1 Introduction to the Principle of Transesterification

Oils and fats transesterification is the process of ester groups interchange or intermolecular rearrangement as triglyceride in oils and fats reacts with fatty acids, alcohols, itself or other esters. It is a process technology to change the characteristics of oils and fats without chemically altering the composition of fatty acids.

Catalyzed by acids, bases or fatty acid syntheses, the following reactions can take place between oils and fats and methanol:

$$
\begin{array}{l} CH_2OCOR_1 \\ | \\ CHOCOR_2 \\ | \\ CH_2OCOR_3 \end{array} + 3(CH_3OH) \longrightarrow \begin{array}{l} CH_2OH \\ | \\ CHOH \\ | \\ CH_2OH \end{array} + \begin{array}{l} R_1COOCH_3 \\ R_2COOCH_3 \\ R_3COOCH_3 \end{array}
$$

In the transesterification step between oils and fats and methanol, 1 mol of oils and fats react with 3 mols of methanol to produce 3 mols of methyl esters and 1 mol of glycerol. In the esterification step between fatty acids and methanol, 1 mol of fatty acids react with 1 mol of methanol to produce 1 mol of methyl esters and 1 mol of water.

Side reactions:
Reactions between oils and fats and bases:

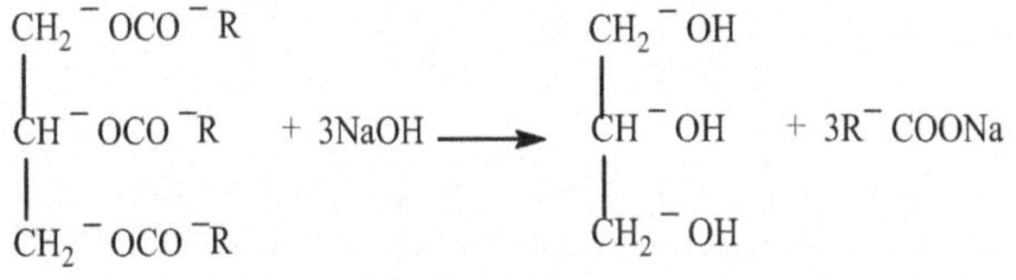

If oils and fats contain fatty acids, esterification takes place between fatty acids and methanol. The reaction equation is described as follows:

$$R{-}COOH + CH_3OH \longrightarrow R{-}COOCH_3 + H_2O$$

In terms of the complexity of the reactions, esterification is relatively simple as it does not produce water. In contrast, water is produced as fatty acids react with methanol. In this reaction system, the concentration of the reactant, methanol, is diluted. In methyl esterefication, methanol concentration is the key to maintaining the reaction speed. A lower methanol concentration level will result in a significantly smaller methyl ester output and a much slower reaction speed, thus leading to a protracted reaction time. Therefore, it is crucial to maintain a methanol concentration above 98%. In the biodiesel production process, fatty acids in the raw material oils should be removed to maintain the methanol concentration. The process for removing fatty acids in oils and fats is elaborated in Chap. 3.

Through transesterification described above, the molecular weight of natural oils and fats (triglycerides) is reduced by 2/3 and their viscosity is reduced by 7/8. This process also improves the volatility of the end product. The biodiesel produced has a similar viscosity compared to diesel with a cetane number of 50.

5.1.2 Reaction Mechanism of Transesterification (Alcoholysis) [14, 15]

The reaction mechanism can be applied to a wide range of interchange reactions between oils and fats and other reactants. This includes (1) alcoholysis, where methyl esters are produced in reactions with monohydric alcohols and where monyl glycerol is produced in reactions with polyhydric alcohols; (2) acidolysis, interchange of fatty acids to produce targeted fatty acids products; and (3) transesterification, a rearrangement process to produce desirable semi-natural oils and fats, monoglyceride, diglyceride or other esters. Transesterification, along with hydrogenation and fractionation, has become a major process technology for oils and fats modification. Particularly, alcoholysis of oils and fats and alcohols is widely used in biodiesel production. In the following section, unless otherwise noted, the transesterification reaction in biodiesel production refers to alcoholysis reaction.

Natural vegetable oils, animal fats, and waste oils of the food industry can be used as raw materials for the production of biodiesel by transesterification. Alcohols that can be used for transesterification include methanol, ethanol, propanol, butanol, and pentanol. Among them, methanol is the most common because it is an inexpensive, short-chain, and strong polar raw material that can readily react with fatty acid glycerides. Moreover, base catalysts are freely soluble in methanol. In this reaction, acid, base or enzyme can be used as a catalytic agent. Base catalysts include NaOH, KOH, carbonate, and sodium and potassium alkoxides. Acid catalysts include sulfuric acid, hydrochloric acid or phosphoric acid. Lipase is an ideal catalyst for transesterification between alcohols and fatty acid glycerides.

Transesterification reactions include the following continuous reversible reactions. Each reaction produces a type of alcohol:

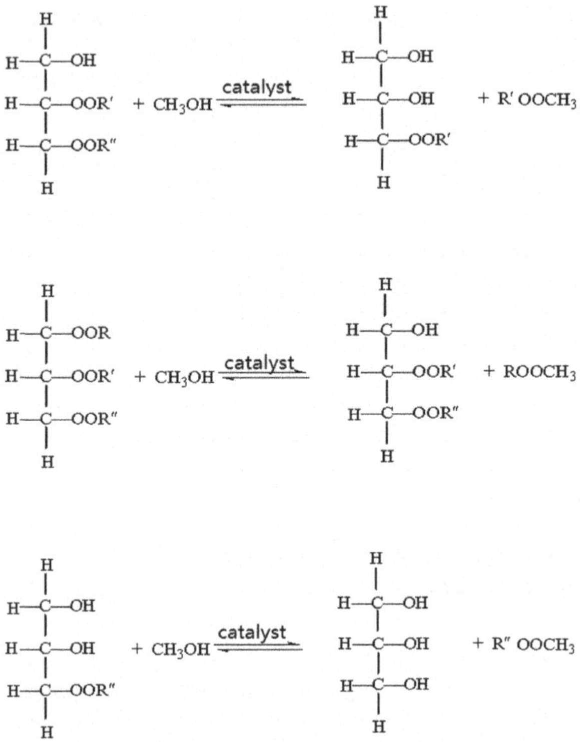

(1) Reaction mechanism of alcoholysis with acid catalysis

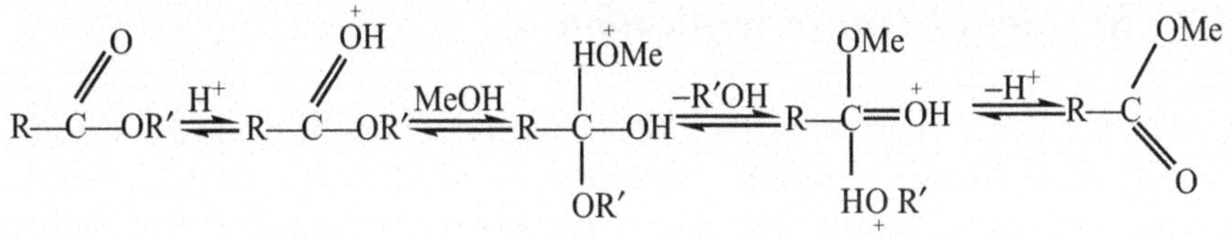

(2) Reaction mechanism of alcoholysis with base catalysis

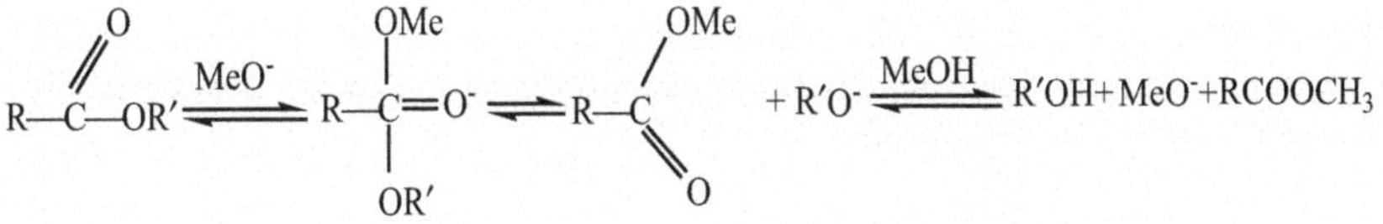

(3) Reaction mechanism of alcoholysis with enzyme catalysis

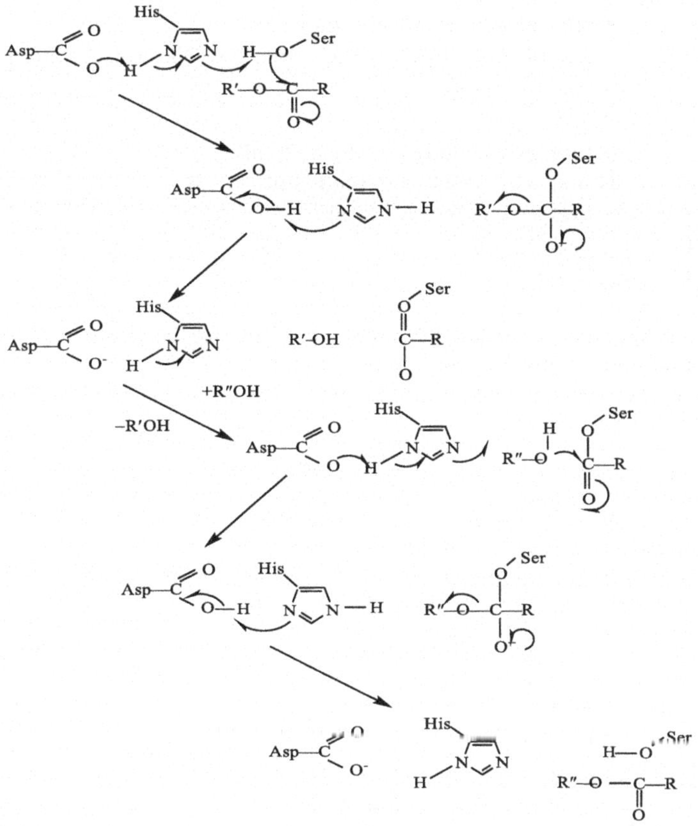

5.1.3 Kinetics of Transesterification

Chemical kinetics is a method to describe chemical reactions by mathematical equations. Chemical kinetics researches also allow for understanding in the control of reaction condition to increase the speed of primary reaction and improve biodiesel output. They also provide insights into the ways to inhibit or slow side reactions to reduce raw material consumption, lessen the burden of detach operation, and enhance end product output [16]. Additionally, quantitative researches on reaction speed could shed light on optimal designs of the biodiesel production process, thus serving as a theoretical basis for optimal control.

Transesterification reactions in biodiesel production can be described by the following reaction-rate equation:

$$-\frac{dC_{Me}}{dt} = kC_o^{\alpha} c_{Me}^{\beta}$$

In this above equation

C_{Me}	concentration of methanol, mol/L
C_O	concentration of triglyceride, in this case, the concentration of edible soybean oil, mol/L
t	time, s
α	reaction order relative to soybean oil
β	reaction order relative to methanol
k	reaction speed constant $(\text{mol/L})^{1-\alpha-\beta}\ S^{-1}$
Subscript O	triglyceride
Subscript Me	methanol

In cases when there is an excessive amount of triglyceride, it can be assumed that the concentration remains the same.

That is, C_o is a constant. Accordingly, the equation can be simplified into

$$-\frac{dC_{Me}}{dt} = kC_{Me}^{\beta}$$

In this equation $K = kc_{Me}^{\alpha} \cdot (\text{mol/L})^{1-\beta} \cdot S^{-1}$.

Take the logarithm on both sides of the equation to get the following result:

$$\log\left(-\frac{dc_{Me}}{dt}\right) = \log K + \beta l \log c_{Me}$$

For a given temperature, K and β are constants. Thus, $\log\left(-\frac{dc_{Me}}{dt}\right)$ 与 $\log C_{Me}$displays a linear relation. Therefore, by measuring the methanol concentration at different time points, we will get a diagram describing the relationship between C_{Me} and t. By solving for the corresponding dC_{Me}/dt at different points along C_{Me} and taking the logarithm of C_M and dC_{Me}/dt, we will get a graph $\log\left(-\frac{dc_{Me}}{dt}\right) \sim \log(C_{Me})$. Accordingly, α (reaction order of methanol) and K (reaction speed constant) can be calculated based on slope and intercept, respectively.

According to Arrhenius equation

$$k = \mathrm{k}_0 e^{\frac{E_a}{RT}}$$

k reaction speed constant
k_0 frequency factor
E_α activation energy.

Take the logarithm on both sides of the equation to get the following result:

$$\log k = \frac{E_\alpha}{2.303R} \times \frac{1}{T} + \log K_0$$

According to this equation, if we draw a graph showing the relationship between lg$k$ and $1/T$, we will get a straight line, its slope representing activation energy E_α and its intercept representing frequency factor k_0.

(1) Measuring methanol

Gas chromatography and colorimetry are most widely used to measure methanol. Gas chromatographic analysis is easy to operate, but it has a low sensitivity and is not suitable for measuring methanol in aqueous solutions. Colorimetric analysis, however, is more complicated. It has high sensitivity and is particularly suited to measure methanol in aqueous solutions. The methods and principles of colorimetry are presented as follows:

In acid solutions, methanol is oxidized into formaldehyde by potassium permanganate, and then forms a purple compound by reaction with chromotropic acid in sulfuric acid medium. At this point, the colorimetry method can be applied at a wavelength of 580 nm.

The equation is described as follows:

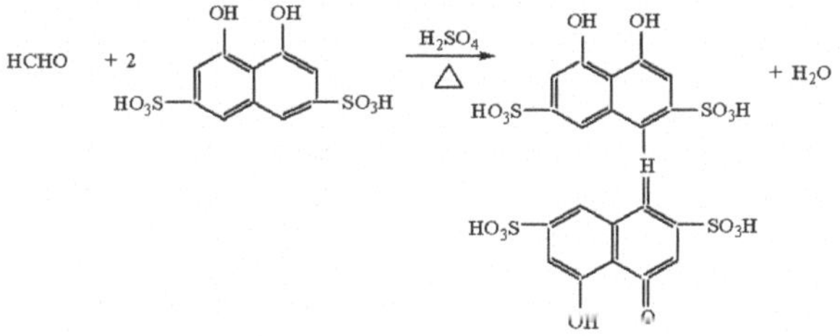

At different time points and for different reaction time, reaction speed $-dC_{Me}/dt$ and methanol concentration C_{Me} can be described as follows (Table 5.1):

Table 5.1 Reaction speed varies with reaction time

Temperature °C	Reaction time (min)	Methanol concentration (mol/L)	$-dC_{Me}/dt$ (mol/L min^{-1})	$lg(C_{Me})$	$lg(-dC_{Me}/dt)$
40	1	0.102	−9.92E−04	−0.99	−3.00
40	10	0.093	−9.20E−04	−1.03	−3.04
40	20	0.084	−8.40E−04	−1.07	−3.08
40	30	0.076	−7.60E−04	−1.12	−3.12
40	40	0.069	−6.90E−04	−1.16	−3.17
40	50	0.063	−6.00E−04	−1.20	−3.22
40	60	0.057	−5.20E−04	−1.24	−3.28
45	1	0.075	−9.90E−04	−1.12	−3.00
45	10	0.066	−9.00E−04	−1.18	−3.05
45	20	0.058	−8.00E−04	−1.24	−3.10
45	30	0.050	−7.00E−04	−1.30	−3.15
45	40	0.044	−6.00E−04	−1.36	−3.22
45	50	0.038	−5.00E−04	−1.42	−3.30
45	60	0.034	−4.00E−04	−1.47	−3.40
50	1	0.077	−6.94E−04	−1.11	−3.16
50	10	0.071	−6.40E−04	−1.15	−3.19
50	20	0.065	−5.80E−04	−1.19	−3.24
50	30	0.060	−5.20E−04	−1.23	−3.28
50	40	0.055	−4.60E−04	−1.26	−3.34
50	50	0.050	−4.00E−04	−1.30	−3.40
50	60	0.047	−3.40E−04	−1.33	−3.47
55	1	0.068	−1.28E−03	−1.17	−2.89
55	10	0.057	−1.10E−03	−1.24	−2.96
55	20	0.047	−9.00E−04	−1.33	−3.05
55	30	0.039	−7.00E−04	−1.41	−3.15
55	40	0.033	−5.00E−04	−1.48	−3.30
55	50	0.029	−3.00E−04	−1.53	−3.52
55	60	0.027	−1.00E−04	−1.56	−4.00
60	1	0.068	−6.92E−04	−1.17	−3.16
60	10	0.062	−6.20E−04	−1.21	−3.21
60	20	0.056	−5.40E−04	−1.25	−3.27
60	30	0.051	−4.60E−04	−1.29	−3.34
60	40	0.047	−3.80E−04	−1.33	−3.42
60	50	0.044	−3.00E−04	−1.36	−3.52
60	60	0.042	−2.20E−04	−1.39	−3.66

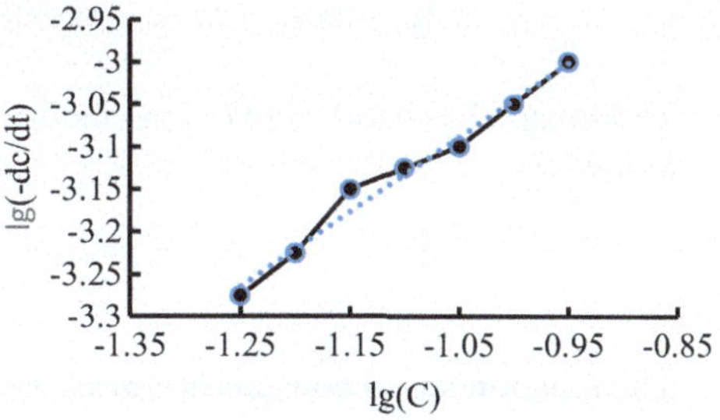

Fig. 5.1 Graph describing the relationship between $\lg(-dC_{Me}/dt)$ and $\lg(C_{Me})$ at 40 °C

By taking $\lg(C_{Me})$ calculated at different temperatures as horizontal axis, $\lg(-dC_{Me}/dt)$ as vertical axis, we can get the following graph (Fig. 5.1).

By linear fitting of the graph above, we will get the following equation:

$$\lg(-dC_{Me}/dt) = -1.9687 + 1.0342 \cdot \lg(C_{Me})$$

Its correlation coefficient $R = 0.9975$.

According to the equation, the reaction at 40 °C is pseudo-first order, its reaction speed constant K being 1.07×10^{-2} min^{-1}.

Similarly, by linear fitting at 45, 50, 55, and 60 °C, we will get the following results:

At 45 °C:

$$\lg(-dC_{Me}/dt) = -1.853 + 1.0106 \cdot \lg(C_{Me})$$

Its correlation coefficient $R = 0.9955$.

Accordingly, the reaction at 45 °C is pseudo-first order, its reaction speed constant being 1.40×10^{-2} min^{-1}.

At 50 °C:

$$\lg(-dC_{Me}/dt) = -1.7219 + 1.2806 \cdot \lg(C_{Me})$$

Its correlation coefficient $R = 0.9958$.

Accordingly, the reaction at 50 °C deviates to some extent from pseudo-first order, its reaction speed constant being 1.90×10^{-2} min^{-1}.

At 55 °C:

$$\lg(-dC_{Me}/dt) = -1.2984 + 1.3358 \cdot \lg(C_{Me})$$

Its correlation coefficient $R = 0.9847$.

Accordingly, the reaction at 55 °C also deviates to some extent from pseudo-first order, its reaction speed constant being 5.03×10^{-2} min^{-1}.

At 60 °C

$$\lg(-\mathrm{d}C_{\mathrm{Me}}/\mathrm{d}t) = -1.2085 + 1.6547 \cdot \lg(C_{\mathrm{Me}})$$

To sum up, the chemical reaction kinetics of transesterification can be summarized as follows:

$$-\frac{\mathrm{d}c_{\mathrm{Me}}}{\mathrm{d}t} = Kc_{\mathrm{Me}}^{1.3}$$

The reaction speed constant K varies with temperature.

(2) Calculating activation energy of reactions

According to the relationship between reaction speed constant and temperature change shown above, activation energy can be calculated as presented in Table 5.2.

According to the above data, and by taking $1/T$ as the horizontal axis and lg (K) as vertical axis, we will get the following graph (Fig. 5.2).

Based on the above graph, the relationship between reactions speed constant K and temperature is as follows:

$$\log K = -4320.7 \times \frac{1}{T} + 11.767$$

Its correlation coefficient $R = 0.9671$.

According to the above equation, the activation energy of transesterification is

Table 5.2 Activation energy calculation datasheet

Temperature (K)	$1/T$ ($\times 10^{-3}$)	Speed constant ($\times 10^{-2}$ min^{-1})	Lg(K) (min^{-1})
313.15	3.19	1.07	−1.9687
318.15	3.14	1.40	−1.853
323.15	3.09	1.90	−1.7219
328.15	3.05	5.03	−1.2984
333.15	3.00	6.19	−1.2085

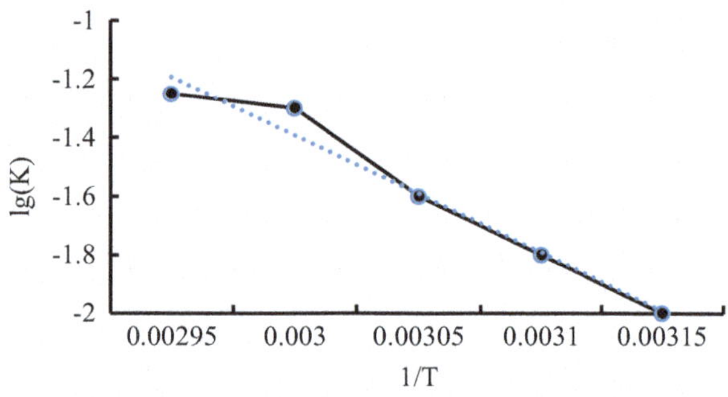

Fig. 5.2 Graph describing the relationship between lg(K) and 1/T

$$E_{\alpha} = 4320.7 \times R \times 2.303 = 82.7\text{kJ/mol}$$

Its frequency factor $K_0 = 5.85 \times 10^{-11}\ \text{mol}^{-1}\ \text{L}\ \text{min}^{-1}$.

The above conclusions show that this reaction has low activation energy and fast reaction speed, as equilibrium is reached within a short period of time. This reaction occurs easily, and its yield only changes slightly with time, and even in a decreasing trend. Therefore, prolonging reaction time alone contributes little to the improvement of reaction yield.

Since transesterification is a reversible reaction, the equilibrium of reaction will shift to the resultant side with the presence of excessive alcohols. Therefore, the actual amount of alcohols used is much larger than that in the chemical equation (alcohol:oil = 3:1). Catalysts can improve yield by accelerating the reaction speed. They can be base catalyst, acid catalyst or biocatalyst. With the help of acid or base catalyst, the transesterification of animal and vegetable oils consists of a series of reversible reactions. Triglyceride is first converted to diglyceride, then monoglyceride, and finally glyceride. In each step, an ester is produced. Freedam et al. have elaborated on the transesterification of soybean oil and other vegetable oils with alcohols. They studied how the type of alcohol, molar ratio, the type and amount of catalyst, and reaction temperature impact on the reaction speed constant and reaction order. With the help of acid or base catalyst, when butanol and soybean oil react at a ratio of 30:1, the reaction is a first-order kinetic process. In contrast, with the help of base catalyst, when butanol and soybean oil react at a ratio of 6, the reaction kinetics follows the rules of a second-order serial process. At 20–60 °C, when the molar ratio of methanol and soybean oil is 6:1 with 0.5% of sodium methylate, the reaction is a combination of a second-order serial process and a fourth-order parallel process. The reaction speed constant in a base-catalyzed reaction is much larger than that in an acid-catalyzed reaction. And the constant grows with the amount of catalyst used. The reaction activation energy, or *Ea*, is 33–83.7 kJ/mol.

5.2 Catalyst Options for Biodiesel Production

Catalysts for transesterification include acid catalyst, base catalyst, and enzyme catalyst [17–19]. Catalysts or biodiesel catalysts can accelerate the reaction speed to improve yield. In transesterification of oils and fats, acid catalysts either have a low catalytic activity or substantially break down oils and fats (inorganic acid). Base catalysts mainly include alkali metal, sodium alkoxides, and bases such as NaOH and KOH. Alkoxides react strongly with H_2O, CO_2, inorganic acid, organic acid, and peroxides, among others. Therefore, oils and fats used for methoxide-catalyzed transesterification must be refined. Additionally, methoxide adsorbed with water vapor is inflammable, making it difficult to be applied on an industrial scale. Alkali metals, on the other hand, are highly chemically active, costly, and difficult to store. They are not suitable for industrial use either. Enzymes tend to be denatured and deactivated in methanol. Their industrial application is also limited. Bases including NaOH and

KOH do not require strict reaction conditions. They are widely used in industrial production because of their affordable price.

5.2.1 Acid-Catalyzed Transesterification

Acid catalysts for transesterification include sulfuric, phosphoric, hydrochloric, and sulfonic acids [20, 21]. Acid-catalyzed transesterification is much slower than base-catalyzed transesterification, but in cases when glycerides have a high level of free fatty acid and water, acid-catalyzed transesterification is the better option. The transesterification process can be sufficiently completed in an acidic condition of the given vegetable oil is low-grade oil (such as sulfatized olive oil). On-site transesterification is different from traditional reaction methods with which vegetable oils are extracted from the mixture. Instead, the raw material containing vegetable oils is in direct contact with acidified alcohols. That is, extraction and transesterification take place simultaneously. Alcohols serve as both extractant and reactant of esterification. By on-site transesterification, sunflower oil reacts with acidified methanol to form fatty acids methyl ester. Its reaction yield is larger than the traditional method with which vegetable oils are extracted in advance.

5.2.2 Base-Catalyzed Transesterification

Base catalysts for transesterification include strong bases (for example NaOH and KOH for unrefined oils and fats), carbonate, and alkyl oxides [22–25] (for example sodium methoxide, sodium ethoxide, sodium isopropoxide, and sodium n-butoxide for refined oils and fats). Generally, NaOMe has a higher catalytic efficiency than NaOH because the mixture of NaOH and MeOH will produce a small amount of H_2O, which can lead to saponification. However, Ma et al. came up with an opposite conclusion that transesterification of cattle tallow requires 0.3% and 0.5% (W/W) of NaOH and MeOH, respectively, as catalysts to maximize the level of activity. Freeman et al. found that if the molar ratio of alcohol/oil is 6:1, the reactions catalyzed by 1% NaOH or 0.5% MeOH will produce almost identical resultants after one-hour reaction time. In particular, NaOH has become the first-choice catalyst for transesterification because of its low price, making it widely applicable in large-scale industrial production.

5.2.3 Enzyme-Catalyzed Transesterification

Despite that base-catalyzed transesterification is capable of producing a high yield in a short period of time, it has the following shortcomings: the reaction consumes a significant amount of energy; it is difficult to recycle glycerol; the reaction has a high alcohol consumption; catalysts must be separated from the resultants, and the basic wastewater produced must be properly treated; the amount of free fatty acids and water has a significant impact on the reaction; the process is complicated. Enzyme-catalyzed transesterification overcomes the shortcomings mentioned above. Particularly, it is convenient to recycle the by-product glycerol. Moreover, the free fatty acids in waste oils and fats can be sufficiently transformed into methyl esters. Therefore, biodiesel production by the enzyme catalysis method is drawing more attention. Researches have shown that with catalysis by Ryzopus oryzae lipase, transesterification of triglycerides and methanol is characterized by a sequential reaction mechanism. Triglycerides and part of the glycerides are hydrolyzed to produce a certain amount of glycerides and free fatty acids, and then free fatty acids react with methanol to produce methyl esters. This process is different from a base-catalyzed one. In enzyme-catalyzed reactions, all free fatty acids in vegetable oils can be transformed into methyl esters. Besides, biodiesel production by enzyme catalysis does not require very strict conditions. Yet, enzymes are expensive. The main obstacle for enzyme catalysis to be applied in industrial production is the cost factor. Immobilized lipase [26], however, can be recycled and used for reaction, which lowers the cost. Many researchers outside China are employing immobilized lipase for transesterification reactions. Furthermore, as enzymes are easily deactivated in high concentration methanol, a new operational method has been developed. That is, enzymes are added in a stepwise fashion to avoid deactivation and achieve a higher conversion rate of methyl esters.

5.2.4 Production Process of Producing Biodiesel by Transesterification

There are various processes for the industrial production of biodiesel. Their end products have developed into various brands. These include Novamont and Ballestra from Italy, IFP from France, and Henkel and ATT from Germany. Today, most factories apply the traditional two-stage transesterification operation–reaction and purification. The most relevant factor in the reaction is the amount of methanol and catalyst. More methanol means higher yield and more difficult separation operation. At present, most of the factories that are using the two-stage operation have a production capacity of 500–10,000 t/a. This type of operation requires a smaller investment to reach a certain output level. However, problems exist in the continuity and security of the process, which could be solved through modern control technologies [27, 28]. The techniques developed by Henkel have become a widely used process for biodiesel manufacturers. In this process, the facilities and equipment used are not exclusive. The distillation operation in the process determines the quality of the diesel produced. The operating pressure is 0.4–0.5 MPa, and the operating temperature is 70–80 °C. The diesel produced through this process has high quality, light color, and high purity, and the purity of its by-product glycerol can reach 92% as well. The main

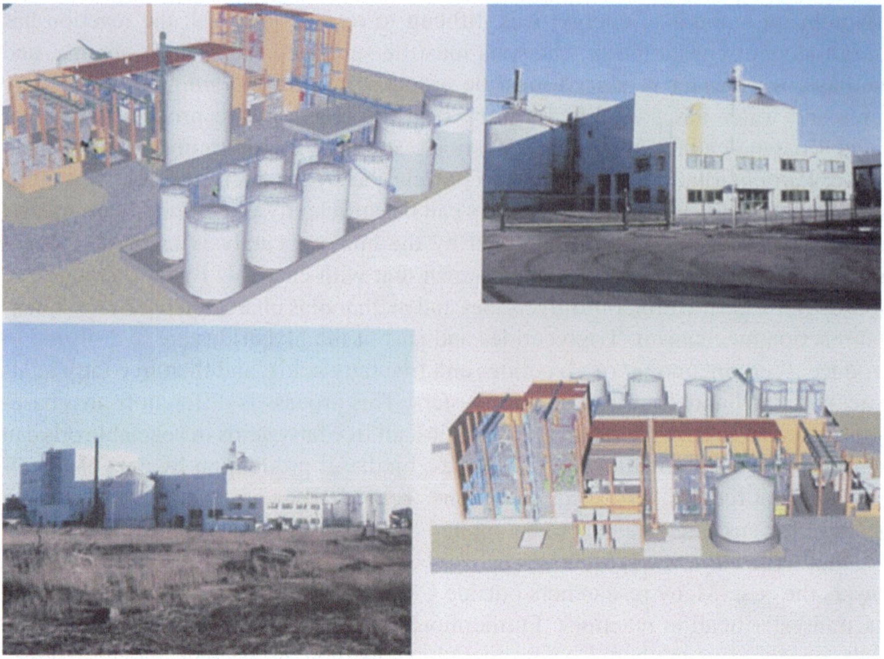

Fig. 5.3 German CIMBRIA SKET GmbH's large-scale biodiesel production line and factory building

disadvantage, however, is its demand for a large amount of investment and energy. At present, some of the European factories that adopt this process have reached an output of 170 kt/a (Fig. 5.3).

5.2.5 Intermittent Transesterification Process for Biodiesel Production

Intermittent transesterification process for biodiesel production includes four steps: transesterification, separation, recycling of methanol, and recycling of glycerol. The process flow diagram is presented below (Fig. 5.4):

(1) Transesterification

A calculated amount of oils and fats, methanol, and catalyst is added to the reaction tank. The weight of methanol is 12–14% of that of oils and fats, and catalyst about 0.8%. The reactants are heated with stirring at a pressure of 0.5–0.6 MPa. The duration of the reaction is 2–3 h after the temperature reaches 160 °C.

(2) Separation

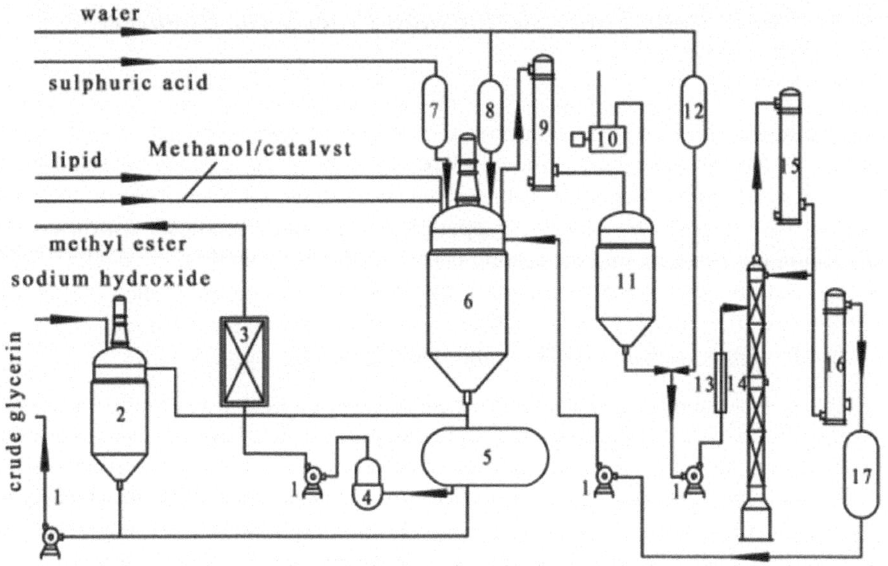

Fig. 5.4 Process flow diagram of intermittent transesterification process for biodiesel production. 1—Pump, 2—neutralization tank, 3—filter, 4—filter, 5—precipitation tank, 6—reaction tank, 7—metering tank, 8—metering tank, 9—metering tank, 10—vacuum pump, 11—methanol tank, 12—metering tank, 13—heater, 14—rectification column, 15—condenser, 16—condenser, 17—methanol tank

Lower the temperature to 90–100 °C and depressurize the reaction system. Add sulfuric acid equivalent to 0.8–0.9% the weight of the oils and fats, and separate the soap produced.

clearpageIn order to separate glycerol and methyl ester, add a proper amount of water because glycerol dissolves in water and methyl ester does not. After the addition of water, allow the solution to stand for some time to ensure precipitation. Drain the glycerol solution at the bottom of the reaction tank, and move the methyl esters to the precipitation tank.

(3) Recycling of Methanol

Lower the pressure to evaporate methanol. Methanol vapor is condensed in the condenser before being recycled in the methanol tank. Add NaOH equivalent to 0.3–0.4% the weight of the oils and fats to neutralize the acidic methanol, and pump the methanol to the rectification column for distillation. Once methanol vapor comes out from the rectification column, part of the vapor is condensed and flows back, and the rest enters the methanol tank to be reused. The water separated from the solution drains from the bottom of the rectification column.

(4) Recycling of Glycerol

The details of glycerol recycling are elaborated in Sect. 5.2.6 of this chapter.

5.2.6 *Continuous Transesterification Process for Biodiesel Production*

The continuous transesterification process is a typical operation that includes vegetable oils as raw material for biodiesel production. It has a high yield and produces biodiesel in a continuous fashion. The continuous transesterification process involves two stages. The first is the operation to produce biodiesel with transesterification (workshop), and the second is re-esterification, methanol recycling, and glycerol refining operation (workshop). The basic flow diagram of the continuous transesterification process for biodiesel production is presented as Fig. 5.5.

The key points of operation are highlighted as follows.

a. Transesterification reaction and separation. First, preheated vegetable oil (about 35 °C), methanol (molar ratio of methanol and oil is 6:1 to 4:1), and catalysts (KOH or NaOH, 1–2%) are continuously added in a given proportion to the first esterification tank for reaction with stirring. The reaction time is determined based on the solution level (10–20 min). This process adopts a two-stage reaction operation. Following the reactions in the first reaction tank, the reactants enter a specially designed separation column, so that the rapeseed oil methyl ester (RME) produced can be continuously separated from glycerol and water. RME then continues to enter the second reaction tank for further and sufficient reaction. After that, the reactions take place in the second separation column to separate the glycerin phase (which flows into the storage tank for recycling).
b. Continuous water rinse, acid cleaning, and separation. The RME separated from the glycerin phase enters the water rinse reactor to be rinsed with hot water. The residues of catalyst, potassium soap, and glycerol in RME are rinsed and then separated in the third separation column. Lastly, RME enters the acid cleaning reactor to be washed with dilute acid, before entering the fourth separation column to separate acid solution from the relatively pure biodiesel.
c. Vacuum drying and desolventizing. Since RME has a high viscosity, the remaining methanol and water needs to go through a two-stage diaphragm vacuum drying process. This is to ensure that the flash point (higher than 110 °C) and moisture content (300×10^{-6}) meet the required standard.

(2) Recycling of glycerol and methanol. This is a four-stage operation including the neutralization of crude glycerol, re-alcoholysis of oil phase, recycling and evaporation of methanol, and rectification of methanol.

a. Neutralization process of crude glycerol. The by-products from transesterification (glycerol, acid solution, KOH the catalyst, methanol) are placed in the mixture tank for neutralization. If more acid is needed for neutralization, add 96% sulfuric acid to ensure that all KOH is transformed into K_2SO_4. Following that, separate potassium sulfate by pressure leaf filters (wash the filter cake, and dry and recycle methanol, as well as potassium sulfate powder to be used as

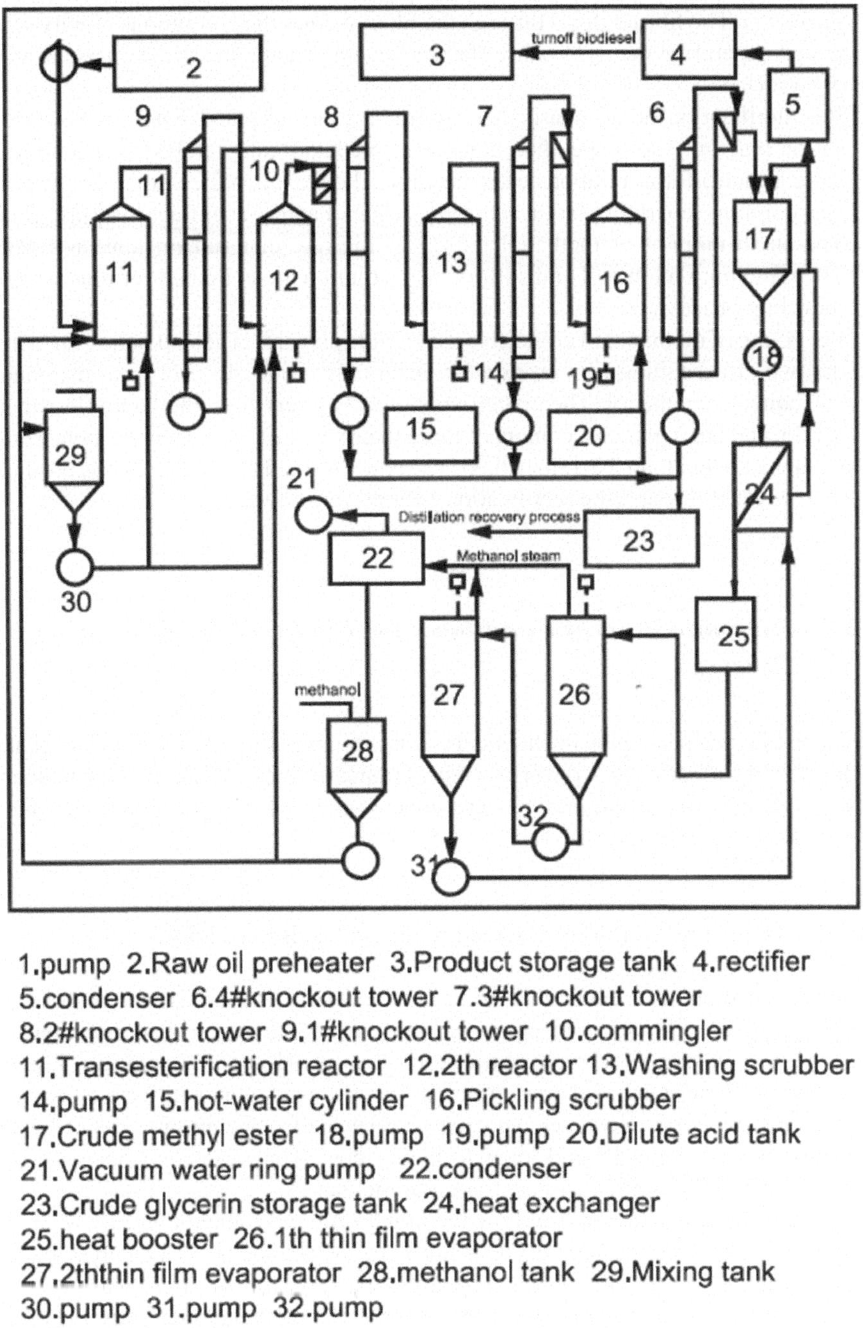

Fig. 5.5 Basic flow diagram of continuous transesterification process for biodiesel production (Excerpt from "Oils and Fats Processing Technology")

fertilizer). The filtrate flows through the filter to enter the continuous gravity separator, separating the light phase (fatty acid and oil) and the heavy phase (crude glycerol).

b. Re-esterification of oil phase. After filtration, the oil phase enters the esterification reactor. Add a specified amount of methanol and catalyst (concentrated sulfuric acid) for continuous esterification. The reactants then enter the separation column, separating biodiesel from methanol and concentrated sulfuric acid. The small amount of methanol returns to the raw material oil tank as RME. The remaining methanol and sulfuric acid solution are collected for reuse in the recycling process.
c. Recycling of methanol in crude glycerol. Crude glycerol is heated before entering the two-effect falling film evaporator, separating crude glycerol from methanol and aqueous methanol. The glycerol obtained has very little methanol (less than 0.1%), and has a glycerol content of more than 80% (13.3% water, 2% potassium sulfate, and less than 3% impurities). The aqueous methanol separated is purified and recycled through the rectification column for reuse.

5.2.7 Determination of Conditions for Biodiesel Production Process

Based on kinetics principles in chemical reactions and researches by Peterson et al., the key factors affecting the conversion rate of transesterification are reaction temperature, methanol concentration, types and concentration of catalyst, fluid mechanics conditions, and reaction time.

5.2.7.1 Impact of Methanol Concentration on Biodiesel Yield

Transesterification is a reversible reaction. Normally, an excessive amount of methanol is used to drive the reaction equilibrium towards the forward direction, so as to enhance the conversion rate of transesterification. However, here is an amount issue [28]. A higher methanol concentration drives the reaction equilibrium towards the forward direction. However, an excessive level of concentration has a lower performance in this regard and leads to higher cost. Therefore, it is necessary to determine the initial optimal concentration of methanol.

Below is the research result of how the molar ratio of methanol and soybean oil affects the yield of biodiesel at 60 °C using 1.0% (wt%) NaOH as a catalyst with a reaction time of 45 min. It is presented as follows:

Figure 5.6 shows that as the initial concentration of methanol increases, the reaction yield of biodiesel also sharply increases. At the methanol/soybean oil molar ratio of 6:1, it reaches the highest point of 85.0%. But if the initial methanol concentration further increases, the biodiesel yield remains almost the same. This demonstrates

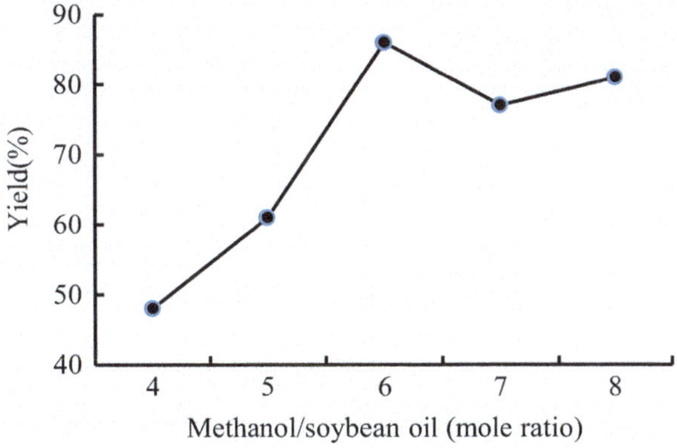

Fig. 5.6 Impact of methanol concentration on reaction yield

that at 6:1, methanol concentration has an optimal impact on the reaction yield. If the concentration keeps increasing, it has little impact on the reaction equilibrium shift. Furthermore, as mentioned above, transesterification is mostly a reversible reaction. Higher methanol concentration will increase the polarity of the reaction system so as to reduce the reaction speed. More specifically, the biodiesel yield will decrease slightly. To sum up, at the methanol/soybean oil molar ratio of 6:1, the biodiesel yield reaches its highest point. Therefore, in the real-world scenario, the methanol/soybean oil molar ratio should be kept at 6:1.

5.2.7.2 Impact of Reaction Time on Biodiesel Yield

Below is the table showing how reaction time affects biodiesel yield at 60 °C with a methanol/soybean oil molar ratio of 6:1 at different catalyst (NaOH) concentration levels.

Figure 5.7 demonstrates that in the initial phase, as reaction time increases, the yield increases significantly. But after 20 min, the reaction yield begins to fall. As the reaction continues, the yield gradually approaches a stable value. All this shows

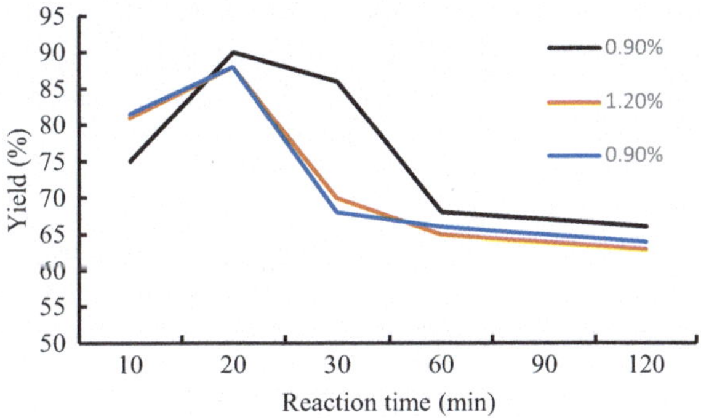

Fig. 5.7 Yield changes over reaction time at different catalyst concentration levels

Fig. 5.8 Reaction yield changes with catalyst (NaOH) concentration

that this is a fast reaction as the transesterification process is generally completed within 20 min. However, side reactions exist

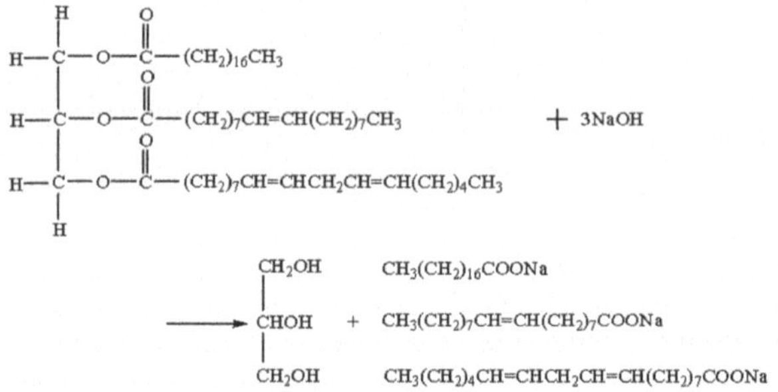

These above side reactions deduce not only the yield, but also the quality of fuel, and thus should be avoided. The above analysis shows that longer reaction time does not necessarily improve the biodiesel yield, but leads to the side reaction–saponification. Therefore, the reaction time should not exceed 30 min.

5.2.7.3 Impact of Catalyst Concentration on Biodiesel Yield

Below is the result showing how different catalyst (NaOH) concentration levels affect biodiesel yield at 60 °C with a methanol/soybean oil molar ratio of 6:1 and with a reaction time of 30 min.

Figure 5.8 shows that when the catalyst (NaOH) concentration level varies between 0.8 and 1.0%, biodiesel yield increases significantly with the catalyst concentration. Yet once the catalyst concentration exceeds 1.1% (wt%), the yield slightly drops. This is caused by saponification between NaOH and fatty acids methyl ester.

The above analysis shows that the catalyst (NaOH) concentration level enhances the yield. However, an excessive level of NaOH concentration leads to a side reaction–saponification, hence a slight drop in the reaction yield. Therefore, in real-world scenarios, the concentration level of the catalyst (NaOH) should be maintained between 0.95 and 1.05% for higher yield.

5.2.7.4 Determination of Reaction Temperature

The temperature of the washing water has a huge impact on the biodiesel yield [29, 30]. At 50 °C, the yield reaches its maximum level. Any rise or fall will reduce the yield. In particular, for a lower temperature, the reaction yield is much smaller than

Table 5.3 Different washing effects at different water temperatures

Temperature (°C)	Biodiesel yield (%)
20	65.3
50	85.0
65	80.8
80	79.2

Table 5.4 Time needed for the emulsified phase to disappear

Temperature (°C)	Biodiesel yield (%)
20	65.3
50	85.0
65	80.8
80	79.2

that at 50 °C. The reason is that lower water temperature tends to cause emulsification, and the emulsified phase is especially stable that it won't demulsify easily. Therefore, during the solution separation process, a significant amount of biodiesel is entrained in the emulsified phase, hence the lower yield. If the temperature of washing water is high, emulsification does not easily occur; and even if it does occur, the emulsified phase is not stable and quickly demulsifies. Accordingly, the reaction yield is higher. But with the rising temperature, the solubility of vegetable oils in water also increases. This leads to a higher amount of oil that is entrained in water, hence a slightly lower biodiesel yield. When the temperature of washing water exceeds 50 °C, the rising temperature has little impact on the time of the emulsification process. Therefore, the temperature of washing water is best maintained at 50 °C (Tables 5.3 and 5.4).

5.2.7.5 Effect of Stirring Intensity on Biodiesel Yield

According to the reaction mechanism of transesterification, methanol first reacts with OH^- to dissociate and then forms the intermediate methoxy. Afterward, the reactive intermediate, methoxy targets carbon atoms in carbonyl groups for nucleophilic substitution reaction [31–33]. Glyceryl in triglyceride is replaced by methoxy, producing the resultant fatty acids methyl ester. According to this reaction mechanism, to control reaction speed, we should control the step where methoxy targets carbon atoms to form the intermediate tetrahedron resultants. Given that methanol and triglyceride are immiscible, increasing the stirring intensity can accelerate the methoxy transfer process from methanol to triglyceride, which, in turn, enhances the reaction between methoxy and triglyceride. Without stirring, there is hardly any reaction between methanol and triglyceride. This further demonstrates that this reaction is a mass transfer control process. Therefore, stirring intensity should be increased as best as possible to improve mass transfer performance.

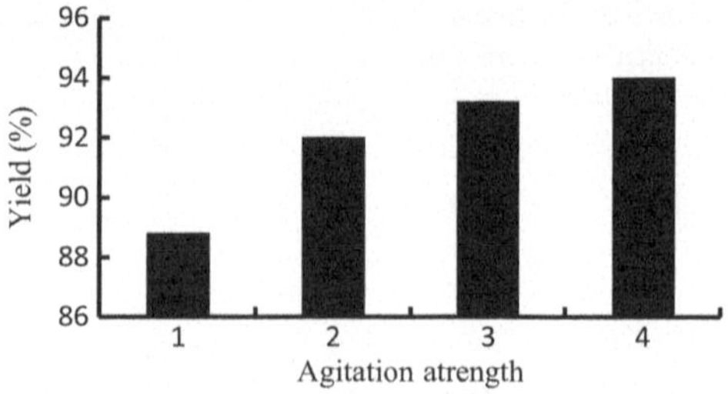

Fig. 5.9 Effect of stirring intensity on biodiesel yield

The mass transfer performance increases with the stirring intensity, and the reaction yield also increases remarkably. It should also be noted that as the stirring intensity increases, the temperature of the reaction system rises as well. It is thus necessary to keep close tabs on the temperature change (Fig. 5.9).

Reaction temperature, catalyst concentration, methanol concentration, reaction time, and stirring intensity all affect the biodiesel yield, but to a different degree. Researchers from East China University of Science and Technology conducted a comparison analysis of those factors through orthogonal experiment and range analysis. The results are shown as follows (Table 5.5):

Particularly, K_{ij} = sum of results on column j, row i,

$$\bar{K}_{ij} = \frac{1}{S} K_{ij}$$

and "s" represents how many times "i" appears in column j.

$$R_j = \max\{K_{ij}\} - \min\{K_{ij}\}$$

By conducting a range analysis on the table above, we will get.

Figure 5.10 indicates that among the five factors listed above, stirring intensity has the greatest range, hence the greatest effect. Therefore, the determination of optimal process conditions should prioritize stirring intensity. The analysis above proves that transesterification as a mass transfer control reaction is mostly affected by the performance of mass transfer. The second most important factor is the amount of catalyst added. Following that are methanol concentration, reaction time, and reaction temperature. Therefore, in choosing the best process conditions, we should first employ the highest possible stirring intensity–level four. The second factor is the catalyst concentration. The analysis shows that concentration level three and level four have almost the same influence on the yield. Considering the catalyst's corrosion effect on equipment and reaction cost, the lower level—1.0–1.2% (wt%)—should be adopted. As to methanol concentration, it has a relatively weak influence on the yield. Furthermore, the higher the concentration level, the less its improvement on the yield. In this regard, the lower methanol concentration level, or a methanol/soybean oil molar ratio of 5:1 to 6:1, should be used. The best reaction time is between 20

Table 5.5 Orthogonal experiment and range analysis on five factors affecting transesterification

Number	Factor					
	Temperature	Amount of catalyst wt%	Reactant ratio	Reaction time (min)	Stirring speed	Yield (%)
	A	B	C	D	E	
1	50	0.8	1:4	10	1	82.1
2	50	1.0	1:5	20	2	93.4
3	50	1.2	1:6	30	3	94.5
4	50	1.4	1:7	40	4	97.0
5	60	0.8	1:5	30	4	90.4
6	60	1.0	1:4	40	3	90.2
7	60	1.2	1:7	10	2	92.1
8	60	1.4	1:6	20	1	91.6
9	70	0.8	1:6	40	2	92.1
10	70	1.0	1:7	30	1	90.4
11	70	1.2	1:4	20	4	96.0
12	70	1.4	1:5	10	3	94.9
13	80	0.8	1:7	20	3	92.6
14	80	1.0	1:6	10	4	92.6
15	80	1.2	1:5	40	1	91.1
16	80	1.4	1:4	30	2	90.1
K_{1j}	367.0	357.2	358.4	361.7	355.2	$T = \sum_{i=1}^{16} y_i = 1471.1$
K_{2j}	364.3	366.6	369.8	373.6	367.7	
K_{3j}	373.4	373.7	370.8	365.4	372.2	
K_{4j}	366.4	373.6	372.1	370.4	376.0	
K_{1j}	91.8	89.3	89.6	90.4	88.8	$y = 91.9$
K_{2j}	91.1	91.7	92.5	93.4	91.9	
K_{3j}	93.4	93.4	92.7	91.4	93.1	
K_{4j}	91.6	93.4	93.0	92.6	94.0	
MAX	373.4	373.7	372.1	373.6	376.0	
MIN	364.3	357.2	358.4	361.7	355.2	
R_j	9.1	16.5	13.7	11.9	20.8	

Data source East China University of Science and Technology

and 30 min. As to reaction temperature that has the least influence on the yield, it should preferably be maintained at 50–60 °C to reduce cost.

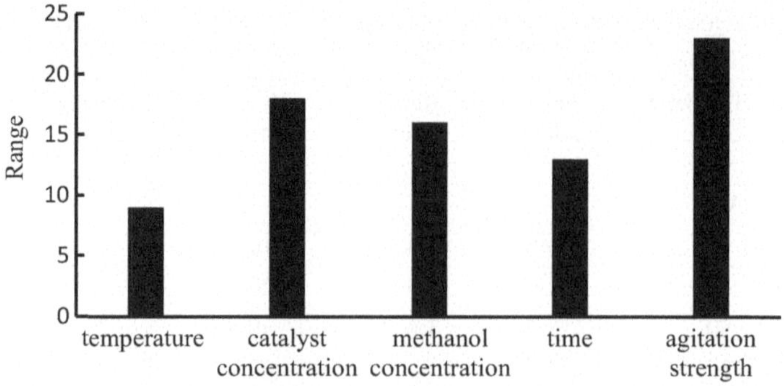

Fig. 5.10 Determination of major affecting factors (*Data source* East China University of Science and Technology)

5.2.8 *Biotechnology for Biodiesel Production*

Despite that base-catalyzed transesterification is capable of producing a high yield in a short period of time, it has the following shortcomings: the reaction consumes a significant amount of energy; it is difficult to recycle glycerol; the reaction has a high alcohol consumption [33, 34]; catalysts must be separated from the resultants, and the basic wastewater produced must be properly treated; the amount of free fatty acids and water has a significant impact on the reaction; the process is complicated. Enzyme-catalyzed transesterification overcomes the shortcomings mentioned above. Particularly, it is convenient to recycle the by-product glycerol. Moreover, the free fatty acids in waste oils and fats can be sufficiently transformed into methyl esters. Therefore, biodiesel production by the enzyme catalysis method is drawing more attention. Researches have shown that with catalysis by *Ryzopus oryzae* lipase, transesterification of triglycerides and methanol is characterized by a sequential reaction mechanism. Triglycerides and part of the glycerides are hydrolyzed to produce a certain amount of glycerides and free fatty acids, and then free fatty acids react with methanol to produce methyl esters. This process is different from a base-catalyzed one. In enzyme-catalyzed reactions, all free fatty acids in vegetable oils can be transformed into methyl esters. Besides, biodiesel production by enzyme catalysis does not require very strict conditions. Yet, enzymes are expensive. The main obstacle for enzyme catalysis to be applied in industrial production is the cost factor. Immobilized lipase, however, can be recycled and used for reaction, which lowers the cost. Many researchers outside China are employing immobilized lipase for transesterification reactions. Furthermore, as enzymes are easily deactivated in high concentration methanol, a new operational method has been developed. That is, enzymes are added in a stepwise fashion to avoid deactivation and achieve a higher conversion rate of methyl esters.

5.2.8.1 Inhibition Effect of Substrates

Enzyme-catalyzed transesterification for biodiesel production uses methanol, ethanol, propanol, and butanol, among others, as substrates. These short-chain alcohols can cause protein denaturalization [35–38]. An excessive amount of these substrates will have a strong inhibition effect on the catalysis of lipase. For reactions in the methanol-oleic acid system (solvent system) with a theoretical optimal molar ratio of substrates at 1:1, the equivalent substrates inhibition weight of methanol is 1 mol (0.2406 ml). However, if 1 molar equivalent of methanol is added all at once, the reaction efficiency will be substantially reduced [39]. This is because the excessive methanol concentration has a strong inhibition effect on lipase. A lower methanol concentration can significantly reduce the activity inhibition effect on lipase. If methanol of less than 1/2 molar equivalent is added in portions, the esterification yield can be greatly improved. But a higher number of portions does not necessarily improve the esterification yield. The inhibition effect of methanol on enzyme-catalyzed transesterification is exhibited in the following figure (Fig. 5.11).

In order to lessen the impact of substrates on enzyme catalysis efficiency, we may add solvents such as petroleum ether (distillation range 60–90 °C), isooctane, n-heptane, n-hexane or cyclohexene to the reaction system. With solvents in the reaction system, the substrate concentrate is reduced and its contact surface area with enzyme is increased, which boosts esterification yield. Various types of solvents have different impacts on the esterification of lipase. This is mainly because solvents have different hydrophobicity levels. The higher the hydrophobicity, the higher the lipase activity and the better the esterification yield. However, in this reaction system, methanol has an exceptionally strong hydrophilicity, which serves as a controlling factor. In this regard, the polarity of the solvent system has a very small impact on the lipase activity and esterification efficiency. From cost-benefit considerations, petroleum ether is a preferred option.

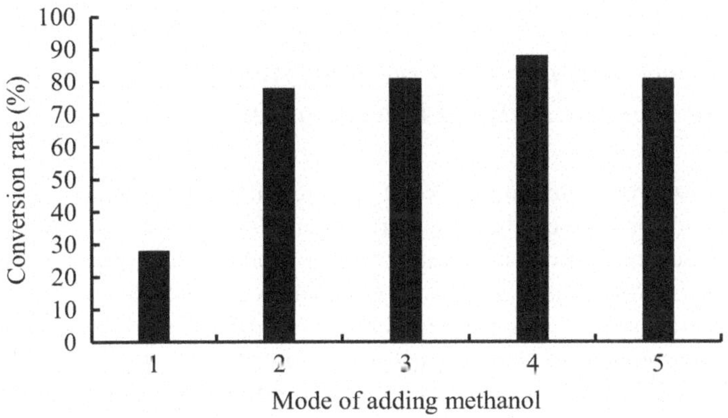

Fig. 5.11 Effect of mode of adding methanol on biodiesel yield

Table 5.6 Effect of various lipase on esterification

Lipase	Conversion rate/%
Procine Pancreas (from Sigma Co.)	22.45
Immobilized Porcine Pancreas	13.79
Lipolase 100T (from Novo Co.) (immobilized lipase)	42.17
Free lipase from *Rhizoups arrhizus* (from our lab)	66.39
Immobilized lipase from *Rhizoups arrhizus*	26.31
Free lipase from *Rhizoups usamil* (from our lab)	61.18
Immobilized lipase from *Rhizoups usamil*	20.60
Free *Candida Cylindracea* (from Sigma)	19.72
Immobilized *Candida Cylindracea*	17.20
Free lipase from *Candida* sp. 99–125 (from our lab)	80.50
Immobilized lipase from *Candida* sp. 99–125	81.51

Amount of lipase; 3% (free), 5% (IM); other conditions are same as 1.3.2; to add methanol stepwise

5.2.9 *Effect of Various Lipase on Reaction*

Different lipase have different catalytic effects on esterification. Below is a table showing the catalytic efficiency of various lipase on esterification [32, 33, 35, 40, 41] (Table 5.6).

5.2.10 *Effect of Water and Free Fatty Acid on Lipase-Catalyzed Esterification*

Lipase-catalyzed esterification does not require such a strict water-free condition as that of acid- or base-catalyzed esterification. Nevertheless, esterification is the reverse reaction of hydrolysis reaction, and it produces an equivalent amount of water in moles. Since water reduces enzyme catalytic efficiency, it is necessary to remove water from the reaction system. Otherwise, the reaction will be inhibited and the micro-environment of enzymes will be disrupted, leading to a drop in enzyme activity and stability. We can add water absorbent to the reaction system to reduce the effect of water on esterification.

Different substrates have specific effects on enzymes. For long-chain saturated fatty acids, the larger the number of carbon atoms, the higher the esterification yield.

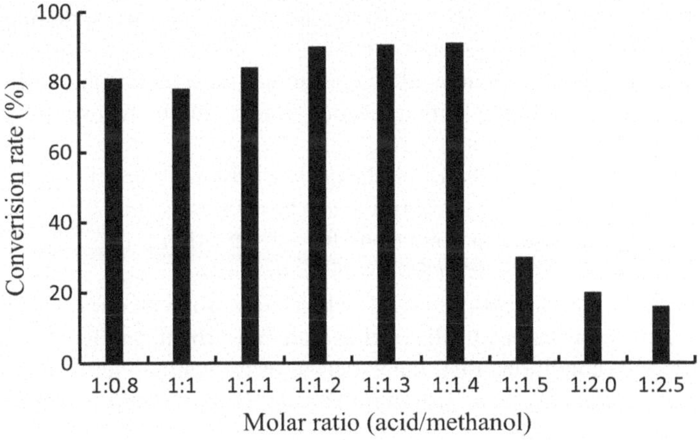

Fig. 5.12 Effect of acid/methanol on biodiesel yield

For a given carbon atoms number, the esterification yield of unsaturated fatty acids is much lower than that of saturated fatty acids.

5.2.11 Effect of Substrate Molar Ratio on Reaction

In a basic reaction system, the theoretical optimal molar ratio of the two substrates is 1:1. But in reality, this is not always the best option. The acid/alcohol ratio should be optimized based on real-world conditions. In practice, the optimal molar ratio of oleic acid and methanol is 1:1.4, at which the esterification yield reaches 92%. Below is a figure showing the effect of substrate molar ratio on transesterification catalyzed by lipase (Fig. 5.12).

5.2.12 Amount and Purity of Enzyme and Lifespan of Immobilized Enzyme

The amount of enzyme directly affects the speed and yield of esterification. Different levels of immobilized enzyme concentration have different catalytic efficiency on transesterification. The optimal enzyme amount is 5%wt (enzyme:acid). Further increase of enzyme amount has little effect on reaction yield. At this point, the enzyme activity is 5000 u/g. The lifespan and half-life of the immobilized enzyme are crucial indicators to determine its activity. Deng et al. used diatomite and a type of textile (scraps of a nonwoven) as support for immobilized enzyme. They believe that inexpensive textiles are easy to recycle and reuse. The corresponding immobilized enzyme has a long lifespan and can be used continuously. For esterification of oleic acid and methanol with an acid/alcohol molar ratio of 1:1, the half-life of the immobilized enzyme is 360 h. With a reaction time of 240 h, the esterification yield decreases by 12% (Figs. 5.13 and 5.14).

To sum up, enzyme catalysis for biodiesel production is a promising biocatalysis process. Lipase immobilized by absorption can be used as a catalyst for esterification to produce biodiesel. The best conditions for esterification process are as follows: in a petroleum ether system, with 5% (wt%) immobilized lipase, at 40 °C, with an oleic acid/methanol molar ratio of 1:1.4, when low-carbon alcohol is added in two molar equivalent portions, with silica gel as water absorbent, and for a reaction time of 24 h. All these conditions met, the expected esterification yield can reach 92%. Experimental results have demonstrated that esterification yield increases with the

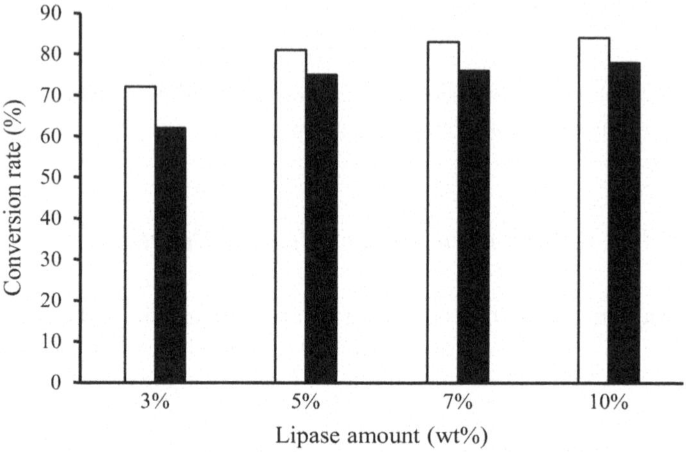

Fig. 5.13 Effect of enzyme concentration on esterification

Fig. 5.14 Lifespan of immobilized enzyme

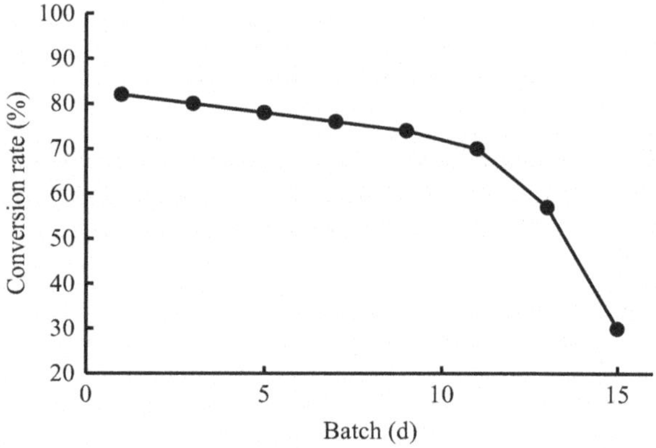

number of carbon atoms of fatty acid and low-carbon alcohol. For a stearic acid system with an acid/alcohol molar ratio of 1:1, the esterification yield can exceed 95%.

References

1. Davis SC, Dohleman FG, Long SP (2011) The global potential for Agave as a biofuel feedstock. Glob Change Biol Bioenergy 3(1):68–78
2. Demirbas A (2002) Biodiesel from vegetable oils via transesterification in supercritical methanol. Energy Convers Manag 43(17):2349–2356
3. Demirbaş A (2003) Biodiesel fuels from vegetable oils via catalytic and non-catalytic supercritical alcohol transesterifications and other methods: a survey. Energy Convers Manag 44(13):2093–2109
4. Demirbas A (2009) Progress and recent trends in biodiesel fuels. Energy Convers Manag 50(1):14–34
5. Diwekar B (2013) Studying various optimal control problems in biodiesel production in a batch reactor under uncertainty. Fuel 103(5):585–592
6. Drapcho CM, Nhuan NP, Walker TH (2011) Biofuels engineering process technology
7. Fangrui MA, Hanna MA (1999) Biodiesel production: a review. Bioresour Technol 70(1):1–15
8. Gui MM, Lee KT, Bhatia S (2008) Feasibility of edible oil vs. non-edible oil vs. waste edible oil as biodiesel feedstock. Energy 33(11):1646–1653
9. Gui MM, Lee KT, Bhatia S (2009) Supercritical ethanol technology for the production of biodiesel: process optimization studies. J Supercrit Fluids 49(2):286–292
10. Haas MJ, Mcaloon AJ, Yee WC, Foglia TA (2006) A process model to estimate biodiesel production costs. Bioresour Technol 97(4):671–678
11. Hill J, Nelson E, Tilman D, Polasky S, Tiffany D (2006) Environmental, economic, and energetic costs and benefits of biodiesel and ethanol biofuels. Proc Natl Acad Sci USA 103(30):11206–11210
12. Hiremath A, Kannabiran M, Rangaswamy V (2011) 1,3-Propanediol production from crude glycerol from jatropha biodiesel process. N Biotechnol 28(1):19–23
13. Hromádko J, Miler P (2011) Principle of monitoring the implementation of sustainability criteria for biofuels. Listy Cukrovarnické A Řepařské 127(7–8):256–260
14. Islam MJ (2015) Experimental and statistical investigation of Australian native plants for second-generation biodiesel production
15. Kang X, Guo X, You H (2015) Biodiesel development in the global scene. Energy Sources Part B 10(2):155–161
16. Knothe G, Krahl J, Gerpen J (2005) The biodiesel handbook. CRC Press (2005)
17. Kumar A (2011) Potential non-edible oil resources as biodiesel feedstock: an Indian perspective. Renew Sustain Energy Rev 15(4):1791–1800
18. Lappas AA, Bezergianni S, Vasalos IA (2009) Production of biofuels via co-processing in conventional refining processes. Catal Today 145(1–2):55–62
19. Leung DYC, Wu X, Leung MKH (2010) A review on biodiesel production using catalyzed transesterification. Appl Energy 87(4):1083–1095
20. Liu X, He H, Wang Y, Zhu S, Piao X (2008) Transesterification of soybean oil to biodiesel using CaO as a solid base catalyst. Fuel 87(2):216–221
21. Marchetti JM, Miguel VU, Errazu AF (2007) Possible methods for biodiesel production. Renew Sustain Energy Rev 11(6):1300–1311
22. Meher LC, Sagar DV, Naik SN (2006) Technical aspects of biodiesel production by transesterification—a review. Renew Sustain Energy Rev 10(3):248–268
23. Qiu Z, Zhao L, Weatherley L (2010) Process intensification technologies in continuous biodiesel production. Chem Eng Process 49(4):323–330

24. Ramadhas AS, Jayaraj S, Muraleedharan C (2005) Biodiesel production from high FFA rubber seed oil. Fuel 84(4):335–340
25. Ramos M, Fernandez C, Casas A, Rodriguez L, Perez A (2009) Influence of fatty acid composition of raw materials on biodiesel properties. Biores Technol 99(1):261–268
26. Saka S, Kusdiana D (2001) Biodiesel fuel from rapeseed oil as prepared in supercritical methanol. Fuel 80(2):225–231
27. Schenk PM, Thomashall SR, Stephens E, Marx UC, Mussgnug JH, Posten C, Kruse O, Hankamer B (2008) Second generation biofuels: high-efficiency microalgae for biodiesel production. Bioenergy Res 1(1):20–43
28. Scott SA, Davey MP, Dennis JS, Horst I, Howe CJ, Lea-Smith DJ, Smith AG (2010) Biodiesel from algae: challenges and prospects. Curr Opin Biotechnol 21(3):277–286
29. Sharma YC, Singh B, Upadhyay SN (2008) Advancements in development and characterization of biodiesel: a review. Fuel 87(12):2355–2373
30. Singh SP, Singh D (2010) Biodiesel production through the use of different sources and characterization of oils and their esters as the substitute of diesel: a review. Renew Sustain Energy Rev 14(1):200–216
31. Sotoft LF, Rong BG, Christensen KV, Norddahl B (2010) Process simulation and economical evaluation of enzymatic biodiesel production plant. Biores Technol 101(14):5266–5274
32. Tang H, Chen M, Garcia ME, Abunasser N, Ng KY, Salley SO (2011) Culture of microalgae Chlorella minutissima for biodiesel feedstock production. Biotechnol Bioeng 108(10):2280–2287
33. Taufiqyap YH (2015) Advanced technologies in biodiesel: introduction to principles and emerging trends
34. Tyndall JC, Berg EJ, Colletti JP, Halvorsen KE, Selfa TL, Becker DR, Hinrichs C (2011) Corn stover as a biofuel feedstock in Iowa's bio-economy: an Iowa farmer survey. Biomass Bioenergy 35(4):1485–1495
35. Vasudevan PT, Briggs M (2008) Biodiesel production–current state of the art and challenges. J Ind Microbiol Biotechnol 35(5):421
36. West AH, Posarac D, Ellis N (2008) Assessment of four biodiesel production processes using HYSYS. Plant Bioresour Technol 99(14):6587–6601
37. Woertz I, Feffer A, Lundquist T, Nelson Y, Safferman S, Lioa W, Saffron C (2009) Algae grown on dairy and municipal wastewater for simultaneous nutrient removal and lipid production for biofuel feedstock. J Environ Eng 135(11):1115–1122
38. Xu H, Miao X, Wu Q (2006) High quality biodiesel production from a microalga Chlorella protothecoides by heterotrophic growth in fermenters. J Biotechnol 126(4):499–507
39. Xue J, Grift TE, Hansen AC (2011) Effect of biodiesel on engine performances and emissions. Renew Sustain Energy Rev 15(2):1098–1116
40. Yang F, Hanna MA, Sun R (2012) Value-added uses for crude glycerol–a byproduct of biodiesel production. Biotechnol Biofuels 5(1):13
41. Zhang Y, Dubé MA, Mclean DD, Kates M (2003) Biodiesel production from waste cooking oil: 1. Process design and technological assessment. Bioresour Technol 89(1):1–16

Chapter 6
By-Products of Biodiesel and Their Recycling

Lin Lin, Mohamed Abdel Shafi Abdel Samie, and Haiying Cui

Abstract After transesterification reaction for biodiesel production, the glycerol-rich phases have to be separated. These phases are traditionally separated by settling, requiring long residence time, especially when soaps and gels are formed. This chapter presents the use of glycerol, showing the quantity and diversity of work related to the search for alternatives to add value to glycerol. It was found that 85.7% of the filed patents fell into four categories of applications: (1) manufacture of chemical products; (2) production of polymer compounds; (3) production of biofuels and biogas; and ((4) purification and use of glycerol. In this regard, the development of new technologies for the use of glycerol, without the need for traditional purification steps, is fundamental. This chapter portrays the efforts that have been made in this direction and the obstacles that still have to be overcome.

Keywords Biodiesel · By-product · Glycerol · Recycling

6.1 Introduction

To enhance reaction yield in biodiesel production, an excessive amount of alcohol is often needed. Therefore, the recycling of methanol is a significant approach to cut production cost. Besides, since glycerol is produced in the reaction, we should consider its recycling and refining as well. Given that methanol recycling is relatively simple, we will focus on glycerol.

L. Lin · M. A. S. Abdel Samie · H. Cui (✉)
School of Food and Biological Engineering, Jiangsu University, Zhenjiang, Jiangsu, China
e-mail: cuihaiying@ujs.edu.cn

M. A. S. Abdel Samie
Faculty of Environmental Agricultural Sciences, Arish University, El-Arish, North Sinai, Arish, Egypt

C. Li et al. (eds.), *Industrial Oil Plant*,
https://doi.org/10.1007/978-981-15-4920-5_6

6.1.1 Physicochemical Characteristics and Uses of Glycerol

Pure glycerol [$C_3H_5(OH)_3$] is a colorless, sweet, and viscous liquid with a relative density of 1.261/l(20 °C). By measuring the relative density of glycerol solution, we can calculate the amount of glycerol. At normal pressure, the boiling point of glycerol is 290 °C. Glycerol has a high viscosity. Pure glycerol has a viscosity 777 times more than that of water, and 50% glycerol solution has a viscosity 5.41 times more than that of water. Glycerol can be mixed with water, methanol, and aniline in any proportion. It is soluble in acetone, alcohol and ether mixtures, or chloroform and ethanol mixtures of a certain mass ratio [1, 2]. Meanwhile, it is hardly soluble in anhydrous ether, and slightly soluble in oils and fats, gasoline, benzene, chloroform, carbon disulfide, and other organic solvents. Glycerol is a preferable solution for many organic compounds, inorganic salts, caustic soda, and heavy metal soaps.

Glycerol is a trihydric alcohol with typical chemical characteristics of that type. It can be involved in many chemical reactions and produce many derivatives.

Since glycerol has a large number of important physicochemical properties, it has become a crucial chemical raw material [3–5] Today in China, glycerol is mainly used in the production of paint, food, medicines, dynamite, toothpaste, cellophane, and insulating materials.

6.2 Production of Crude Glycerol

Since biodiesel production is yet to form a large-scale industry in China, recycling by-products of the production is not a meaningful approach [6]. But with the developing oils and fats manufacture industry, biodiesel is steadily heading toward a mass-production industry. In this sense, the recycling and reuse of by-products are becoming more important [7, 8]. In fact, the recycling and reuse of glycerol in biodiesel are similar to that in oils and fats hydrolysis wastewater and in saponification wastewater [9–11]. We may adapt the glycerol recycling approach in both cases to biodiesel production. Below is a brief introduction to this glycerol recycling approach:

(1) Raw materials of glycerol and their origins

Glycerol products can be divided into chemically synthesized glycerol and natural glycerol. Most synthesized glycerols are made through chemical synthesis of petrochemical products. Natural glycerol widely exists in nature mainly in the form of triglycerides (vegetable oils and fats and animal fats). It is produced by saponification, hydrolysis or alcoholysis of oils and fats. Biodiesel production by transesterification falls under the alcoholysis category. With oils and fats as industrial raw material, a huge amount of wastewater is produced in the production of fatty acids, soaps, and biodiesel. Such wastewater contains useful substances including glycerol. If directly emitted, they not only contaminate the environment, but also waste resources. When recycled and reused, they can create significant social and economic benefits.

At present, the main raw materials of natural glycerol originate from the oils and fats hydrolysis wastewater and saponification wastewater. The production process is as follows: raw material wastewater → wastewater purification → concentrated dilute glycerol solution → crude glycerol → refining → finished glycerol product.

(2) Purification of raw material wastewater

In the oils and fats hydrolysis wastewater and saponification wastewater, there are many types of impurities in addition to glycerol. Raw material wastewater must be purified before concentration to remove the impurities. This approach can help avoid glycerol losses due to solution fermentation, reduce corrosion of glycerol solution on evaporation equipment, reduce fouling on equipment, improve heat transfer efficiency, facilitate evaporation and distillation, and enhance the quality and recovery percentage of glycerol.

For different glycerol production techniques [12–15], the impurities in wastewater have various properties and amounts. Therefore, different purification methods are needed, including chemical purification, ion-exchange purification, and electric purification.

① Chemical purification

a. Purification of oils and fats saponification wastewater. Oils and fats are saponified and salted out to produce wastewater that contains 10–15% of salt, 5–10% of glycerol, 0.1–0.5% of total alkali (sodium hydroxide and sodium carbonate), less than 0.25% of soap and other trace components including protein, mucus substances, resin, hydrocarbons, pigment, fermentation products, and sediments [16, 17]. The wastewater must first be purified to ensure the quality of glycerol, and the operation conditions of evaporation and distillation must be improved as much as possible.

The purification of oils and fats saponification wastewater includes three steps—acid treatment, degumming, and base treatment. The process is presented as follows:

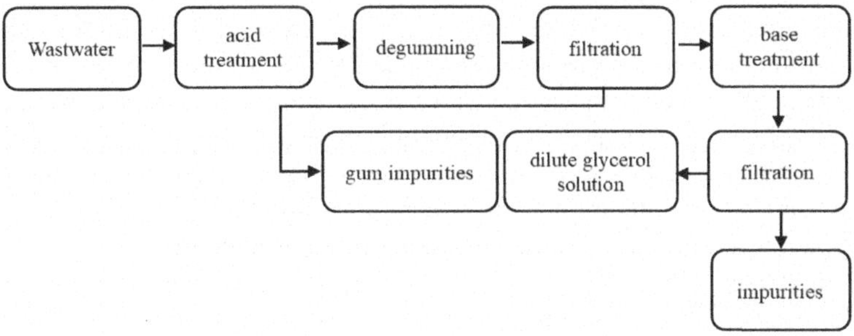

I. Acid treatment. Add inorganic acid to saponification wastewater. Acid is added to neutralize free bases and decompose the small amount of soap into fatty acids so that they can float on the solution surface to be removed (but some of the fatty acids precipitated are difficult to remove because of their colloidal state and dispersion in the solution). This step helps reduce the amount of slime residues to be filtered in the following process.

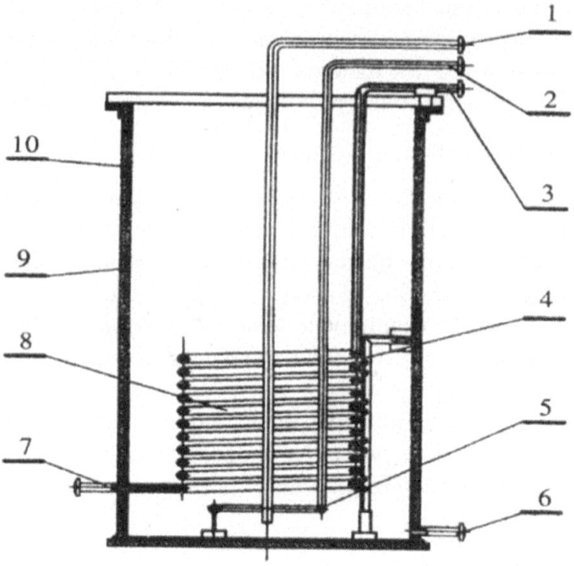

Fig. 6.1 Reaction tank for saponification wastewater purification
1-Compressed air nozzle 2-Direct steam connection pipe 3-Indirect steam connection pipe 4-Coil support 5-Direct steam nozzle 6-Material discharge pipe 7-Exhaust connection pipe 8-Heating coil 9-Lining 10-Tank main structure

The acid treatment equipment for saponification wastewater is normally made of ordinary carbon steel plates of 8–10 mm. Due to corrosion by acid and salt, they can only last for three to four years. To solve this problem and extend the equipment lifespan, we may use stainless steel plates that are acid- and salt-resistant as raw material. But this will cost a lot more. Therefore, we can use carbon steel plates as the main material and apply epoxy fiber-reinforced plastic to the inner surface. This approach lowers the equipment cost and meets the process requirements. The structure of the treatment tank is described in Fig. 6.1.

II. Degumming. Add flocculant to the crude waste solution to turn the water-soluble sodium soap into insoluble metallic soap. The metallic soap and flocculant are hydrolyzed in solution to form hydroxide sol, and absorb other impurities that precipitate. Meanwhile, under the influence of flocculant metal ions, electric neutralization takes place between a small amount of soap in a colloidal dispersion state and charged impurities, resulting in their coagulation. The waste solution contains protein and its decomposition product—amino acid. As amphoteric compound, they precipitate in acidic solutions when they reach the isoelectric point.

The flocculant used can be aluminum salts (aluminum sulfate and alum) or ferric chloride. Aluminum salt turns sodium soap into insoluble aluminum soap, and it is hydrolyzed to form aluminum hydroxide sol. Aluminum salt is inexpensive, and its colloids have a strong precipitation effect than colloids with iron ions. However, it tends to produce sodium sulfate. Just like acid treatment, this will have negative impacts on the evaporation and recycling of salt for reuse and lead to more glycerol losses. In general, ferric chloride is a preferred option. At present, most factories use ferric chloride for degumming.

The main degumming equipment is the same as acid treatment equipment.

III. Base treatment [18]. After degumming and filtration, the acidic filtrate still contains an excessive amount of ferric chloride. Base treatment is applied to neutralize acid and reduce corrosion on the equipment. It also transforms ferric chloride into ferric hydroxide precipitate and absorbs impurities that can be removed through filtration.

Normally, caustic soda solution is used for this purpose. The amount of base solution used in base treatment is crucial to glycerol quality, recycling, and evaporation procedures. Since acidic solution also contains a little lower fatty acid soap, the amount of base added should be controlled to avoid foam during evaporation and facilitate the hydrolysis process of soap solution as it turns into fatty acids. Besides, part of the lower fatty acids will escape during the evaporation process, which enhances the glycerol quality. Otherwise, soap tends to foam during the evaporation, which will cause unnecessary loss of substances including glycerol, and lead to salt coagulation in the heating pipe, which decreases the efficiency of evaporation equipment and adds to the burden of equipment washing. Moreover, excessive bases will lead to extra losses due to glycerol polymerization during distillation. However, insufficient base will cause the solution to become acidic. For base treatment, the pH value should be strictly maintained between 8 and 9, particularly not exceeding 9.

b. Purification of oils and fats hydrolysis wastewater [19]. The glycerol concentration in oils and fats hydrolysis wastewater is usually between 10 and 15%. The composition and amount of its impurities are relevant to the hydrolysis method. Whatever method is used, the impurities in wastewater affects the concentration and distillation operations, as well as the quality and recycling percentage of glycerol. Therefore, oils and fats hydrolysis wastewater should also be purified.

Catalytic hydrolysis of oils and fats produces more impurities in its wastewater [20], including sulfuric acid, sulphonic acid, low molecular weight organic acids, and colloidal impurities. Its process equipment flow diagram is described in Fig. 6.2, and the purification process flow diagram is presented below:

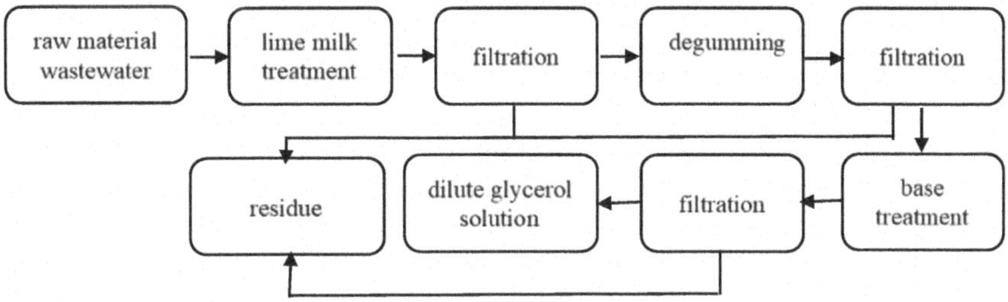

I. Lime milk treatment. Lime milk is added in order to neutralize the acidic solution, and allow calcium hydroxide to react with fatty acids and sulfuric acid in the waste solution to form precipitates such as calcium soaps and calcium salts. In the meantime, with calcium hydroxide as electrolyte, part of the colloidal impurities are flocculated and precipitated. This will help reduce the corrosion on the evaporation equipment, facilitate evaporation operation, and ensure the quality of glycerol.

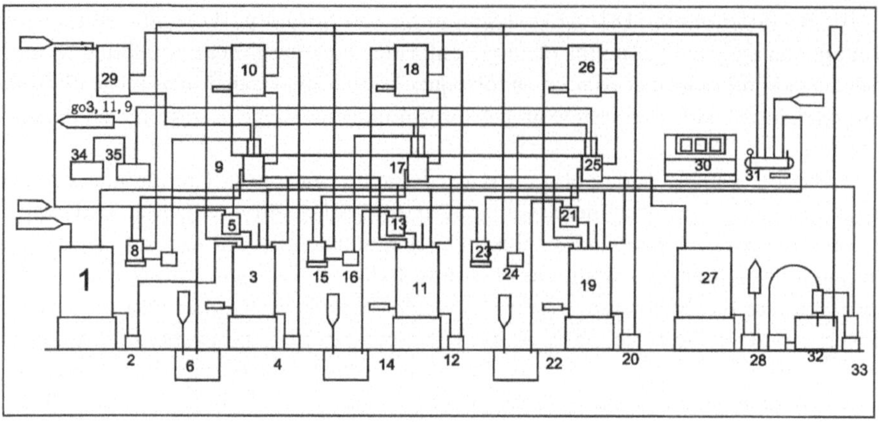

1. Crude waste liquid storage tank 2 . tank 3. Deacidification tank 4. High-level metering tank for deacidifier
5. High-level metering tank for deacidifier 6. Deacidification tank 8. Alkaline recycling tank 9 .Filter press;
10.High storage tank 11.Degumming tank 12.pump 13.Flocculant High-level Quantitative Tank 14.Flocculation tank
15.Acid return tank 16.pump 17. Filter press 18. High storage tank 19 .Soda tank 20.pump 21.High level metering tank
22.Caustic soda tank 23.Return water tank 24.pump 25.Filter press 26.High storage tank
27.Net waste liquid storage tank 28.pump 29.High water storage tank 30.Industrial washing machine 31.Steam package
32.Water jet vacuum pump 33.Vacuum balance pump 34.Air compressor 35.Gas storage tank

Fig. 6.2 Basic flow diagram of wastewater treatment from oil hydrolysis process

II. Degumming [21]. Degumming is to add flocculant to the clear filtrate after lime milk treatment. The purpose is to allow the colloidal impurities to be flocculated and precipitated, just as in the degumming process in saponification wastewater treatment.

The flocculant used is also similar. It can be aluminum salts (aluminum sulfate and alum) or ferric chloride. The principles of degumming in this case is similar to saponification wastewater degumming. The difference, however, is that the hydrolysis agent used for oils and fats hydrolysis contains low molecular weight petroleum sulfonate [22]. Accordingly, lime milk treatment produces calcium salts that are soluble in glycerol solution and that cannot be removed. Therefore, ferric chloride is the preferred option for degumming because it helps to produce precipitable sulfonic acid iron. Despite having a slightly weaker precipitation effect than aluminum salts, ferric chloride generally does better in terms of the overall purification performance.

III. Base treatment. The glycerol waste solution after degumming still contains calcium sulfate (0.15–0.40%) and iron salt residues. They are precipitated from the solution during concentration and form deposit on the inner wall, which reduces heat transfer and evaporation efficiency and negatively affects the quality and recycling of glycerol. They need to be further purified.

The last step of glycerol purification can be completed with any one of the following compounds: soda, sodium oxalate, and barium carbonate.

The procedure is to add soda ($NaCO_3$) or sodium oxalate ($Na_2C_2O_4$) to the waste glycerol solution for neutralization. As more soda or sodium oxalate is added, the waste solution will appear neutral or slightly alkaline. At this point, soda or sodium oxalate will react with calcium sulfate to form calcium carbonate or calcium oxalate, respectively, as well as soluble sodium sulfate. Meanwhile, in a slightly alkaline environment, ferric chloride is transformed into ferric hydroxide or ferric oxalate precipitate. Ferric hydroxide further absorbs impurities to be removed by filtration.

The aqueous solution containing barium carbonate ($BaCO_3$) is alkaline [23]. Adding barium carbonate in the first step will neutralize the acidic glycerol waste solution and forms barium sulfate by reacting with calcium sulfate. As more barium carbonate is added, the solution turns neutral or slightly alkaline. Barium carbonate then reacts with calcium ions in the solution to form calcium carbonate, while ferric chloride is transformed into ferric hydroxide, which further absorbs impurities to be removed by filtration.

In the final step, soda or sodium oxalate is used for purification. The purified waste solution contains water-soluble sodium sulfate. Sodium sulfate will not lead to precipitate in the evaporation equipment, but it does not reduce the ash content of glycerol. Therefore, barium carbonate is a better option to remove calcium sulfate and reduce the ash content. However, due to the high cost of barium carbonate, many factories use soda for the final purification step.

② Purification by ion exchange

Ion-exchange resin is a good option to purify dilute glycerol solution, and it is widely used in the industry [18, 22].

Ion-exchange resin is a complicated, network-structured organic substance. With many exchangeable active genes, it is stable in common acid, basic, or other solutions. Ion-exchange resins can be categorized into strong acid, weak acid, strong base an weak base types. It depends on whether it contains acidic or alkaline genes and how strong the acidity or alkalinity is. Ion-exchange resins can also be categorized into cation and anion types. The former is represented by the strong sulfonic acid type, which is widely applicable in acidic, alkaline, and neutral solutions for exchanging simple or complicated inorganic and organic cations. Common cation exchange resins include sulfonated coal, sulfonated styrene-divinylbenzene copolymers, and sulfonated phenolic resin. The latter is strongly alkaline, and can be applied in acid, basic, or neutral solutions. It can react with strong or weak acid for anion exchange. Common anion exchange resins are those contain $-N(CH_3)_3OH^-$, including styrene-divinylbenzene-chloromethyl-trimethylamine resin.

The basic principle for the purification of dilute glycerol solution using ion-exchange resin is as follows. Ionizing impurities such as sodium chloride, sodium sulfate, other salts and fatty acid, soap and pigment solutions are removed by exchange and absorption through a set of cations and anions exchange equipment. Glycerol is nonelectrolyte and will not transform in this process. Some impurities including polyglycerol and esters are difficult to remove, but they are rare in glycerol. Therefore, with proper treatment, no other process except evaporation and concentration is needed to produce refined glycerol.

After ion exchange purification, the glycerol solution is concentrated to a certain level. If the evaporation equipment is made of stainless steel, no bleaching process is needed as the finishing step. For any other material, the solution should be discolored.

③ Electric purification [24, 25]

Chemical purification does not necessarily disrupt the emulsion in glycerol solution. The addition of extra chemicals normally increases the ash content of the solution and leads to glycerol loss.

By using direct current to purify glycerol waste solution, the emulsion of the solution is disrupted with the presence of electric field. The particles in disperse phase float on the surface layer of the solution in the state of "bubble cap", while the clear, purified glycerol solution is left on the lower layer.

Lab experiments have been made to study the effect of the following factors on glycerol solution purification. The parameters have been determined as follows:

Electric field strength: no less than 30 V/cm
Electrode material: aluminum
Electrode spacing: the spacing should be as small as possible and avoid danger
Temperature: not lower than 70 °C
Glycerol liquid layer thickness: 70 cm.

On-site experiments using these parameters in factories function well. One ton of 10–12% glycerol wastewater has an average electricity consumption level of 18.6 kW·h and 0.23 kg aluminum consumption. On the anode attach aluminum paste and other residues that need to be cleaned regularly. This is a major weakness of this approach. There must be accompanying methods for precipitation prevention or cleaning.

6.3 Concentration of Purified Glycerol Wastewater

After purification, most impurities in glycerol solution have been removed [26]. But the glycerol concentration does not increase as much water still remains. As to the dilute glycerol solution made from wastewater treated by oils and fats saponification and hydrolysis, it still contains much salts and thus has to be dehydrated and concentrated. In this way, the salts are gradually crystallized and precipitated to the point that the glycerol level in solution reaches about 80%, known as crude glycerol.

(1) Concentration theories [27, 28]

The theoretical basis of dilute glycerol solution concentration is the evaporation theory. As to the evaporation process, the glycerol solution is first heated to its boiling point so that part of the water is gasified and escapes, and the glycerol concentration increases. In fact, during the evaporation of dilute glycerol solution, a tiny amount of glycerol is gasified along with water. Since the boiling point of glycerol is way higher than that of water, the amount is too tiny to be taken into serious consideration. Therefore, glycerol is regarded as nonvolatile during the evaporation and concentration of dilute glycerol solution. As water keeps vaporizing, salts are crystallized and precipitated, so that the dilute glycerol solution is concentrated.

Fig. 6.3 Equipment for evaporation at normal pressure

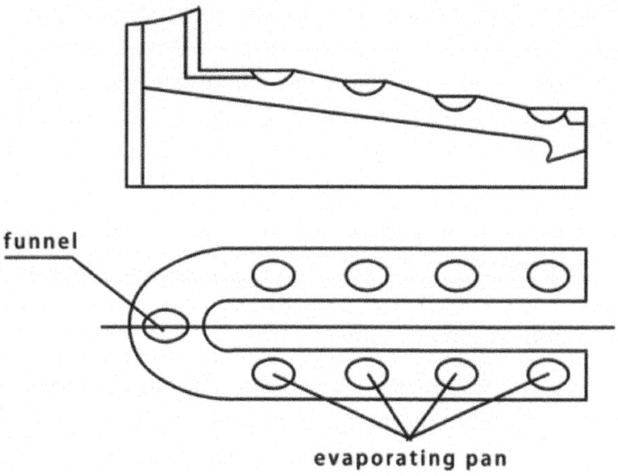

(2) Concentration methods

The concentration process of glycerol can be operated by evaporation under normal pressure or in a vacuum state.

④ Evaporation under normal pressure

During the evaporation process under normal pressure, the solution is usually placed in a huge iron pot to be heated by the fire in the stove or by steam. The set of iron pots are preferably assembled as illustrated in Fig. 6.3, so that they can save energy and achieve continuous production. The addition of glycerol solution starts from the pot that is closest to the chimney. When the concentration reaches a certain level, use a spoon to transfer the glycerol to the next pot. Repeat this process to the last pot to get the final crude glycerol. There should be a level difference between each pot to facilitate glycerol transfer.

During the operation, the half-finished products in the pots are controlled by temperature [29]. The temperature in the last pot should be controlled at about 125 °C, at which point the glycerol concentration stands at 65–70%. The temperatures in the preceding ones decrease by 4–5 °C for each pot. Increasing the number of pots used can improve thermal efficiency. However, the temperature of the flue gas should be maintained at or above 100 °C.

During glycerol evaporation, the salts in the waste solution will be gradually crystallized and precipitated. To minimize the effects of crystallized salts on evaporation, discs made of iron sheet are suspended in the evaporation pot. It allows salts to crystallize on it as the solution is boiling, so that they can be easily removed.

Evaporation at normal pressure requires only basic equipment and is easy to operate [30]. However, evaporation in open air will lead to glycerol loss, especially in the case of glycerol of high concentration. 3% of the glycerol is lost at a concentration level of 60, and 6.5% at 88%. Additionally, a higher concentration requires a higher temperature, which will cause the glycerol to polymerize, thus negatively affecting its color. Furthermore, it also consumes more fuel (to increase the concentration from 6 to 65%, 1.3–1.7 tons of coal is needed for every ton of glycerol waste solution) and the process takes a long time. Therefore, this approach is applicable only in small factories with limited conditions for the pre-concentration of glycerol solution. Due to the high value of glycerol, every bit of loss is costly. Therefore, vacuum evaporation is a more common method.

- Vacuum evaporation. Vacuum evaporation can be categorized into single-effect evaporation and multi-effect evaporation. The glycerol concentration process can be categorized into single-effect evaporation, multi-effect evaporation, and the combination of the two.

For single-effect evaporation, it requires a smaller equipment investment. But it consumes more thermal energy because the secondary steam cannot be reused. The condensation of the secondary steam requires a large amount of cooling water, which increases the production cost.

Multi-effect evaporation takes advantage of secondary evaporation to improve thermal efficiency, saving fuel as well as cooling water. Multi-effect evaporation is similar to single-effect evaporation in terms of evaporation efficiency. For a given temperature difference, the multi-effect evaporator efficiency, or the amount of steam generated from the heated area per square meter is smaller than that of single-effect evaporator. However, single-effect evaporation costs less.

Multi-effect evaporation requires several evaporators that increase the equipment cost. Meanwhile, it is not feasible to increase the concentration of crude glycerol using multi-effect evaporator. This is because the boiling point of glycerol solution increases with its concentration, so does the amount of heat required for evaporation, which exceeds the heat provided by secondary steam. Therefore, vacuum state is required for the process to accelerate the evaporation speed to match the first-effect evaporation operation (using forward feeding).

At present, glycerol factories utilize double-effect evaporation in the first stage. When the glycerol concentration reaches a certain level, they continue to use single-effect evaporation (Fig. 6.4).

(3) Main equipment for dilute glycerol solution concentration

The vacuum evaporation system consists of three major components: vacuum evaporator with a trap, condenser, and vacuum pump.

(1) Vacuum evaporator. Circulating evaporator is used for glycerol solution evaporation. For this type of evaporator, the solution is circulating in the evaporator, which improves heat transfer efficiency. At present, most glycerol factories are using central circulating evaporator (standard-type), suspending-basket-type evaporator, and the most commonly used external heating evaporator.

Central circulating evaporator (standard-type). Its structure is shown in Fig. 6.5. The heating chamber contains a vertical heating tube (or "boiling tube") bundle. In the center of the bundle there is a tube with a large diameter, known as the central circulating tube. Since the central circulating tube has a large cross section area, the heat transfer area of the solution per unit volume is smaller than that in the other

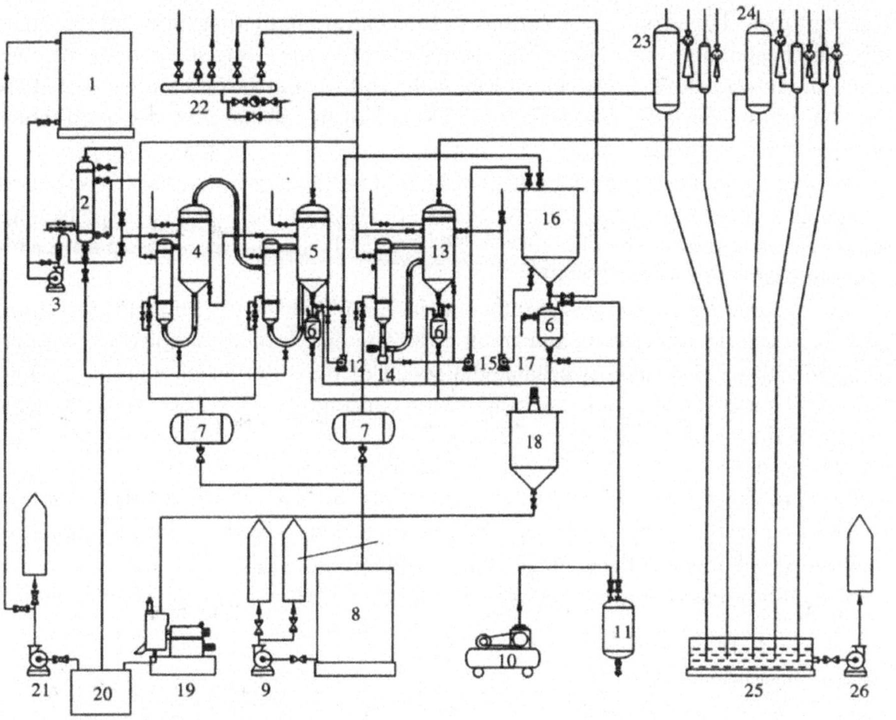

Fig. 6.4 Process flow diagram of dilute glycerol solution evaporation
1-Waste solution header tank 2-Evaporation preheater 3-Feeding pump 4-First evaporator 5-Second evaporation 6-Salt collection tank 7-Exhaust steam drum 8-Centrifugal pump 9-Condensate storage tank 10-Air compressor 11-Gas storage tank 12-Pump 13-Evaporator 14-Circulating pump 15-Pump 16-Salt precipitation tank 17-Pump 18-Salt washing tank 19-Peeler centrifuge 20-Cleaning liquid recovery pool 21-pump 22-Steam distributor 23-Vacuum system 24-Vacuum system 25-Water sealing pool 26-Circulating water pump

Fig. 6.5 Standard-type evaporator

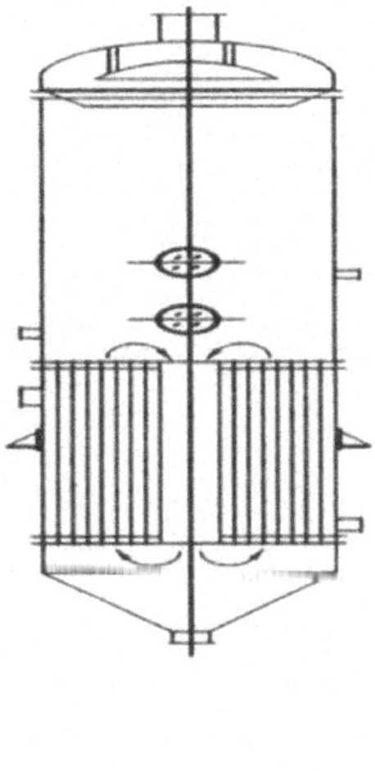

heating tubes. Therefore, the solutions in the central circulating tube and the other tubes are heated unevenly. Due to the suction effect by the rising steam in the heating tubes, the solution level in the central tube falls while that in the other tubes rises. This circulation improves the heat transfer efficiency of the evaporator, thus facilitating the evaporation process.

In order to facilitate the solution's circulation in the evaporator, the cross section area of the central circulating tube is usually 40–100% of the total area of the heating tubes. The height of the heating tube is 0.6–2 m, and its diameter is 25–75 mm, with a height-diameter ratio of 20–40.

The advantages of this evaporator are its simple structure, reliable operation, good heat transfer efficiency, and small cost. The disadvantages, however, are that it is difficult to clean and overhaul, the speed of the solution circulation is slow (normally under 0.5 m/s), and the evaporation concentration level is almost identical to that of the resultant solution with a high viscosity and high boiling point, thus decreasing the temperature difference during heat transfer and obstructing the heat transfer efficiency.

The evaporator featuring a central circulating tube is widely used in chemical industry. It is known as the standard-type evaporator.

Suspending-basket-type evaporator. Its structure is shown in Fig. 6.6. It is named

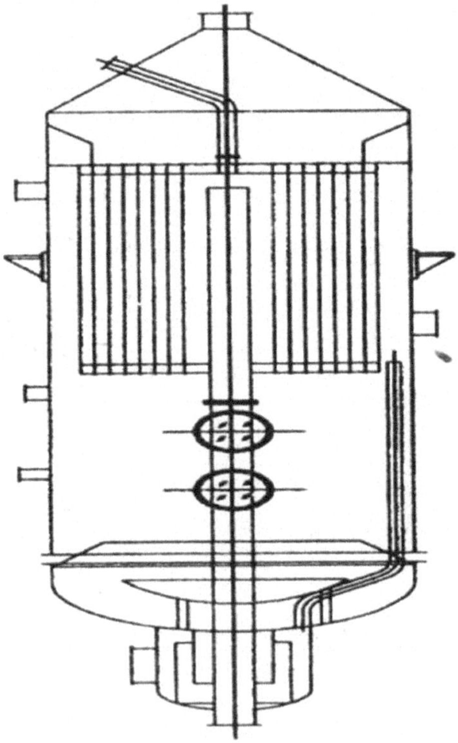

Fig. 6.6 Suspending-basket-type evaporator

as such because the heating chamber looks like a basket hanging in the lower section of the evaporator casing. The solution circulation in this evaporator is similar to that in a standard-type evaporator. But the circulating route is descending along the crevice between the heating chamber and the evaporator casing, and ascending along the heating tube. The circular cross section area of this evaporator is 100–150% of the heating tube. Therefore, the circulating speed of the solution is faster than that in a standard-type evaporator at 1–1.5 m/s. Since the evaporator casing is in direct contact with the boiling liquid of a relatively low temperature, the heat loss is small. Additionally, since the heating chamber is removable from the top of the evaporator, it is easy to wash, overhaul, and replace. The disadvantages of this evaporator lie in its complicated structure and higher metal consumption for per unit heat transfer area. This design is applicable to solutions that tend to form fouling or crystallize during evaporation.

External heating evaporator. Its structure is shown in Fig. 6.7. The heating chamber is installed outside the separation chamber. Therefore, it is easy to clean and replace, and reduces the height of the evaporator. This evaporator has a long heating tube, and its circulating tube is not heated by steam. Therefore, the circulation speed is fast. As the glycerol solution evaporates, the vapor enters the heating chamber through its

Fig. 6.7 Self-circulation external heating evaporator

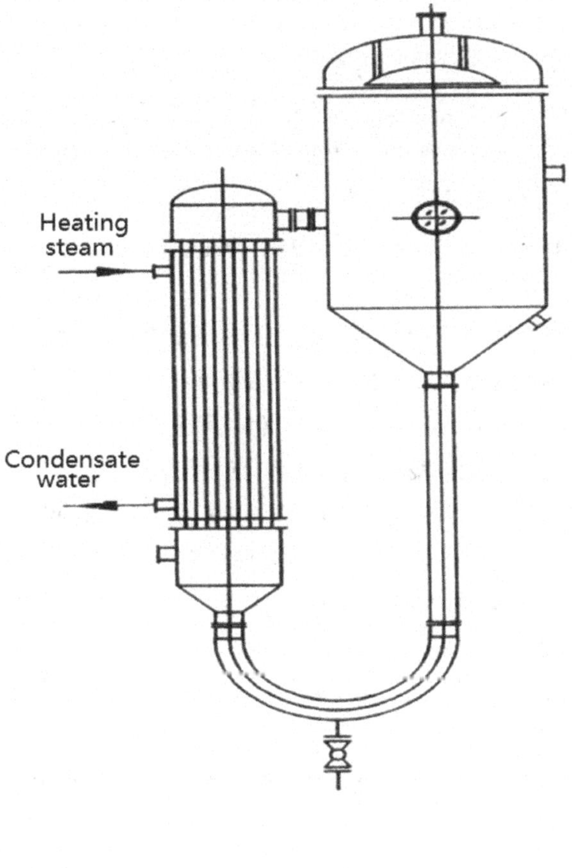

bottom. Its density decreases after heating, so the vapor rises along the heating tube into the separation chamber. In a vacuum and suddenly expanding state, the water in the solution evaporates rapidly. The solution becomes more condensate and has a high density. It then enters the heating chamber through the connection pipe, draining the hotter, lighter solution into the separation chamber. This process is repeated as above.

Since the solution circulates fast in this evaporator, the overall production efficiency is better than the other two evaporators mentioned above, and it is also easy to wash. Therefore, it is widely adopted by glycerol factories.

However, when processing thermally unstable substances like glycerol, if they stay in the heating zone for an excessive amount of time, there will be negative impacts on the end products, which is the main disadvantage of this evaporator. In particular, in cases where the concentrated glycerol has been refined by ion-exchange resins, this disadvantage stands out. This is because this refinement process can produce products better than common distilled glycerol. If they are treated by the above three evaporators, their quality may be affected. Centrifugal evaporator and membrane rotary evaporator can reduce the time required for the evaporation process and minimize thermal conversion of the production.

② Auxiliary equipment for vacuum evaporation

Collector. It is used to collect the liquid drops carried by the steam from the separation chamber, in order to reduce glycerol loss. The collector manages to change the velocity and direction of the

steam movement, or effects the centrifugal force to separate steam from liquid.

A simple form of collector consists of a cylinder equipped with one or several baffles. The incoming steam hits the baffles and changes direction and speed, leaving the drops it carries in the collector. If the steam enters the collector in a tangent direction (Fig. 6.8) and travels through the spiral plate where it is under the centrifugal

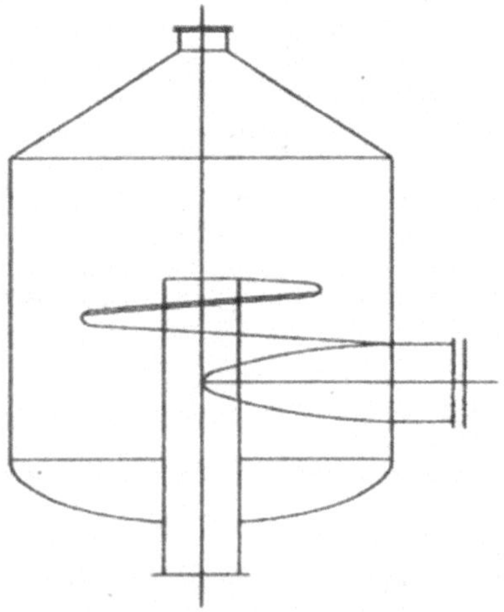

Fig. 6.8 Collector

force, the separation process will become more efficient.

② Vacuum pump. The vacuum state as in vacuum evaporation is achieved through vacuum pump. The vacuum pump applicable in glycerol solution evaporation can be a mechanical vacuum pump, a water jet pump, or a steam jet pump. The selection of the proper equipment is based on the vacuum level and air displacement capacity required, referring to specific on-site conditions.

A. Mechanical vacuum pump. This pump has a limited vacuum level and a low air extraction capacity. If the evaporation process does not require a strict vacuum level, mechanical vacuum pump can be a proper option. However, glycerol evaporation normally generates a huge amount of steam, which requires a high air extraction capacity of the vacuum pump. If a mechanical vacuum pump is selected, a condenser of a certain capacity should be installed between the pump and the evaporation equipment, in order to ensure the required vacuum level for the pump to function properly.

B. Water jet pump. This pump has a limited vacuum level but a high air extraction capacity. If the glycerol evaporation process does not require a strict vacuum level, water jet pump can be a proper option. Furthermore, this pump has a low energy consumption and is resistant to damage, which lowers the glycerol production cost.

C. Steam jet pump. This pump can be categorized into single-stage and multi-stage types. Multi-stage jet pump has a high vacuum level threshold and a high air extraction capacity. It is applicable in the glycerol concentration process that demands a strict vacuum level and a high air extraction capacity. The glycerol evaporation process mostly adopts a two-stage or three-stage jet pump. Multi-stage pumps have strict steam pressure requirements, larger steam consumption, more equipment investment, and higher production cost. But it ensures high glycerol quality and recycling rate.

(3) Other equipment. In addition to the above equipment, the glycerol concentration process also requires auxiliary equipment including salt precipitation tank, salt washing tank, separator, and steam condensate storage tank.

References

1. Chen J, Jiang JC, Xu JM et al (2011) Production of renewable liquid fuel by catalytic cracking of soybean oil. Chem Indus For Prod 31(4):25–30
2. Chen J, Jiang JC, Xu JM (2010) Process research on the production of renewable liquid fuel from catalytic cracking of soybean oil. In: International conference on chemical and biological utilization of biomass resources
3. Choi JH, Jeong KJ, Dong Y et al (2006) Electro-oxidation of methanol and formic acid on PtRu and PtAu for direct liquid fuel cells. J Power Sources 163(1):71–75
4. Consultant DIScmp (2006) Converting waste plastics into liquid fuel by pyrolysis: developments in China. John Wiley & Sons, Ltd 28
5. Demirbaş A (2000) Conversion of biomass using glycerin to liquid fuel for blending gasoline as alternative engine fuel. Energy Convers Manag 41(16):1741–1748
6. Dote Y, Sawayama S, Inoue S et al (1994) Recovery of liquid fuel from hydrocarbon-rich microalgae by thermochemical liquefaction. Fuel 73(12):1855–1857

7. Fallah RN, Azizian S (2012) Removal of thiophenic compounds from liquid fuel by different modified activated carbon cloths. Fuel Process Technol 93(1):45–52
8. Huang DM, Jiang WW, Lin L et al (2013) Study on the catalyst performance on cornus wisoniana oil catalytic cracking prepared biological fuel oil. Appl Mech Mater 477–478(477–478):1457–1463
9. Kunkes EL, Simonetti DA, West RM et al (2008) Catalytic conversion of biomass to monofunctional hydrocarbons and targeted liquid-fuel classes. Science 322(5900):417–421
10. Li CZ, Zhang AH, Xiao ZH et al (2011) Preparation of bio-diesel from Cornus wilsoniana fruit oil with basic ionic liquids as catalyst. J CentL South Univ For Technol 31(3):38–43
11. Lin L, Cui H, Vittayapadung S et al (2015) Synthesis of recyclable hollow Fe/C—SO3H fiber as a catalyst for the production of biodiesel. Environ Prog Sustain Energy 33(4):1432–1437
12. Liu CW, Yang Z, Dai XA et al (2005) Study on the preparation of bio-diesel by vegetable oil. Henan Chemical Industry
13. Lu L, Quan K, Xu J et al (2014) Liquid hydrocarbon fuels from catalytic cracking of rubber seed oil using USY as catalyst. Fuel 123(123):189–193
14. Luo Z, Wang S, Liao Y et al (2004) Research on biomass fast pyrolysis for liquid fuel. Biomass Bioenerg 26(5):455–462
15. Nian LI, Zhong SA (2011) Preparation of biodiesel from Cornus wilsonian oil catalyzed by solid cesium phosphotungstates. J Cent South Univ 42(5):1226–1231
16. Oasmaa A, Kuoppala E (2003) Fast pyrolysis of forestry residue. 3. Storage stability of liquid fuel. Energy Fuels 17(4):1075–1084
17. Ooi YS, Zakaria R, Mohamed AR et al (2004) Catalytic cracking of used palm oil and palm oil fatty acids mixture for the production of liquid fuel: kinetic modeling. Energy Fuels 18(5):1555–1561
18. Ooi YS, Zakaria R, Mohamed AR et al (2004) Catalytic conversion of palm oil-based fatty acid mixture to liquid fuel. Biomass Bioenerg 27(5):477–484
19. Ooi YS, Zakaria R, Mohamed AR et al (2005) Catalytic conversion of fatty acids mixture to liquid fuel and chemicals over composite microporous/mesoporous catalysts. Energy Fuels 19(3):736–743
20. Qiana W, Wilkinsona DP, Shena J et al (2006) Architecture for portable direct liquid fuel cells. J Power Sources 154(1):202–213
21. Ramya G, Sudhakar R, Joice JAI et al (2012) Liquid hydrocarbon fuels from jatropha oil through catalytic cracking technology using AlMCM-41/ZSM-5 composite catalysts. Appl Catal A 433–434(31):170–178
22. Todsapon T, Rauchfuss TB (2010) Efficient production of the liquid fuel 2,5-dimethylfuran from fructose using formic acid as a reagent. Angew Chem Int Ed 122(37):6766–6768
23. Twaiq FAA, Mohamad AR, Bhatia S (2004) Performance of composite catalysts in palm oil cracking for the production of liquid fuels and chemicals. Fuel Process Technol 85(11):1283–1300
24. Wang Z, Yu ST (2016) The production of liquid hydrocarbon fuel from catalytic cracking of rubber seed oil using new mesoporous molecular sieves. Acs Sustain Chem Eng 4(10)
25. Zhang A, Li C (2014) Swida wilsoniana oil based biodiesel catalyzed by 1-(4-sulfonic acid) butyl-pyridinium hydrosulfate. In: International conference on materials for renewable energy and environment, pp 355–359
26. Zhang A, Yuandong XU, Zhang Y et al (2010) Biodiesel production from swida wilsoniana fruit oil using 1-(4-sulfonic acid) butyl-pyridinium hydrosulfate ionic liquid as catalyst. In: 2010 China renewable energy science and technology development conference
27. Zhang AH, Liang-Hou LI, Lin L et al (2015) Preparation of biodiesel by magnetic solid acid fiber catalytic Swida wilsoniana oil. Applied Chemical Industry
28. Zhang AH, Xiao ZH, Zhang LB et al (2013) Study on bio-fuels drop acid by [BSPy]HSO4 catalytic esterification. Adv Mater Res 781–784:2433–2437
29. Zhang LB, Xiao ZH, Li PW et al (2015) Research on Swida wilsoniana for biodiesel production. J Biobased Mater Bioenergy 9(1)

30. Zhang SP, Yan YJ, Li TC et al (2005) Upgrading of liquid fuel from the pyrolysis of biomass. Biores Technol 96(5):545–550

Chapter 7
Technologies for Conversion Bio-Lubricant Production in Fatty Acids

Yu Cao, Ning Wang, Hongchen Fu, Fei You, and Liangnian He

Abstract In response to the increasing environmental pollution concern and depleting petroleum reserves, bio-lubricants have received wide attention in replacement of mineral oil. Vegetable oils can be used as potential source of environmentally friendly bio-lubricants because of their biodegradability, renewability, and excellent lubrication performance. Present chapter introduces properties of several raw oils as well as oil extraction methods. Besides, vegetable oils also have some limitations such as poor oxidative stability and low-temperature fluidity, which can be improved through various modification methods such as direct transesterification or selective hydrogenation. Technology for modification of raw oils has also been described. Additionally, additives have a significant influence on the performance of the lubricating oil. The characteristic and applicability of different commonly used additives in bio-lubricant have been summarized in this chapter. Castor oil has a higher viscosity and a lower viscosity index compared with other vegetable oils. This chapter in particular describes the development of production process for the castor-based biodegradable lubricant.

Keywords Bio-based lubricant · Vegetable oil · Chemical modification · Renewable resources · Castor oil

7.1 Introduction

With rapid development of machinery industry, the demand and consumption of lubricants rise continuously. Almost all lubricants are composed of base oils and additives. At present, the vast majority of base oil of lubricants is mineral oil. In consideration of the progressive depletion of the world reserves of fossil fuels, it is very necessary to develop alternatives of petroleum-based products. Furthermore,

---

Y. Cao · N. Wang · H. Fu · F. You · L. He (✉)
State Key Laboratory, Institute of Elemento Organic Chemistry, College of Chemistry, Nankai University, Tianjin 300071, China
e-mail: heln@nankai.edu.cn

H. Fu
College of Pharmacy, Nankai University, Tongyan Road 38, Tianjin 300353, China

C. Li et al. (eds.), *Industrial Oil Plant*,
https://doi.org/10.1007/978-981-15-4920-5_7

lubricants are inevitably discharged into the environment due to transportation, leakage, and natural replacement. Mineral oil-based lubricants which have high ecological toxicity and poor biodegradability may result in pollution to the environment. Such problems have aroused the increasing concern of the government and public. Therefore, the development and utilization of environmentally friendly lubricants is an important issue in the current development of lubricants. In this regard, bio-lubricants may play an important role as substitutes of mineral oil-based lubricants, resulting in significant economic and social sustainable benefits.

7.2 Development of Bio-Lubricants

The present consumption of fossil fuels has led to significant levels of environmental pollution and is rapidly diminishing petrochemical and energy reserves. Biomass could be regarded as a promising carbon-based alternative energy source and a sustainable chemical feedstock.

In this context, vegetable oil, which is extracted from the seeds or fruits of plants, is the most commonly used feedstock for developing lubricants. The base oils of bio-lubricants are mainly highly unsaturated or High Oleic Vegetable Oils (HOVOs), including soybean oil, rapeseed oil, olive oil, peanut oil, castor oil, palm oil, etc. Compared with mineral oil-based lubricants, bio-lubricants are environmentally friendly alternatives. Bio-lubricants have a number of advantages over mineral lubricants, such as

1. Renewable ability: very low or almost negligible aquatic toxicity; good lubricating properties.
2. Excellent biodegradability: in general, the biodegradation rate of vegetable oils is between 90 and 98%, whereas in the case of mineral oils this value is around 20% [4].
3. High Viscosity Index (VI): because of the strong intermolecular interactions of the double bonds and the molecule linearity.
4. Low evaporation rates: they have an approximately 20% lower evaporation rate than mineral oil-based fluids [28].

Molecularly, vegetable oil is an unsaturated fatty acid glyceride with an average relative molecular weight of 800–1000. The chain length of the fatty acids is usually in the range of C_{12}–C_{24}. The saturated fatty acids in vegetable oils mainly contain soft fatty acids (such as palmitic acid), stearic acid, etc. The unsaturated fatty acids mainly include oleic acid, erucic acid, linoleic acid, linolenic acid, and ricinoleic acid. Accordingly, the composition of fatty acids varies between feedstocks, thus determining the properties of vegetable oil.

A low melting point and good oxidative stability are essential to high-performance lubricants. The number of the carbon–carbon double bond and the chain length of fatty acids are two main factors to affect the oil properties. In general, increasing the number of double bonds, i.e., increasing the degree of unsaturation of the oil usually results in better low-temperature properties. However, high unsaturation in the molecule weight causes poor oxidative stability. A longer chain length results in a higher melting point and viscosity. Monounsaturated fatty acids, such as oleic acid and palmitoleic acid, have low melting point and good oxidative stability. Therefore, plant oil feedstocks which possess high contents of oleic or palmitoleic acid are generally preferred in this purpose.

Among the various crops with oil-bearing seeds, non-edible crops are more desirable than edible crops because they grow on land that is largely unproductive and do not in direct competition with existing edible crops resources. As a non-edible crop, castor is an excellent renewable feedstock for biomass-based energy products, such as biodegradable lubricants, bio-polymers, and biodiesel. It is easy to cultivate under drought, salt, and barren conditions. Besides, castor oil raw material is easy to scale up. At present, among the bio-energy crops, castor produces the highest yield. Castor is a key industrial crop cultivated in tropical and subtropical regions of India, China, and Brazil. Globally, around one million tons of castor seeds are produced every year, with a worldwide market of over $400 million of castor oil. Currently, US imports about $41 million worth of castor oil annually.

In recent years, the development and utilization of castor oil has gained wide attention. Castor is one of the top ten oil crops with high oil content, which is 45–60% [25]. And it is unique because main of its fatty acid content is ricinoleic acid, a monounsaturated, 18-carbon fatty acid with a hydroxyl function at 12-position. It is the only major commercial source of hydroxy fatty acid. Due to the unique molecular structure and facilely processing of castor oil, it can be considered as a great alternative to petroleum to produce a wide range of products including lubricants, coatings, surfactants, plasticizers, cosmetics, pharmaceuticals, polymers, cosmetics, soaps, inks, fungicides, and others.

With the development of castor technology and industry, green high-performance lubricants have been developed. Castor-based biodegradable lubricate is a kind of promising product with high viscosity and nice performance characteristics at low temperature. Castor oil is the only one which has high content of hydroxy fatty acid among vegetable oils. Its viscosity at ambient temperature is five times more than that of most vegetable oils and it also has a high viscosity index. The advantages of castor oil as lubricant base oil covers (1) renewable, non-food biomass; (2) biodegradable, eco-compatible; (3) outstanding anti-oxidation, anti-wear properties, low-temperature fluidity, and long life-time; (4) the properties controlled by easily modifying their structures.

Bio-lubricants are potential alternative lubricants. This chapter gives an overview of a bio-lubricant based on vegetable oil, especially castor oil. In Sect. 7.2.1, the source as well as oil extraction methods of the bio-lubricant are discussed. Section 7.2.2 introduces the main modification methods of raw oil, including biological modification and chemical modification. Additives have a significant influence on the performance of the lubricating oil. Section 7.2.3 describes the commonly used additives in bio-lubricants.

7.2.1 *Raw Oil for Bio-Lubricants Production*

Lubricants consist of base oil and additives. Base oil can be divided into mineral oil, synthetic oil, and bio-lubricant base oil. Unlike mineral oil and synthetic oil, bio-oil is generally obtained from the modification of plant raw oil. For the bio-lubricants, the raw oil is the oil extracted from seeds via a series of procedures. Therefore the properties of raw oil are of great importance to bio-lubricant. In this section we will introduce production methods of raw oil.

As vegetable oils are being engaged for industrial lubricants usage, the competition between using oil as food or as industrial lubricants have prompted the search for non-edible oils as suitable raw material. Castor oil, a non-edible vegetable oil produced from the seeds of castor plant, is one of the largest investigated oils for industrial and automotive lubrication [3]. Relative to other vegetable oils, it has a good shelf life and does not deteriorate unless subjected to excessive heat [27]. The castor oil is characterized by high viscosity although this is unusual for a natural vegetable oil. This behavior is largely due to hydrogen bond of its hydroxyl groups. With these advantages, castor is mainly produced as a promising resource for bio-lubricants, and the industrial production technology is developed. So we choose castor oil as an example to illustrate the process of oil manufacture.

The extraction of oil from castor seed is by one or a combination of mechanical pressing and solvent extraction. In mechanical pressing, the seeds are crushed and then adjusted to low moisture content by warming in a steam-jacketed vessel. Thereafter, the crushed seeds are loaded into hydraulic presses and they are pressed by mechanical means to extract oil. The oil from mechanical pressing has light color and low content of free fatty acids. However, mechanical pressing will only remove about 45% of the oil while the remaining oil in the cake can be recovered only by solvent extraction. In the solvent extraction method, the crushed seeds are extracted with a solvent in a Soxhlet extractor or commercial extractor. Solvents used for extraction include heptane, hexane, and petroleum ethers [27]. The seeds can be extracted after pretreatment. Conventional pretreatment may include dehulling, size reduction, breaking, grinding, and thermal treatment (cooking) with the purpose of debilitating the cell coats and preparing the material for optimal oil extraction.

7.2.1.1 Transportation

Castor seeds have a thin and crisp crust which makes seeds easy to break. Meanwhile, the nutlet of castor is soft. Once the curst is broken, the nutlet will cohere with the dust and it is difficult to separate them. Taking these into consideration, the seeds should be transported into the factory after cleaning. To avoid the breaking of seeds and tedious procedure, cleaning process is operated with the combination of hull beater and double-layer aspirated plansifter. The aspirated plansifter provides both high cleaning efficiency and low damage percentage. The size of castor seeds is generally around 13, 8.5, and 6 mm, so the holes' diameter of first layer should be around 15 mm. For second layer, the holes' diameter should be 5–6 mm, less than the minimum size of seeds. This kind of design can simplify the cleaning process. After cleaning, the seeds are transported into factory by chain-type bucket elevator, which will not skid and block. Additionally, using chain-type bucket elevator as transport implement can keep the system steady with the help of frequency control and e-metrics.

7.2.1.2 Crushing

The nuts of castor are easy to adhere with each other, so rolled compact procedure is necessary for the production of oil. To simplify the steam-roasted process, the seeds should be crashed as small as possible. The thickness of the castor seeds should be rolled between 2.2 and 2.5 mm. Furthermore, the seeds of castor are large and have a smooth shell, which makes it difficult to get into the machine, so there should be a wide gap between two rollers

7.2.1.3 Steam-Roasted

During steam-roasted process the seeds produced from crushing are humidified, heated, and dehydrated, to make it fit for pressing. The time of steam-roasted should be no longer than 1 h. The temperature and moisture should between 115–120 °C and 4–5%, respectively. Additionally, the cooker should be designed with a shallower bottom, a higher rotate speed, a larger entrance, and a stronger draft fan. Thirdly, the drive device should have a large power.

7.2.1.4 Oil Manufacture

The castor seeds are soft and contain plenty of oil, so the seeds can be squeezed with shell. Compared with the ordinary seeds, the oil press for castor needs stronger power, higher rate of feeding, and narrower gap between the cage bar and pressed cake. The time for squeeze should also be prolonged. The residual oil ratio of the castor cake, which is produced by the oil press device, should be no more than 13% constantly. Only in this way can provide a prerequisite for the low residual oil ratio for the castor flake.

7.2.1.5 Filtration

The crude oil should be filtered at high temperature, generally between 80 and 85 °C, as a result of the presence of high viscosity and adhesive impurities in the unrefined rough oil. Among several types of filtration equipments, the leaf filter with stainless wire net is the most effective, as it has a less adhesive force for the rough oil and residue than the filter cloth. Before the filtration, there is no need to remove the dregs thoroughly from the oil, as appropriate amount of big-diameter residues can prevent the filtration equipment from being bunged up. After the filtration, the filtrate can go on next process without further refined filtration. Dregs are beneficial to the separation during the alkali refining process.

7.2.1.6 Extraction

There are two difficulties in the extraction process. The first one is that compared with other oils, the castor oil is more difficult to lixiviate when using ordinary conditions, as the existence of hydroxyl group in the molecule results in high polarity and viscosity. The other one is that castor cake has a strong tackiness and poor permeability, which lead to the difficulties in the solvent leaching and desolventizing. The key to solve these two problems is increasing extraction temperature and unchocking the blockings. To ensure the temperature of the cake is above 60 °C, the delivery system should be equipped with a heat insulation layer. The solvent preheater should preheat to 55 °C. The pipelines that transport miscella should be equipped with heating jacket, ensuring the extraction temperature is 57–60 °C at which the castor oil has a good solubility and low viscosity. Meanwhile, countermeasures should be taken to prevent the blocking of the extractor and desolventizer. Combining these measures with other technical means, such as prolonging time of extraction and reducing the height of the castor cake, the residual oil ratio of castor could be controlled under 1%.

7.2.1.7 Desolvation

To assure the airflow channel unobstructed, there should be a pre-desolvation in desolventizer toaster and the thickness of the bed material should be decreased. It is of great importance for desolvation to intensify internal heat transfer, as this can avoid the oilseed meal absorbing water and clotting with each other.

7.2.1.8 Refinement

During vegetable oils' production, it is usual to refine the crude oil obtained from either mechanical pressing or solvent extraction. The main aim of refining is to remove impurities (e.g., colloidal matter, free fatty acid, and coloring matter) and other undesirable constituents, thus making the oil more resistant to deterioration during storage. Refining includes (a) removing solid and colloidal matter by settling and filtration, (b) neutralizing the free fatty acid by alkali, (c) removing colored matter by bleaching, and (d) deodorizing by treatment with steam at high temperature and low pressure. The general method of refining used for edible oils is applicable to castor oil [27]. But the condition for the refinement is harsher than other oils as a result of the high viscosity of castor oil. The ricinoleic acid is gadoleic acid with a hydroxy group, so it has a strong polarity and strong affinity to water, which means ricinoleic acid is easy to emulsify and difficult to demulsify. Although castor oil is stable at the room temperature, the acid value will decrease after a long time storage, normally below 10 mg KOH/g. Castor oil is easy to dehydrate and be oxidized under high temperature, deepening the color, reducing the transparency, increasing the viscosity as well as reducing oil quality. According to these features of the castor oil, semi-continuous refining technics, the combination of batch caustic refining and continuous desiccation, decoloration and deodorization (desolvation), is the best choice, as it can prevent the emulsification and simplify the separation steps. What's more, semi-continuous can also keep the system stable, preventing the production line breaking off.

The proportion of phospholipids in castor oil is no more than 0.3%, so the castor oil can undergo caustic refining process directly. The temperature for alkali refining is normally 60 °C. To avoid the formation of loose nigre during this process, it is necessary to use lye with high concentration. Meanwhile, adding hot water during alkali refining process can prevent the emulsification. Oil refining caldron should have enough space to add a large amount of water as well as demulsifier. After washing, alkali process is completed.

The castor oil produced from alkali refined process is deep in color as a result of the oxidation at high temperature. Therefore attapulgite, sometimes combined with a small amount activated carbon, is always used to decolor castor oil. To meet the demand of the national standard, castor oil is stirred with atlapulgite at 95 °C for 1 h. After separating castor oil and atlapulgite, decolored castor oil is produced.

To increase the efficiency of the deodorization, the vacuum degree of equipment should achieve 667 Pa. The temperature and time should also be controlled precisely for the purpose of preventing the intermolecular dehydration of the castor oil. The temperature should between 130 and 140 °C and the time should be no more than 30 min. After decreasing the temperature to 90 °C and removing vacuum, castor oil can be obtained with high quality.

7.2.1.9 Mixed Oil Refining

Although the refining process introduced before has been used widely, there are some drawbacks for this method as a result of the unique properties of castor oil. Firstly, castor oil has a high polarity and strong affinity to water, which results in the emulsification during the alkali refining process and difficulties in separation. Adding sodium chloride solution is an efficient way to demulsify, but this will cause the waste of the castor oil and the increase in cost. Additionally, the density of castor oil and water is close, so they are difficult to separate from each other after the washing process. During the desiccation, ricinoleic acid is easy to dehydrate under high temperature, yielding unsaturated diolefinic acid, which will decrease the transparency and quality of the oil. The temperature of desolvation and deodorization should be controlled strictly as castor oil is liable to be oxidized [34].

Mixed oil refining technique means that the mixed oil produced from extraction is directly refined without desolvation [8]. In mixed oil, there are lots of solvent molecules around the glyceride molecule. These solvent molecules can prevent alkali contacting with the glyceride, so the oil molecule will not be saponified. However, the free fatty acid can react with alkali as a result of a smaller steric hindrance. The purified castor oil will be produced after a combination of alkali refining process, decoloration, evaporation, and desolvation.

Compared with the first refining method, mixed oil refining process has several advantages [34]. Firstly, the viscosity and density of mixed oil are much smaller than castor oil. These features make the alkali refining process, separation and filtration simplified. Secondly, the temperature is low during the whole process, thus the pigments will not be oxidized and is easy to be adsorbed by atlapulgite. Thirdly, the impurities have been removed before evaporation, so the quality of oil and efficiency of the equipment are improved.

The production of raw oil needs a series of complicated procedures, and the quality of raw oil is influenced by the parameters of these processes. Although refined raw oil has much better properties, it cannot be used as lubricants directly. The modification and additives are needed to further improve the properties of raw oil.

7.2.2 Preparation of Bio-Lubricant Base Oil

Plant oils are produced by photosynthesis with the characteristics of cleanness, richness, and renewability. Due to the mature cultivation, refining and processing technology, there are a wide range of sources and relatively low-cost plant oils. The structure of plant oil is very different from that of mineral oil. The main component is ester, whose the molecular weight is large, and it contains unsaturated Triacylglycerol (TAG). The differences among various plant oils lie mainly on the content and species of fatty acids. According to the differences of fatty acids, it can be divided into saturated fatty acids, monounsaturated fatty acids, and polyunsaturated fatty acids. Most oilseed crops produce TAG containing a mixture of fatty acids with a chain length of 18 carbons and 1–3 double bonds. The structures and species of fatty acids play a decisive role in various properties [14].

7.2.2.1 Principles of Preparation of Bio-Lubricant Base Oil

Because of the good biodegradability and excellent lubrication performances of plant oils, they can be applied to the production of lubricants. Due to the limitation of the structure and performance of plant oil, it is difficult to apply the raw oil to the field of lubrication directly. The most important performances such as high-temperature oxidation stability and low-temperature fluidity of plant oil must be improved. Because the molecular structure of plant oil determines that it is vulnerable to be attacked by oxygen, resulting in oxidation and thermal decomposition. The

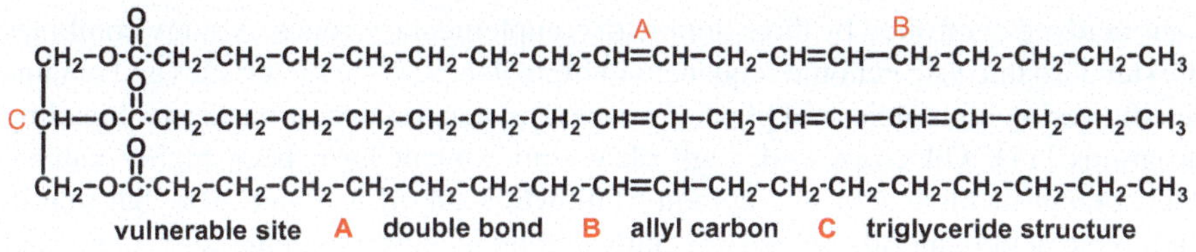

Fig. 7.1 Triacylglycerol (TAG) and the vulnerable sites

vulnerable parts of plant oil molecular, including double bond (A), allyl carbon (B), and triglyceride structure (C) (Fig. 7.1).

Saturated fatty acid glycerides are usually solid under normal temperature conditions and cannot directly be applied to lubricant base oil. The fluidity of plant oil at low temperature is closely related to the degree of unsaturation and the structure of triglyceride. TAG is easy to accumulate and form larger crystals at low temperature, which leads to poor low-temperature performance of plant oil. Different triglycerides in plant oils eventually lead to different melting points (or freezing points). In general, the longer the carbon chain and the higher unsaturation degree of fatty acids, the lower the triglyceride melting point (or freezing point) will be.

The presence of double bonds improves low-temperature fluidity, but the more the double bonds in the plant oil, and the worse the oxidation stability of the oil [14]. The oxidation mechanism of plant oil is mainly characterized by active allyl radical reaction, which is the main reason for its poor oxidation stability. The oleic acid molecule contains only one double bond, which has good thermal oxidation stability and low-temperature performance. In contrast, the oxidation stability of the polyunsaturated compounds such as linoleic acid and linolenic acid is very poor.

Currently, the main modification methods include biological modification and chemical modification:

(1) Plants with high oleic acid content are cultivated by modern biotechnology (mainly genetic engineering technology); the key to biological modification of bio-lubricant lies in increasing the content of oleic acid in plant oils.
(2) Chemical modification of plant oil aims to reduce the content of double bond or adjust the triglyceride structure to achieve the purpose of improving lubricating performance. Through chemical modification, the content of unstable location points (double bond and allyl carbon) in plant oil can be reduced, thereby enhancing the oxidation stability of plant oils.

7.2.2.2 Biological Modification of Bio-Lubricant Base Oil

Improvement of plant oils by genetic modification to obtain high oleic oils has led to better acceptance of these oils as lubricants [14]. By genetic analysis of sunflower seeds with high oleic acid content, it is confirmed that the presence of dominant gene in the sunflower crop dominates oleic acid content. Besides, the characteristics of high oleic acid are controlled by three dominant complementary genes. A study combines the standard intrinsic genome sequences of high oleic acid characteristics and obtains self-flowering high oleic acid plants from several generations of backcross breeding programs [11]. Oil crops with high oleic acid content have been highly valued. The content of oleic acid in common sunflower seed oil is about 30%, but that in high oleic acid plant oils, which are cultivated by modern biotechnology, oleic acid content is more than 90% [24]. Genetic engineering is an effective approach to further improve the lubricity performances of high oleate plant oils. By expressing a fatty acid hydroxylase gene from castor in canola, oil containing up to 16% hydroxy fatty acid can be produced in a high oleate background. Results show that the introduction of hydroxy fatty acids significantly increases the viscosity and oxidation stability of plant oils [14].

However, the biological modification of plant oil will inevitably meet the problems of biological genetic stability and long research cycle. Industrial manufacture requires a stable and effective means to modify plant oil in order to achieve the purpose of applying to the field of lubrication.

7.2.2.3 Chemical Modification of Bio-Lubricant Base Oil

Although biological modification can increase the content of oleic acid and therefore improve the overall oxidation stability and low-temperature fluidity, it has the problems of genetic stability. The chemical modification method has higher biological safety, better industrial processing basis and product performances. The vulnerable parts of plant oil molecules include double bond, allyl carbon, etc. Chemical modification is mainly directed against these vulnerable parts [24].

At present, the research on chemical modification of plant oil is mainly focused on improving its degree of saturation and branching. The main ideas of chemical modification are hydrogenization, polymerization, transesterification, esterification, and isomerization. The selective hydrogenation, oligomerization, and cyclization of plant oil can reduce the unsaturated components in plant oil, that is, to improve the service life of the oil, and have no obvious effects on its low-temperature performance [13].

Selective Hydrogenation

Selective hydrogenation has great application prospects in the synthesis of lubricating oil. In the process of hydrogenation, if the unsaturated fatty acids are completely converted to saturated fatty acids, which are solid under normal conditions, the oxidation stability can be improved, while the low-temperature fluidity becomes worse. It is generally believed that the unsaturated fatty acid triglyceride containing one double bond has good oxidation stability and low-temperature fluidity. Therefore, selective catalytic hydrogenation, to a certain extent, reduces the unsaturation, and is

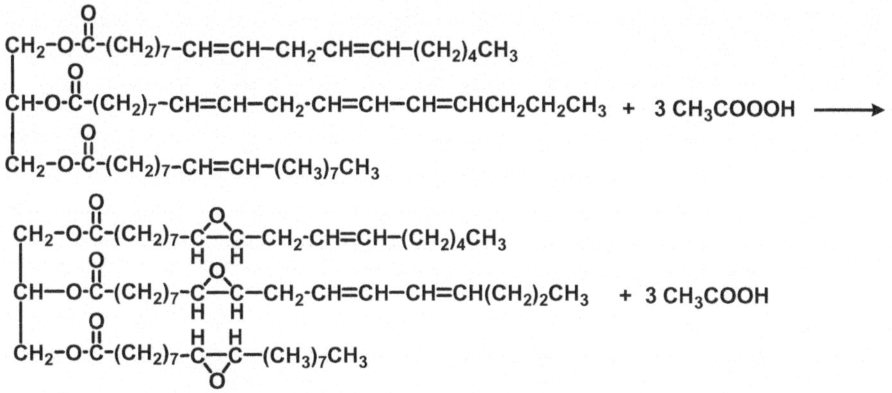

Fig. 7.2 The epoxidation reaction

one of the key points for the preparation of lubricating base oil by chemical modification. The selective hydrogenation can transform polyunsaturated fatty acids, such as linolenic acid and linoleic acid, into monounsaturated fatty acids, which will improve the oxidation stability of plant oil without affecting the low-temperature performance.

Dimerization/Oligomerization

Dimerization and oligomerization are also approaches to modify double bonds. Two or more fatty acid molecules are involved in this process, which contain a carbon chain of 18 carbons with one or more double bonds, to form dimer or trimer-aliphatic product [38]. Polymerized modified plant oil has better viscosity and temperature performances. Polymerization can eliminate the double bonds in lube base oil and obtain base oils with different viscosity. If the polymerization reaction produces too many branched chains, the biodegradability and acidity of the products will be increased.

Epoxidation

The double bonds easily react with peracid. Therefore, the epoxidation is one of the main reactions to reduce the number of double bonds in fatty acids (Fig. 7.2). But the epoxidation oil is not stable. The reason is the epoxy group is a three-membered ring with high tension. It is a highly reactive chemical and tends to react with other substances containing active hydrogen in the presence of acid or alkali [37].

The thermal oxidation stability and tribological behavior of Epoxy Soybean Oil (ESBO), soybean oil, and high oleic acid soybean oil are compared, and it is found that ESBO which has better thermal oxidation stability is a potential lubricant base oil. In boundary lubrication, ESBO shows the most excellent wear resistance [1]. After epoxidation, it is found that the oxidation stability of rapeseed oil is also improved [37]. And biodegradation tests show that epoxidation does not affect its biodegradability [33].

Both selective hydrogenation and epoxidation can improve the oxidation stability of plant oils, but they cannot improve the low-temperature performance. So there is a good idea to change the low-temperature performance and oxidation resistance at the same time by improving the branching degree of plant oils.

Branching

Branched-chain fatty acid esters have good physical properties such as excellent low-temperature performance and stronger hydrolysis stability due to the increase of steric hindrance chain. Low pour point, good fluidity, high stability, and high flash point make them widely used in the fields of lubricating oil, cosmetics, plastics, and so on.

There are some effective methods of branching modification of plant oil. (1) Esterification of hydroxy plant oil. By esterification of castor oil with C1 ~ C3 linear chain or branched chain anhydride, a mixture of monohydric, dihydroxy, and trihydroxy castor oil ester will be produced with good lubricating performances. The pour point can be reduced to - 40 °C, and the viscosity index is more than 135 [32]. (2) The direct branching modification of unsaturated fatty acid bonds in plant oil. Unsaturated double bonds can be removed by using cyclopentadiene and methylcyclopentadiene on the fatty acid chain, and hydrogen reacts with the double bonds on the alkyl ring after addition reaction, the oxidation stability and low-temperature fluidity of the products will be improved [19].

Transesterification or Esterification

The glyceryl group in plant oil has a strong tendency of hydrolysis and thermal degradation. The structure of triglycerides can be replaced by transesterification to yield polyol esters which have good oxidation stability and low-temperature performances [29]. There are two main ways to prepare polyol esters from plant oil: one is to produce fatty acids from oil and then esterified with polyols [21]; another one is to convert oil into methyl ester and then conduct transesterification with polyols [9, 10]. The corresponding polyol esters are prepared by transesterification of plant oil with Trimethylol Propane (TMP), pentaerythritol, and 2,2-dimethyl-1,3-propanediol. The transesterification products have good low-temperature fluidity, oxidation stability, viscosity temperature, and lubricity [20].

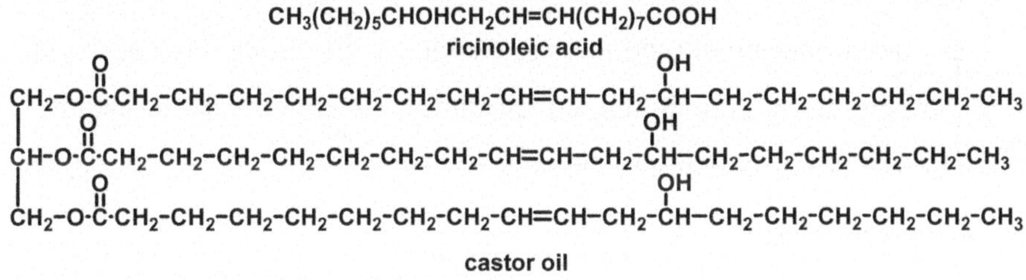

Fig. 7.3 The structure of ricinoleic acid and castor oil

7.2.2.4 Modification of Castor Base Oil

Structure and Composition of Castor Oil

Castor oil, one of the most commonly used vegetables oils, contains a hydroxyl group in addition to an olefinic linkage in its predominating fatty acid. Ricinoleic acid (or 12-hydroxy oleic) accounts for about 80% of the total fatty acid content produced upon hydrolysis of castor oil (Fig. 7.3).

The introduction of this hydroxyl group into the fatty acid chain gives castor oil unique physical and chemical properties. One important physical property is viscosity, of which castor oil is considerably higher than other fatty oils. Since viscosity is a measure of the ease with which the molecules are able to slide past one another, it seems likely that the hydroxyl groups may protrude from the fatty chains in such a way as to impede their free flow. Another quite different physical property is solubility in various types of solvents. Castor oil tends to dissolve in oxygenated solvents like alcohols and not to dissolve in hydrocarbon-type solvents, while the other fatty oil is opposite. This peculiarity is also attributable to the presence of the hydroxyl group [12]. The other typical performance of castor oil is oxidation stability, which is better than other plant oils. It is mainly due to the low content of polyunsaturated acid, such as linoleic acid and linolenic acid.

Most of the plant oils have low viscosity, and castor oil is a nice choice when preparing high viscosity environmentally friendly lubricants, such as gear oil and grease. It has found that castor oil is a promising biodegradable lubricant based on the viscosity, oxidation stability, deposition tendency, volatility, lubricity, and compatibility with additives [2].

Modification and Application of Castor Base Oil

The castor is further modified to produce a series of lubricating oils with better performance. The biological modification of castor is mostly used to increase the oil yield and environmental suitability of the castor oil plant. In 1984, the first species of ricinus parthenogenetic was discovered, and the female inbred line of castor plant was

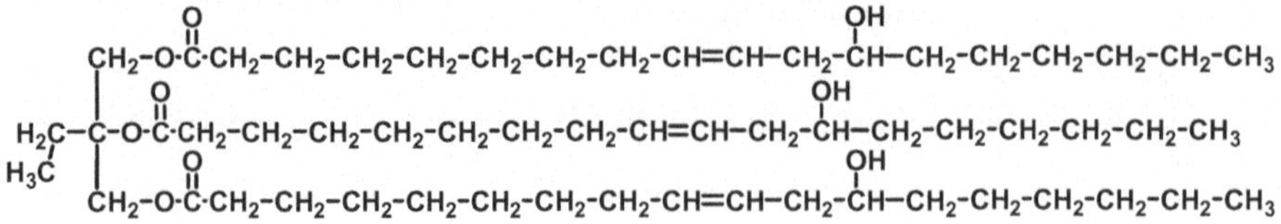

Fig. 7.4 Trimethylolpropane ester

bred by Agronomy Measures. In 1996, the first castor hybrid in China was approved, which improved the oil yield and environmental adaptability of castor [31].

Besides, similar to the chemical modification methods of common plant oils, hydrogenation, branching, and esterification/transesterification can be applied to modification of castor oil as well.

Hydrogenated Castor Oil (HCO), the main component is 12-hydroxy stearic acid triglyceride prepared by catalytic hydrogenation of castor oil, is a powder or flake substance with a melting point of 86 °C. It can be used in cosmetics, hair beauty, and ointments, as paraffin substitute and brightener, and it can also be used in the preparation of hydrogenated stearic acid and its derivatives. The high-pressure lithium-based lubricant prepared through HCO has been used in automotive and military fields because of its gel action, which greatly improves the lubrication performance, with high drop points and excellent water resistance [26].

By isomerization modification of natural castor oil, the length of the molecular side chain is increased, which is able to improve its viscosity index and decrease its pour point.

Ethyl/methyl ricinoleate, produced from the transesterification reaction between castor oil and ethanol/methanol, is the main component of castor biodiesel. A transesterification reaction between this ricinoleate and the superior alcohol TMP results in polyolester (Fig. 7.4). Because of the saturation level and the presence of branching, the product is a synthetic base stock with good oxidative stability and low-temperature properties. After isomerization modification, the rheological properties have been greatly improved, and its anti-wear and antifriction properties are better than those of the same viscosity mineral oil, equivalent to pentaerythritol ester or dioctyl sebacate [30].

The shorter chain derivatives of castor oil are also excellent lubricating base oils. In the modification process, sebacic acid or other medium or short-chain fatty acids are often obtained from cracking castor oil, and then esterification occurs with 2-octanol to produce the lubricating oil with excellent performance. But sebacic acid production would employ phenol or cresol as a thinning agent in current process, thus causing environmental pollution [6, 23]. He's group has developed a novel clear process without phenol for producing sebacic acid from castor oil at high temperature [16] (Fig. 7.5). Firstly, by mixing alkali solution, thinning agent, and castor oil at sufficient temperature, sodium ricinate can be generated by saponification. Volatiles such as 2-octanol can be distilled off. After dehydrogenation and isomerization, pyrolysis reaction happened and formed disodium sebacate. After acidified, sebacic

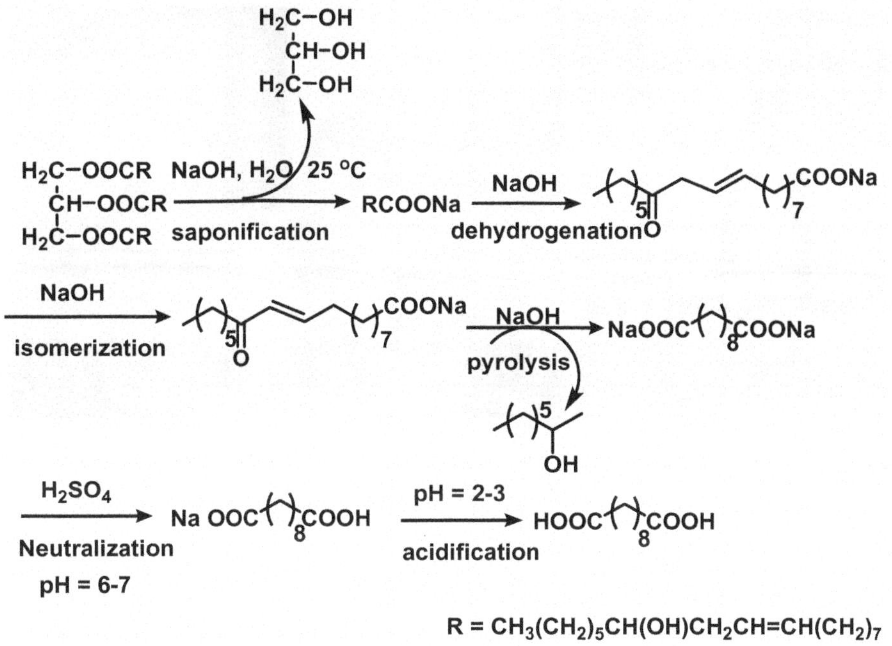

Fig. 7.5 Sebacic acid production starting from castor oil

Fig. 7.6 Synthetic lubricant from castor oil

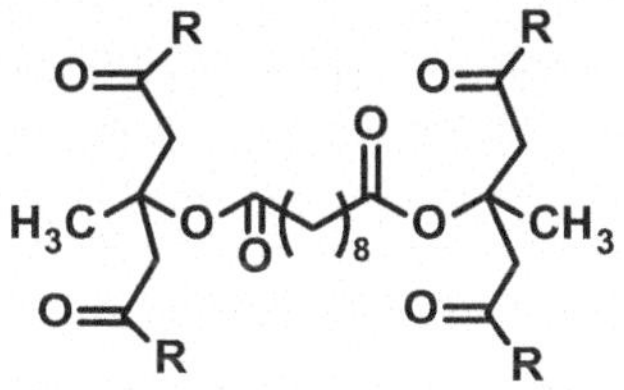

acid and thinning agent can be effectively recovered from reaction mixture. The multi-component thinning agent used in this process is relatively inexpensive, non-foaming, non-volatile, and resistant to decomposition under the reaction conditions.

In order to improve its lubricating and anti-wear performances as well as environmental friendliness, a novel lubricant—ester derivative of sebacic acid has been developed (Fig. 7.6). The lubricant does not include any of the contaminating components used to manufacture traditional industrial lubricants [17].

Sebacic acid is produced in laboratory. After success at laboratory level, production plant in Tianjin (Fig. 7.7) has been set up for manufacturing castor-based lubricate with annual capacity of 5000 tons. This technology was given as a golden award for patents in Tianjin, and recognized nationwide supported by Department of Agriculture, China [17]. The method has obvious advantages over traditional method. Products operate smoothly in extremely cold conditions and can reduce the use of

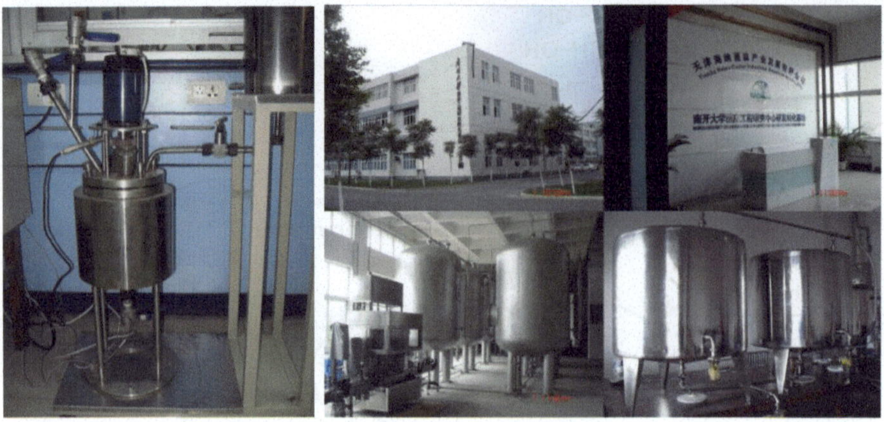

Fig. 7.7 Bench scale facility for sebacic acid preparation (left) and Pilot plant for production of environmentally friendly castor-based lubricate with annual capacity of 5000 tones (right)

fuel by 8% without the decrease of lubrication performance. Besides, the sulfur and nitrogen content in our products is negligible thus leading to a low sulfur and nitrogen emission. With these advantages, they have been certified as bio-based labeled product by United States Department of Agriculture (USDA), which is the authority certification of the biomass industry. Engine oil Energy-saving Bench is the lubrication industry certification authority, which proves that the product can save fuel. This product passed the certification of Sinopec in 2016 and is the first of its kind.

Poor oxidative stability and low-temperature performance are the bottlenecks that affects the application of plant oil. The main purpose for modification is to reduce the double bond content of plant oil. Biological modification can be used to increase the content of oleic acid in plant oils to improve oxidative stability. Furthermore, it is feasible to improve its oxidation stability by chemical modifications, such as hydrogenation, transesterification, isomerization, and esterification. By modifying the plant oil, the oxidation stability and low-temperature fluidity are improved, which makes the plant oil exhibit better lubrication performance. The bio-lubricant products are then prepared by blending additives.

7.2.3 Bio-Lubricant Blending Technology

The plant-based oil obtained after the modification cannot be directly used, and it is also necessary to add additives for blending. Lubricating oils can be prepared by mixing base oils and additives in intermittent or continuous blending process. Additives have a significant influence on the performance of the lubricating oil. Different uses of additives make lubricants differ in their physical and chemical properties.

7.2.3.1 Bio-Lubricant Additives

Plant oils used as base oils need to be selected from the following aspects: good dispersion and dissolution of additives, excellent low-temperature properties, fine lubricating properties, and high flash point. The inadequacies of plant oils such as substandard oxidation stability, poor hydrolytic stability, deficient low-temperature fluidity, blistering compared to mineral oil, and poor filterability can be improved by adding appropriate additives [35].

Additives are chemical compounds that give the lubricants some special properties. Some additives make lubricants have new beneficial properties, for example, cleanliness and dispersion, abrasion resistance, corrosion resistance, and emulsifying ability. Other additives can improve the existing performance of lubricants, such as viscosity, viscosity index, pour point, antirust properties, and anti-oxidation properties. Firstly, when additives are added to the lubricants, the real contact area increases, the contact stress decreases, and non-smooth friction surface can be filled and repaired. Secondly, because most additives are polar substances, they can react with metal surfaces to form a chemisorbed film which has extreme pressure protection. Finally, when the friction surface is under local high temperature and high-pressure conditions, additives can be decomposed to produce new substances. These new substances can react with metals and yield a film which can prevent the surface from gluing. These are the main processes by which additives play a protective role.

Lubricant additives have a variety of functions being divided into viscosity index modifiers, detergent and dispersant additives, extreme pressure and anti-wear additives, antioxidants, and other additives.

Viscosity Index Modifiers

Viscosity index modifiers are linear polymer compounds, which are soluble in hydrocarbon base oils. They are based on the difference in the viscosity of different forms at different temperatures and improve the viscosity performance of the oil [15]. When the polymer shrinks and curls at low temperatures, they have a relatively small effect on the viscosity. However, polymer swells at high temperatures, and the internal friction of the base oil increases significantly, resulting in an obvious increase in the viscosity of the base oil. The addition of the viscosity index modifiers compensates for the disadvantages of decreasing the viscosity of lubricants caused by the rise in temperature. The most commonly used viscosity index improvers are hydrogenated styrene diene polymers, polymethacrylates, ethylene–propylene copolymers, polyisobutylene, polyvinyl *n*-butyl ether, and so on.

Detergents and Dispersants

Lubricating oil can be oxidized at high temperatures. Oxidation products form deposits, thus reducing the efficiency of the internal combustion engine. Therefore, in order to neutralize these substances, detergents are added to the lubricating oil. Detergents are mostly alkaline which can neutralize the acids produced by the oxidation of lubricating oils and prevent them from further oxidative condensation. Detergent has a role in promoting dissolution. Because they can dissolve and wash the materials such as colloidal substances and carbon deposits produced by the oxidation of lubricating oils. Detergents are rarely used singly in oils. They are often used in combination with dispersants [41].

Carbonaceous sludge forms in the combustion chamber due to incomplete combustion. This sludge accumulates to form an oil-insoluble substance. Dispersant additives can disperse the sludge particles and keep them suspended in the oil [39]. The main function of the dispersant is similar to that of the detergent. Both dispersants and detergents play a part in dispersing and solubilizing, but the performance of dispersant is better than that of detergent. This is because the dispersant provides a larger oil-soluble group, which makes carbon deposits and colloidal substances harder to aggregate. The dispersant can directly interact with the carbonyl, carboxyl, hydroxyl, nitro, sulfate of the sludge. These polar groups are dissolved and dispersed in oil in the form of small particles thus preventing the clogging of the screen.

Commonly used dispersants and detergents include metal sulfonate, ash-less sulfonate, overbased sulfonate, salicylates, alkyl phenolates, overbased carboxylate, polyisobutylene succinimides, glycidol modified succinimides, Mannich adducts, polyethylene glycol esters, polyol poly- (12-hydroxy stearic acid), phosphates and phosphonates [39].

Antioxidants

Lubricants inevitably undergo oxidation during use and storage. Oxidation is the main cause of deterioration of the lubricating oil. In order to solve this problem, it is necessary to add a certain amount of antioxidant to extend the working time of the lubricating oil. Antioxidants have two mechanisms: one is the capture of free radicals generated during the oxidation process to terminate the oxidation process; the other one is decomposing the peroxide and yielding a stable compound. Therefore, according to these two mechanisms, antioxidants can be divided into two categories: free radical trapping agents and peroxide decomposition agents. Among the commonly used antioxidants, the radical capturing agents are mainly phenol antioxidants and amine antioxidants. And the peroxide decomposition antioxidants are mainly dialkyldithiophosphates e.g., dialkyl zinc dithiophosphate and dialkyl dithiocarbamate [36], etc.

Extreme Pressure and Anti-Wear Additives

Extreme pressure and anti-wear additives play a role in preventing metal surface from abrasion, scratching, and sintering under high load on the metal surface. Chlorine, phosphorus, and sulfur are typically contained in extreme pressure and anti-wear additives [5]. These elements protect the metal surface with easily sheared layers of chlorines, phosphides, or sulfides, which forbid severe seizure and wear. However, it should be noted that the use of these elements is controlled due to environmental concerns—since these elements lead to corrosion of metal specimens as well as pollution [22]. In the past ten years, nitrogen-containing heterocyclic compounds and their derivatives have become research hotspots for green lubricating oil additives. The reported multifunctional lubricating oil additives containing nitrogen heterocyclic derivatives can be summarized as thiadiazole, thiazole, benzothiazole, benzotriazole, morpholine, imidazoline, and triazine [40, 42]. Kalam investigated that waste vegetable oil-contaminated lubricants with amine phosphate as anti-wear additive reduced wear and friction coefficient as the same time increased viscosity [18, 25].

Other Additives

Plant oils and synthetic esters are easily hydrolyzed to produce acidic substances that can cause metal rust. To achieve a good rust-proof effect, rust inhibitors need to be added. The main types of lubricating oil rust inhibitors are sulfonates, carboxylic acids/carboxylates, alcohols, esters, organic amines, phosphoric acid, and phosphates.

Thickeners that are commonly known for increasing viscosity can be used for biodegradable greases. Although inorganic thickeners such as clay and silica cannot be biodegraded in the environment, they do not pose a hazard to the environment therefore still be used. Because soap-based greases such as lithium/calcium-based mixed soaps, $C_{18}H_{35}O_3Li$ or $(C_{18}H_{35}O_3)_2Ca$, and composite aluminum soaps have good heat resistance and water resistance, they can also be used as thickeners. Of course, the choice of thickener can also be determined by the properties such as viscosity, toxicity, and degradability of the grease.

Other lubricant additives include corrosion inhibitors such as fatty acid derivatives, amines, imidazolines, and triazoles for preventing corrosion of equipment; anti-foaming agent such as polysiloxanes and methacrylates to prevent foaming; biodegradation accelerators such as phosphor-nitrogen modified fatty acid [7], etc.

7.2.3.2 Bio-Lubricant Blending Technology

Bio-Lubricant Blending Process

Lubricant blending is mostly a homogenous phase mixing process. When the lubricant additive is a solid, it is a heterogeneous phase mixing process, during which solid additives will eventually dissolve and form a homogeneous phase. It is generally believed that the liquid-liquid homogenous mixing is a combination of three diffusion mechanisms: molecular diffusion, eddy diffusion, and convection diffusion.

The blending of lubricant has two methods: intermittent and continuous blending. During intermittent blending process, the material is blended under the effect of

Table 7.1 Additives in antirust oil

Ingredients	Quality/g	Effect
Barium petroleum sulfonate	5.00	Corrosion inhibitor
Benzotriazole	0.02	
Diisooctyl sebacate	5.00	Solubilizers
Dibutyl phthalate	0.10	
2,6-di-tert-butyl *p*-cresol	0.20	Antioxidants
Methyl paraben	0.20	Preservatives and antioxidants
Propylparaben	0.10	
Silicone oil	0.10	Defoamer

mechanical stirring or pump circulation stirring. Continuous blending requires that base oils and additives are fed into the mixer in a defined ratio. When the process is completed, the physical and chemical properties of mixed oil can meet the demands, and the oil can be directly filled or entered into the product tank for storage. Compared with continuous blending, the intermittent blending process equipment is relatively simple. This blending process eliminates the need for accurate flow meters, highly reliable automatic controls, and online quality inspection methods. Therefore, the investment required to build such a blending device is low and it is easy to implement. The production capacity of such a blending device is limited by the capacity of the tank. As long as a suitable blending tank is selected, it can meet a certain demand for production capacity, but the labor intensity is huge. Continuous blending requires meters and controls to accurately measure and control the flow of each component, as well as on-line quality analysis instruments and computer control systems. Since this method of blending has advanced equipment and means mentioned above, continuous blending can achieve optimal control, rational use of resources, thereby reducing costs. Continuous blending production capacity depends on the size of the blending tank. The efficiency of equipment and the quality of components in the blending process directly affect the quality of the oil.

Bio-Lubricant Blending Examples

With the development of environment-friendly lubrication antirust oil as the starting point, L-RD-4-2 lubricant antirust oil that has been developed is used for steel corrosion protection.

The base oil and additive are confected using the polar corner design method. The base oil in the lubricant is a mixture of hydraulic oil, rapeseed oil, castor oil, and Span-85, and was mixed at a ratio of 1:1:1:1, accounting for about 80–95% of the total content of bio-lubricant. The remaining part is the functional additives (Table 7.1). This ratio is assigned based on the product's performance requirements. For example, L-RD-4-2 lubricant antirust oil products are required to meet several

Table 7.2 Product quality index

Items		National standard requirements	Result	Testing method
Flash point (°C)		≥170	212	GB/T 3536
Pour point (°C)		≤−25	−30	GB/T 3535
Dynamic viscosity(mm^2/s, 100 °C)		9.3–12.5	10.1	GB/T 265
Low-temperature dynamic viscosity (mPaos, −18 °C)		≤2500	2000	GB/T 6538
Viscosity index		≥75	83	GB/T 1995
Viscosity ratio		≤2.0	1.4	–
Increased total acid number (mg KOH/g)		≤3.0	2.6	–
Volatile content (m/m)		≤2.0%	0.9%	SH/T 0660
Foam volume (mL)	24 °C	≤300	Qualified	GB/T 12579
	93.5 °C	≤25		
Acid neutrality		Qualified	Qualified	SH/T 0660
Removability (wet heating)		Removable	Removable	SH/T 0212
Rust prevention	Hot and humid (Grade A)(h)	≥480	500	SH/T 2361
	Salt spray (Grade A)(h)	–	–	SH/T 0081
	Saline impregnation (Grade A)(h)	≥20	24	SH/T 0025

demands: viscosity in a range of 9.3–12.5 mm^2/s, price is lower than 20 yuan/kg and plant base oil content not less than 50%.

After nearly one year of storage under natural conditions (high temperature and humidity in summer; large temperature difference between day and night in autumn; cold and wet in winter), the testing results showed that the quality index (Table 7.2) of L-RD-4-2 lubricant antirust oil meets the requirements of the SH/T 0692-2000 standard.

The lubricant antirust oil has the characteristics of stable quality, excellent performance, and easy use, and is a product with broad application prospects.

The properties of bio-lubricant can be improved by the introducing additives such as viscosity modifiers, detergent and dispersant additives, extreme pressure and anti-wear additives, antioxidants and other additives, depending on the requirements of the application. The main blending mechanism of lubricating oil is summarized, and the intermittent and continuous blending of lubricating oil is compared. L-RD-4-2 lubricant antirust oil blended by plant-based oils and additives has excellent performance. With people's attention to the environment, bio-lubricants will be widely studied.

Table 7.3 Fire-resistant hydraulic oils-pentaerythritol oleate

Items	National standard requirements	Result	Testing method
Dynamic viscosity (mm^2/s, 40 °C)	62–68	63	GB/T 265
Acid value (mg KOH/g)	≤0.8	0.47	GB/T 7304
Iodine value (g I_2/100 g)	85–100	102	AOCS CD1-25
Hydroxyl value (mg KOH/g)	≤10.0	6.2	GB13482
Flash point (°C)	≥300	338	GB/T 3536
Pour point (°C)	≤–21	−23	GB/T 3535
Demulsibility	≤25	19	GB/T 7305
Chroma	≤2	1	GB/T 1722
Moisture	≤0.05	0.05	07033.053

Table 7.4 Fire-resistant hydraulic oils-trimethylolpropane oleate

Items	National standard requirements	Result	Testing method
Dynamic viscosity (mm^2/s, 40 °C)	≤53	46	GB/T 265
Acid value (mg KOH/g)	≤1.0	0.42	GB/T 7304
Iodine value (g I_2/100 g)	75–90	86.4	AOCS CD1-25
Hydroxyl value (mg KOH/g)	≤7	6.6	GB13482
Flash point (°C)	≥300	324	GB/T 3536
Pour point (°C)	≤−39	–45	GB/T 3535
Demulsibility	≤25	18	GB/T 7305
Chroma	≤2	1	GB/T 1722
Moisture	≤0.1	0.03	07033.053

Different bio-based lubricants have different physical and chemical properties and are used in different occasions. Here we list the detection indicators of fire-resistant hydraulic oils—pentaerythritol oleate (Table 7.3), trimethylolpropane oleate (Table 7.4) and some bio-based lubricants produced by Heda (Table 7.5) for reference.

7.3 Outlook

The requirements for lubricants are varied with the development of the industry and technology. During the 1950s, appropriate viscosity and the absence of acidic components of the base oil was of great importance. While from 1990s, base oil development was influenced not only by the demands of lubricant performance but also by environment, health, and safety criteria. Taking these into consideration, bio-lubricants are more attractive than mineral-based lubricants. Bio-lubricants exceed

Table 7.5 Some bio-based lubricants produced by Heda

	Priolube 1414	Priolube 1446	Priolube 1851	Priolube 1936	Priolube 2089	Priolube 3987
Dynamic viscosity (mm^2/s, 40°C)	6.0	30	495	26	44	138
Dynamic viscosity (mm^2/s, 100°C)	2.3	7.0	49	5.3	8.7	18.2
Viscosity index	229	207	153	139	181	140
Pour point (°C)	−21	−24	−36	−54	−54	−33
Flash point (°C)	195	295	300	244	315	320
Biodegradable OECD 301B (%)	**> 60**	**84**	**65**	**74**	**99**	**70**
Renewable ingredient (%)	79	85	87	0	85	85

the performance of mineral lubricants in terms of viscosity, low carbon-forming tendency, stability, volatility, response to additives, and biodegradability. Bio-lubricants are now widely accepted as offering a number of inherent performance advantages over conventional petroleum-based oils to formulate modern automotive engine oils [25]. Using bio-lubricants as a replacement of mineral oil can reduce the discharging of harmful gases and may remit the greenhouse effect.

Among various bio-lubricants, castor oil-based lubricants are the most promising one as castor oil is non-edible and has excellent physical and chemical properties. By using castor oil-based lubricants instead of mineral-based lubricants, we can achieve the goal of reducing carbon emission, increasing energy efficiency and increasing carbon sink. Besides bio-lubricants, castor oil can also be used to produce other high value-added products, thus has a great economic benefit. For example, Tianjin Nankai University Castor Engineering Technology Co., Ltd. has developed a series of high value-added products based on castor oil, such as biodegradable lubricants, bio-jet fuel, bio-materials, and so on [17]. Currently, industrial uses of castor oil are largely limited by its supply [3]. The problem associated with castor oil production is that its seeds contain the toxin ricin and hyper-allergenic 2S albumins, which are detrimental to growers as well as processors [27]. Furthermore, the current production of castor oil-derivatives generally suffers from high energy consumption and heavy environmental pollution. Therefore, eco-friendly and sustainable technologies are highly desirable in the improvement of castor crop to develop high oil and low toxin or toxin-free castor oil for the production of castor oil-derived lubricants, biodiesel, bio-plastics, and other value-added bio-chemicals.

Although there have been some achievements in bio-lubricants industry, further improvements are inevitable and are already being recorded with an increasing number of studies directed toward such areas. Firstly, new extraction method should be developed for large scale production in order to produce castor oil with high quality effectively. Secondly, lower power-consuming and more efficient modification methods are needed to achieve a better performance of bio-lubricants [13, 14]. And more and more bio-lubricants should be developed to meet various lubrication demands [17]. The cost of modification should also be decreased. Thirdly, additives which can change the chemical and physical properties of bio-lubricants may also have unknown negative effects on bio-lubricants such as a drop in viscosity or a decrease in biodegradability [5, 15, 35]. Hence, development of high-performance, multi-functional, and environmental friendly additives with less side effects will become research hotspot. Additionally, further systematic research of interactions between bio-lubricants and additives should be conducted. Bio-based high-performance green lubricants are good for energy-saving and emission-reducing. By accelerating the application of high-performance green lubricants, improving power, saving fuel, reducing emissions, and preventing gray skies can be achieved.

References

1. Adhvaryu A, Erhan SZ (2002) Epoxidized soybean oil as a potential source of high-temperature lubricants. Ind Crop Prod 15:247–254
2. Asadauskas S, Perez JM, Duda JL (1997) Lubrication properties of castor oil-Potential basestock for biodegradable lubricants. Lube Eng 53:35–40
3. Bongfa B, Peter AA, Barnabas A, Adeoti MO (2015) Comparsion of lubricant properties of castor oil and commercial engine oil. J Tribol 5:1–11
4. Birová A, Pavlovicová A, Cvengros J (2002) Lubricating oils base from chemically modified vegetable oils. J Synth Lubr 18:292–299
5. Canter N (2007) Special report: trends in extreme pressure additives. Tribol Lubr Technol 63:10–12
6. Charles D, WO 9835925 (1998) Method for preparing cleaved products from castor oil or derivatives thereof.
7. Chen B, Fang J, Dong L, Sun X, Wang J (2008) Enhancement of biodegradability of lubricants by biodegradation accelerants. Lubr Sci 20:311–317
8. Chen J (1999) Refining mixed oil. China Oils Fats 24:25–26 (in Chinese)
9. Demirbas A (2005) Biodiesel production from vegetable oils by supercritical methanol. J Sci Ind Res India 64:858–865
10. Demirbas A (2006) Biodiesel production via non-catalytic SCF method and biodiesel fuel characteristics. Energ Convers Manage 47:2271–2282
11. Fernández-Martinez J, Jimenez A, Dominguez J, Garcia JM, Garces R, Mancha M (1989) Genetic analysis of the high oleic acid content in cultivated sunflower (Helianthus annuus, L.). Euphytica 41:39–51
12. Gilbert EE (1941) The unique chemistry of castor oil. J Chem Educ 18:338–341
13. Grant I, Charne DG (1999) Brassica napus plant oil wherein the levels of oleic, alpha-linolenic, and saturated fatty acids are endogenously formed and simultaneously provided in an atypical highly beneficial distribution via genetic control. US Patent 5955623 21 Sep 1999
14. Grushcow J (2005) High oleic plant oils with hydroxy fatty acids for emission reduction. World Tribology Congress III, Washington, D.C. USA, 12–16 Sep 2005

15. Han H (2002) Viscosity index improvers for oil products. Sichuan Chem Ind 5:42–44 (in Chinese)
16. He LN, Dou XY, Ye F (2009) Phenol-free process for Sebacic acid production. CN 101367719 18 Feb 2009
17. He LN, Dou XY, Gao YS, Ye F (2011) Biomass-based energy: castor-derived biodegradable lubricate from laboratory to pilot plant. Prep Pap Am Chem Soc Div Fuel Chem 56:15–16
18. Kalam MA, Masjuki HH, Varman M, Liaquat AM (2011) Friction and wear characteristics of waste vegetable oil contaminated lubricants. Int J Mech Mater Eng 3:431–436
19. Kodali DR (2002) Process for modifying unsaturated triacylglycerol oils: resulting products and uses thereof. US Patent 6420322 16 July 2002
20. Kodali DR, Nivens SC (2001) Transesterified oils. US Patent 6278006 B1 21 Aug 2001
21. Koh MY, Ghazi TIM, Idris A (2014) Synthesis of palm based biolubricant in an oscillatory flow reactor (OFR). Ind Crop Prod 52:567–574
22. Li B, Wang X, Liu W, Xue Q (2006) Tribochemistry and antiwear mechanism of organic-inorganic nanoparticles as lubricant additives. Tribol Lett 22:79–84
23. Logan RL, Udeshi SV (2002) Method for preparing sebacic acid and octanol-2. US Patent 6392074B1 21 May 2002
24. Miller JF, Zimmerman DC, Vick BA (1987) Genetic control of high oleic acid content in sunflower oil 1. Crop Sci 27:923–926
25. Mobarak HM, Niza Mohamad E, Masjuki HH, Kalam MA, Al Mahmud KAH, Habibullah M, Ashraful AM (2014) The prospects of biolubricants as alternatives in automotive applications. Renew Sust Energ Rev 33:34–43
26. Naughton FC (1974) Production, chemistry, and commercial applications of various chemicals from castor oil. J Am Oil Chem Soc 51:65–71
27. Ogunniyi DS (2006) Castor oil: a vital industrial raw material. Bioresour Technol 97:1086–1091
28. Salih N, Salimon J, Yousif E, Abdullah BM (2013) Biolubricant basestocks from chemically modified plant oils: ricinoleic acid based-tetraesters. Chem Cent J 7:128–140
29. Schneider MP (2006) Plant-oil-based lubricants and hydraulic fluids. J Sci Food Agr 86:1769–1780
30. Silva JACD, Habert AC, Freire DMG (2013) A potential biodegradable lubricant from castor biodiesel esters. Lub Sci 25:53–61
31. Sun LJ (2007) High yield cultivation techniques of hybrid castor. J Seed Ind Guid 4:20–21 (in Chinese)
32. Tao DH (2003) Ye B (2003) Modified castor lubricants and preparation methods. CN Pat 1424387:18 (in Chinese)
33. Todd LK, Brajendra KS, Kenneth MD, Sevim ZE (2007) Adsorption behavior of epoxidized fatty esters via boundary lubrication coefficient of friction measurements. Chem Eng Commun 194:1065–1077
34. Tu G (2001) Refining technology of mixed castor oil. China oils and fats 26:40–43 (in Chinese)
35. Wagner H, Luther R, Mang T (2001) Lubricant base fluids based on renewable raw materials: their catalytic manufacture and modification. Appl Catal A-Gen 221:429–442
36. Xiao CL (2000) Synthesis and properties of zinc dialkyl aromatic dithiophosphates. Chem Eng 1:14–15 (in Chinese)
37. Wu X, Zhang X, Yang S, Chen HG, Wang DP (2000) The study of epoxidized rapeseed oil used as a potential biodegradable lubricant. J Am Oil Chem Soc 77:561–563
38. Yan XP, He BL, Zhang J, Liu HF (2005) Modification of transition metal cations to polymer-stabilized platinum colloidal clusters in enantioselective hydrogenation of methyl pyruvate. Chin J Polym Sci 23:393–399
39. Zainal NA, Zulkifli NWM, Gulzar M, Masjuki HH (2018) A review on the chemistry, production, and technological potential of bio-based lubricants. Renew Sust Energ Rev 82:80–102
40. Zeng X, Wu H, Yi H, Ren T (2007) Tribological behavior of three novel triazine dervivative as additives in rapeseed oil. Wear 262:718–726

41. Zhang C, Liu H, Zhang P, Zhang Z (2014) Lubricant additives-colorful condiments. Chin J Nat 36:264–273 (in Chinese)
42. Zhu F, Fan W, Wang A, Zhu Y (2009) Tribological study of novel S-N style 1,3,4-thiadiazole-2-thione derivatives in rapeseed oil. Wear 266:233–238

Chapter 8
Oleochemistry Products

Riccardo Tesser, Rosa Vitiello, Vincenzo Russo, Rosa Turco, Martino Di Serio, Lin Lin, and Changzhu Li

Abstract To overcome the well-known environmental problems, in recent years the attention of many researchers worldwide has been directed towards the replacement of petrochemical raw materials with renewable ones such as, for example, plant oils. Among the reasons that lead to the replacement of petrochemical raw materials with greener ones are the high CO_2 emissions, the contamination of the aquatic environment, the lack of adequacy of production, the greenhouse effects and, above all, because petrochemical raw materials are non-renewable. Oleochemicals, on the contrary, are renewable chemical substances derived from plant oils and other fats, including animal fats, or even synthesized from the metabolism of microorganisms. In some cases, oleochemicals are also produced as by-products from the biofuel production. In this chapter, an attempt has been made to review and discuss this trend in chemicals productions according to which petrochemicals were replaced by oleochemical raw materials derived from plant oils and obtaining, in the meantime, products with a comparable quality. Edible vegetable oils are not well accepted as oleochemicals feedstock because that will have competition with food usage; on the contrary, a vast class of non-edible oil plants (castor, linseed, soybean, palm, rapeseed, camelina, tall, jatropha, mahua, etc.), are preferred in this regard. Many feedstock could be successfully obtained from plant oils, i.e. fatty acids, long-chain fatty acid

R. Tesser (✉) · R. Vitiello · V. Russo · R. Turco · M. Di Serio
Department of Chemical Sciences, University of Naples Federico II, 80126 Naples, Italy
e-mail: riccardo.tesser@unina.it

V. Russo
Åbo Akademi, Laboratory of Industrial Chemistry and Reaction Engineering, 20500 Piispankatu, Turku, Finland

M. Di Serio
IROAST, Kumamoto University, Kumamoto 860-8555, Japan

L. Lin
School of Food and Biological Engineering, Jiangsu University, Zhenjiang 212013, China

L. Lin · C. Li
National Joint Laboratory of Oil Energy Plants High Efficient Conversion, Beijing 410004, China

C. Li
Hunan Academy of Forestry, Changsha 410004, China

C. Li et al. (eds.), *Industrial Oil Plant*,
https://doi.org/10.1007/978-981-15-4920-5_8

esters, including saturated or unsaturated and both could be used in the preparation of functional groups including hydroxyl groups, epoxide groups or glycerol. Four areas were covered by this work: 1—biodegradable surfactants, 2—safer fungicides, 3—bio-based plastics and plasticizers, preferably biodegradable, 4—paintings and coatings that do not lead to the air pollution (release of volatile organic compounds).

Keywords Oleochemicals · Plant oils · Surfactants · Fungicides · Plasticizers · Plastics · Paintings and coatings

8.1 Introduction: An Overview on Oleochemistry Products

Similarities between fossil oil, which is considered as a non-renewable source of energy, and plant oil, that would be further processed to reach quality standard as biofuel, a renewable source of energy, was reported in many studies. Annually, of the world's transport fuels, 4% was biofuels and this percentage is increasing dramatically [237]. The demand for plant oil will probably double by 2030 but the plant oil production may not cover such needs. Therefore, the exploration of novel plant oil sources has become increasingly important in order to make up this future plant oil shortfall [237]. As a general consideration, products derived from oleochemical processes have favourable price, performance and safety criteria and must be seriously considered also for future product development. Renewable resources have often been shown to have many advantages when compared with petrochemical raw materials and can, therefore, be considered ideal candidates for new product synthesis and the ideal raw material basis.

Vegetable oils and fats are important constituents of human and animal foodstuffs and are nowadays gaining more and more relevance as raw materials in the production of different useful industrial chemicals. Certain fats and oils are industrially used and are important renewable resources compared to fossil and mineral raw materials, whose availability will become restricted in the near future. In particular, plant oils are long-chain fatty acid esters that can be obtained from agricultural crops. They may be saturated or unsaturated, and may bear functional groups, such as hydroxyl groups, epoxide groups or glycerol. They are used in many applications 'as is', or as 'supports' or additives in formulations for complex use. However, technological developments have made it possible to transform plant oils directly into a multitude of 'biological' industrial products, such as biosolvents, biodetergents, biolubricants, bio-plastics, biomaterials and biosurfactants [87].

Worldwide results from oleochemistry show that the use of vegetable fats and oils allows the development of competitive, powerful products, which are both consumer and environment friendly.

Recently, new findings in oleochemical industry, brought to new products in different fields such as: surfactants, additive for cosmetic formulations, polymer plasticizers and additives, lubricants components, modified epoxides, etc. In many of the mentioned fields, the products derived from oleochemistry have proven to be powerful alternatives to conventional mineral oil products.

However, the use of edible vegetable oil for industrial oleochemical production adds to the debate over whether land should be farmed for fuel or food. This problem will continue to grow in the near future since the percentage of global vegetable oil production used for biodiesel is projected to increase from 12% in 2012 to 14% in 2023 [152].

Plant and vegetable oils produced to meet the demands for food, feed and industrial applications are largely derived from oil palm (Elaeis guineensis) and several major temperate oilseed crops, including soybean (Glycine max), rapeseed (Brassica napus), sunflower (Helianthus annuus) and peanut (Arachis hypogaea); the existing vegetable oil production platforms have been developed from these sources. But recently the steady increase in population created a remarkable shortage in the plant oil needed for these purposes [81]. The usage of plant oil as an alternative source of bio-based biodiesel and oleochemicals has contributed to widening the gap between production and needs [17].

Oleochemistry studies showed that the use of vegetable fats and oils allows the development of competitive, powerful products, which are both consumer friendly and environment friendly. Recently developed products, which fit this requirement profile, are the anionic surfactants cocomonoglyceride sulfate and the non-ionic sugar surfactant alkyl polyglycoside. These products are used especially as mild surfactants in cosmetic formulations. In polymer applications derivatives of oils and fats, such as epoxides, polyols and dimerizations products based on unsaturated fatty acids [88], are used as plastic additives or components for composites or polymers like polyamides and polyurethanes. In the lubricant sector, fatty-acid-based esters have proven to be powerful alternatives to conventional mineral oil products [85].

World energy demand is expected to increase due to the expanding urbanization, better living standards and increasing population. At a time when society is becoming increasingly aware of the declining reserves of fossil fuels besides the environmental concerns, it has become apparent that biodiesel is destined to make a substantial contribution to the future energy demands of the domestic and industrial economies. There are different potential feedstock for biodiesel production. Non-edible vegetable oils which are known as the second-generation feedstock can be considered as promising substitutions for traditional edible food crops for the production of biodiesel [17].

Biofuels have the potential to alter the transport and agricultural sectors of decarbonizing societies [90]. Yet, the sustainability of these fuels has been questioned in recent years in connection with food versus fuel trade-offs, carbon accounting and land use. Recognizing the complicated playing field for current decision-makers, we examine the technical attributes, policy and global investment activity for biofuels [15]. Two potential solutions include the production of lipids in photoautotrophic algae or conversion of plant biomass using engineered microbes. Building upon the early work of the Aquatic Species Program at the National Renewable Energy Laboratory, significant progress has been made using systems biology, synthetic biology and metabolic engineering to increase the production of algal lipids [174].

Increasing energy use, climate change and carbon dioxide (CO_2) emissions from fossil fuels make switching to low-carbon fuels a high priority. Biofuels are a potential low-carbon energy source, but whether biofuels offer carbon savings depends on how they are produced. Converting rainforests, peatlands, savannas or grasslands to produce food-crop-based biofuels in Brazil, Southeast Asia and the United States creates a 'biofuel carbon debt' by releasing 17 to 420 times more CO_2 than the annual greenhouse gases (GHG) reductions that these biofuels would provide by displacing fossil fuels. In contrast, biofuels made from waste biomass or from biomass grown on degraded and abandoned agricultural lands planted with perennials incur little or no carbon debt and can offer immediate and sustained GHG advantages [63]. Production of vegetable oil is generally considered environmentally unsustainable, due in part to deforestation associated with establishing new palm and soy oil plantations and the long time required to repay the carbon debt associated with the establishing new oil crops [63].

Plant oils are the major feedstock for oleochemical production. As a consequence of growing oleochemical demand, production of plant oils has steadily increased (up 14% from 148.96 Mt in 2010–11 to 169.56 Mt in 2013–14) and is expected to increase further (up 28% by 2023 relative to the 2011–2013 average). Usage of plant oils can be divided into three categories: food, biodiesel and oleochemicals. Of the 157 Mt of plant oil consumed in 2012–13, 77% were used for production of food, 12% were used for production of biodiesel and the remaining 11% for production of oleochemicals [152, 153].

A tremendous geographical and feedstock shift of oleochemical production has taken place from North America and Europe to Southeast Asia and from tallow to palm oil. It will be important to introduce and to cultivate more and new oil plants containing fatty acids with interesting and desired properties for chemical utilization while simultaneously increasing the agricultural biodiversity. The problem of the industrial utilization of food plant oils has become more urgent with the development of the global biodiesel production. The remarkable advances made during the past decade in organic synthesis, catalysis and biotechnology using plant oils and the basic oleochemicals derived from them will be reported, including, for example, w-functionalization of fatty acids containing internal double bonds, application of the olefin metathesis reaction and de novo synthesis of fatty acids from abundantly available renewable carbon sources [25].

Oils and fats of vegetable and animal origin are historically and currently the most important renewable feedstock of the chemical industry. Oleochemical transformations occur preferentially at the ester functionality of the native triglycerides, such as hydrolysis to free fatty acids and glycerol and transesterification to fatty acid methyl esters. Fatty acids are transformed by reactions at the carboxy group to soaps, esters, amides or amines. Hydrogenation of both fatty acids and their methyl esters gives fatty alcohols, which are used for the production of surfactants. Competitive petrochemical processes to produce fatty alcohols, such as the Ziegler Alfol process and hydroformylation of alkenes, exist, but the share of fatty alcohols from renewable resources is steadily increasing, from about 50% in 2000 to just under 65% in 2010 [25].

Microbial synthesis of oleochemicals has advanced significantly in the past decade. Microbes have been engineered to convert renewable substrates to a wide range of molecules that are ordinarily made from plant oils. This approach is attractive because it can reduce a motivation for converting tropical rainforest into farmland while simultaneously enabling access to molecules that are currently expensive to produce from oil crops. In the past decade, enzymes responsible for producing oleochemicals in nature have been identified, strategies to circumvent native regulation have been developed, and high yielding strains have been designed, built and successfully demonstrated. Many studies have discussed the metabolic pathways that lead to the diverse molecular features found in natural oleochemicals [165].

Natural lipid biosynthesis is regulated by enzymatic activity and leads to a broad variety of oils in seeds and fruits in which a restricted subset of fatty acid compositions predominates and accounts for the traditional use of oils and fats in food and oleochemical applications. Both genetic and environmental factors affect triacylglycerol (TAG) biochemistry. Manipulation of fatty acid and TAG synthesis for the purpose of usability for both nutritional and industrial applications by means of natural breeding methods and genetic engineering techniques rely on detailed knowledge of biochemical pathways. The features of various oil recovery and refining techniques are outlined and the applications of unusual fatty acids are described [21].

Figure 8.1 shows an overview of fatty acids structures and position of double bond inside molecule [25].

As it is well known, fatty acids of plant seed oils show a remarkable variety while the fatty acids of bulk oils currently used in oleochemistry are rather uniform. Referring to Fig. 8.1, saturated fatty acids with an even number of carbon atoms (C8–C18) and unsaturated C18 fatty acids, such as oleic and linoleic acid as well as relatively small amounts of linolenic acid, erucic acid and ricinoleic acid and are industrially utilized. The most important oleochemical reactions performed with are the thermal cleavage to 10-undecenoic acid and basic cleavage to sebacic acid (decanedioc acid). Interestingly, the enantiomeric purity of, which makes it an interesting substrate for organic synthesis, has not yet been exploited appropriately (for some examples, see

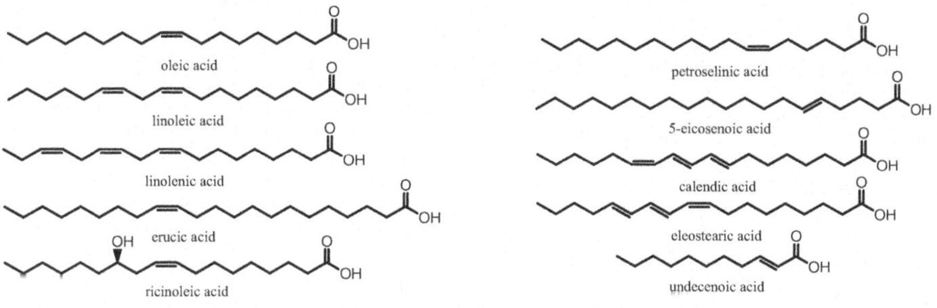

Fig. 8.1 Fatty compounds as starting materials for synthesis

Sects. 8.3.2 and 8.3.3). The latter applies generally to the utilization of the synthetic potential of nature [25].

TAG, as the dominant form of plant oil, has recently attracted immense interest in terms of being produced in plant vegetative tissues via genetic engineering technologies. Multidiscipline-based '-omics' studies are increasingly enhancing our understanding of plant lipid biochemistry and metabolism. As a result, the identification of biochemical pathways and the annotation of key genes contributing to fatty acid biosynthesis and to lipid assembly and turnover have been effectively updated. In recent years, there has been a rapid development in the genetic enhancement of TAG accumulation in high-biomass plant vegetative tissues and oilseeds through the genetic manipulation of the key genes and regulators involved in TAG biosynthesis. Genetic engineering strategies ranging from single-gene manipulation to multigene stacking aimed at increasing plant biomass TAG accumulation [237].

The basic oleochemicals (Scheme 1) are fatty acids (ca. 52%), the respective methyl esters (ca. 11%), amines (ca. 9%) and alcohols (ca. 25%). These are used for the production of important product groups such as surfactants, lubricants and coatings. Fatty acid methyl esters production increased during the past 10 years because of their large-scale utilization as biodiesel. This fact stimulated research on glycerol as a platform chemical for the production of bulk chemicals. Most of the native oils contain unsaturated fatty acids, such as oleic acid (1a) which is applied in the well-known reactions of petrochemical alkenes but only very few reactions across the double bond of unsaturated fatty compounds are currently applied in the chemical industry [25].

Thus, it will be important to introduce and to cultivate more and new oil plants that provide fatty acids with new and interesting properties for chemical utilization, such as petroselinic acid (6a) from the seed oil of Coriandrum sativum, (5Z)-eicosenoic acid (7a) from meadowfoam (Limnanthes alba) seed oil, Calendic acid (8a) from Calendula officinalis, and a-eleostearic acid (9a) and punicic acid (10a) from tung (chinese wood) oil and pomegranate [25].

As we have seen, the scenario of oleochemistry products is rather wide and complex and, in this review, we have restricted our analysis to four classes of oleochemistry fundamental products. These products have been examined from the point of view of synthesis, application and market growth. The considered categories are: (i) surfactants, (ii) essential oils as fungicides and additive for plastics, (iii) paints and coatings and (iv) epoxy-based plasticizers.

8.2 Surfactants

Surfactants: "Surfactants, also referred to as tensides, are active substances whose molecules or ions have the property that, when a characteristic concentration in aqueous solution is exceeded, they associate by reversible aggregation to form larger particles, known as micelles, which impart important colloidal behavior to the solution" (Ullmann's Encyclopedia, [218].

8.2.1 Introduction and Reason to Favour Oleochemicals Surfactants

Surfactants are amphiphilic compounds containing both hydrophobic and hydrophilic moieties, thus conferring them the ability to accumulate at the interface between fluid phases, such as oil/water or air/water, reducing the surface and interfacial tensions to form emulsions. They do their function through their structure, as it contains a hydrophilic head and hydrophobic tail, that enables them to interpose themselves between water and water-insoluble substances. Surfactants are classified according to their ionic properties in water as anionic, cationic, non-ionic or amphoteric surfactants depending on the type and charge of the hydrophilic groups. Surfactants are used in very wide applications such as household detergents, cleansing sector, textile treatment, food, industrial cleanser, cosmetics, personal care products, crop protection in mining and also has its applications in the production of paints, dyes and paper [85, 185, 222].

Traditionally, surfactants are produced by organic chemical reactions from petroleum feedstock on one hand. On the other hand, the growing awareness towards the use of renewable-based products and 'green products' has stimulated the development of alternatives to these chemical surfactants [222]. Environmental protection has led to the phasing out of surfactants having toxicological effects on nature (synthetic surfactants). In addition, biodegradability has become an important factor in the environmental acceptance of a surfactant. Synthetic surfactants, i.e. alkyl phenols, have often shown good functional properties but it has its hazard on the environment. This has led to the development of surfactants based on natural products using millions of components that can be used as raw material. Many natural raw materials incorporate special structures in the surfactant that may reveal new and unexpected functional properties, which can lead to good substitutes for the traditional surfactants. Fatty acids, monoglycerides and glucosides are natural raw materials that have been used for many years in the production of surfactants. Sterols are a more novel class of surfactant raw materials from a natural origin [69].

The toxicological and environmental properties of surfactants based on oleochemicals are not necessarily better than those of surfactants prepared from petrochemical products. However, the life cycle of a surfactant also needs to be studied. For example, the transport cost is an important factor, and it is thus of interest to source raw materials that are locally available. Other life cycle aspects are, for example, the biodegradability and toxicity of the surfactants; how long it takes for a sewage treatment plant to degrade the surfactant; and the eco-toxicity of the degradation products. Furthermore, the synthesis of the surfactant plays a role: how many steps are needed, how much carbon dioxide is produced and how toxic are the by-products. Sterol surfactants have been used as a surfactant for years and it has two sources, synthetic ethoxylated sterols and natural steryl glucosides [69].

Surfactants from renewable resources contribute less to the greenhouse effect if harvested and grown sustainably. When the life cycle of the product is analysed covering the CO_2 emissions from production, use and degradation after disposal, oleochemical surfactants replacing the petrochemical ones would cause a reduction of 34% (reducing absolute emissions to 0.99 Mt) of the CO_2 emission. For surfactant applications in laundry detergents, lowering wash temperatures is also an interesting strategy to reduce CO_2 emissions [160].

Natural oleochemicals sugar-based surfactants (e.g. akyl polyglucosides) and microbial surfactants (biosurfactants) are examples of such environmental-friendly surfactants [222]. Bio-based surfactants can be classified as oleochemical surfactants, saponins and biosurfactants. Oleochemicals comprise vegetable oils and animal fats. Oleochemical surfactants including soaps are mainly made from fatty acids, fatty alcohols or fatty amines using tallow, palm oil, palm kernel oil or coconut oil as feedstock. Saponins are kind of natural soaps. They have antifungal and antibacterial activity and are produced by many plants to form part of the plant's protection against disease [184]. Anionic surfactants like cocomonoglyceride sulfate and the non-ionic sugar surfactant alkyl polyglycoside are surfactants prepared of an oleochemicals origin which fit to be powerful, consumer and environment friendly. These surfactants find its application in cosmetics [85]. Surfactants are the most important oleochemical outlet for fatty acids in Europe as 200 kt of vegetable oil methyl ester are produced annually as raw materials especially for surfactants [21].

Synthetic surfactant solutions containing short polyoxyethylene chains have been shown to be highly toxic and may, therefore, have a major impact on aquatic life due to the discharge of detergents into the environment. Environmentally, removing detergent-like substances from wastewater before its discharge into aquatic systems is a must, especially as legislated in Europe, as concerns the protection of aquatic environments (Ecolabel, Blue Angel, Wasser Gafahrdung Klasse classification, etc.), air quality (Kyoto Protocol, etc.); and toxicological protection (Biocide guideline, European REACH project, etc.). That has enforced research to focus on replacing raw materials extracted from fossil fuels with organic products from forestry and agriculture. These so-called 'agromolecules' are considered to be particularly advantageous, as they are natural, non-toxic, biodegradable and renewable, with a balanced carbon footprint. Considerable efforts are currently being made to promote the use of agromolecules in developed countries, in which there is a clear industrial market for alternatives to fossil fuels and their derivatives [87].

Concluding that, in the purpose of the protection of the environment, reduce the greenhouse effects, reduce CO_2 emission, protect the aquatic environment, oleochemicals surfactants should replace the synthetic surfactants. In the following section, we will present some applications of oleochemicals surfactants in various fields from many resources.

8.2.2 Chemistry of Surfactants

Surfactants are an important class of chemical products capable to reduce surface tension of liquids or the interface of liquid. The molecular structure of a surfactant is characterized by the presence of both hydrophilic head and hydrophobic tails.

The most important hydrophobic and hydrophilic group are listed in the following Table 8.1.

Surfactants do occur as simple monomers but also as more complex oligomers or polymers. They are used in many sectors of industry, agriculture and households. In particular, the industrial sectors in which they are applied in detergents [54, 135, 159, 227, 228], personal care, industrial cleaning, textile, oilfield, emulsion polymerization [30, 41, 68, 82, 175], food, construction, paper and plastics [40, 43, 61, 149, 168, 225] with different functional properties, such as emulsifier, foaming or anti-foaming agent, solubilizer, disperser, lubricant, wetting agent.

Surfactants can be classified in different ways according to the electric charge after dissociation in aqueous solution [159]. In particular, there are different types of surfactants classified by the charge of the head group which can be negative, positive, neutral or zwitterionic.

Anionic surfactants (negative charge): They are amphiphilic compounds in which the hydrophobic residues carry anionic groups with small counterions such as sodium, potassium or ammonium ions which only slightly influence the surface-active properties of the substance. Examples are soaps, alkylbenzene sulfonates, alkyl sulfates and alkyl phosphates [218].

Cationic surfactants (positive charge): They are amphiphilic compounds in which the hydrophobic residues exist as cationic with counterions such as chloride, sulfate or acetate that only slightly influence the active properties of the compound. Examples are tetra-alkyl ammonium chloride and N-alkylpyridinium chloride [218].

Non-ionic surfactant (no charge): are amphiphilic compounds that are unable to dissociate into ions in aqueous solutions, for example, fatty acid alkylolamides, alkyl and alkylphenyl polyethylene glycol ethers, sucrose fatty acid esters, alkyl polyglucosides or tri-alkylamine oxides [218].

Amphoteric surfactants (positive and negative groups): They have zwitterionic hydrophilic groups, for example, aminocarboxylic acids, betaines and sulfobetaines [218].

Commercially, the most important class of surfactants are anionic, but they have recently lost some of their market share gained by non-ionic. Non-ionic surfactants represent the fastest-growing type of surfactant because of their increased use in

Table 8.1 Hydrophilic and hydrophobic groups

Hydrophilic groups	Hydrophobic groups
COO^-K^+	C_nH_{2n+1}
$SO_3^-Na^+$	C_nH_{2n-1}
$PO_3^{2-}2Na^+$	$C_nH_{2n+1}C_6H_4$
$OPO_3^{2-}2Na^+$	C_nF_{2n+1}
$OSO_3^-Na^+$	–
$(OCH_2CH_2)_nOSO_3^-NH_4^+$	–
$(OCH_2CH_2)_nOH$	–
$NR_3^+Cl^-$	–

detergents, where they exhibit synergistic effects with other surfactants, and owing to their suitability for the manufacture of highly concentrated detergents or compact products. Cationic surfactants are used for fabric softening and liquid household products and finally amphoteric surfactants are used for body-care preparations [159].

8.2.3 Biosurfactant as an Alternative to Synthetic Surfactant: Raw Materials

Usually, the choice of feedstock is between petrochemical or oleochemical raw materials [43, 175]. The choice of the raw material depends on many factors, in particular, the availability and its price are very important. Another important aspect is the environmental impact of the final product. Traditional petrochemical surfactants show a low rate of biodegradation and high potential aquatic toxicity [54]. Due to poor biodegradability and the growing surfactants demand new green surfactants are developing. Bio-based surfactants have the same properties as petrochemical surfactants and, at the same time, they are, in most cases, biodegradable and with low toxicity. Consequently, industry and research organizations are active in finding out new ways of producing surfactants that are more environmentally friendly and safer, which entail a minimum health risk, and which can be produced from domestic renewable resources. Most bio-based surfactants are made from vegetable oils and they are named 'oleochemicals surfactants'.

The oleochemical industry uses natural oils and fats as its raw material, and converts these to three major oleochemical building blocks, all of which are relevant for the surfactants industry as a source of hydrophobic compounds. These are fatty acids, fatty methyl esters and fatty alcohols respectively. Oleochemical hydrophobic building blocks for the surfactants industry are based on natural oils and fats. Therefore, to choose the raw material (animals and plants), we should consider some crucial factors regarding this, in particular, we must consider availability, composition and cost-effectiveness. For what concern the availability, natural oils and fats can be derived from a wide range of vegetable and animal source. In particular, the most important vegetable oils used in the surfactants industry are coconut oil, palm kernel oil, corn oil, cottonseed oil, olive oil, sunflower oil and soybean oil, while the most important animal source is tallow.

The most common hydrophobes derived from renewable raw materials are obtained from plant fatty acids or animal tallow triglycerides. Fatty acids obtained usually from oils are stearic acid (C18) and palmitic acid (C16) [68]. The most important raw materials for the surfactants industry are coconut oil and palm kernel oil which have a high amount of fatty acids: about 70% of fatty acid with tails 14 carbons or less in length.

The palm kernel oil is produced from the fruit kernel, the coconut oil, instead, is obtained from the fruit pulp, while the remainder of the coconut is used for biochar (shell) and for food (milk).

8.2.4 *Synthesis of Surfactants*

Biosurfactants are generally categorized by their microbial origin or chemical composition. Starting from renewable raw materials there are several chemical reactions to obtain surfactants or surfactant precursors. In particular, it is possible to use standard oleochemical transformations (transesterification, hydrogenation, hydrolysis, etc.) or more specific reaction to produce surfactants (reduction, sulfonation, chlorination, etc.).

Considering oleochemical raw materials, some chemical transformations to obtain surfactants or surfactant precursors are reported in Fig. 8.2. Esterification, ethoxylation, neutralization or condensation reaction are considered to obtain, starting from fats and oils, the corresponding surfactants such as soaps, mono- and diglycerides, sugar esters, polyglycol esters, fatty amines, fatty alcohols.

Fatty acid methyl esters

First of all, it is possible to consider as starting materials fatty acid methyl esters that can be produced by transesterification of triglycerides [56, 188].

Fatty acids (FAs)

Fatty acids (FAs) and their derivatives could act like a surfactant as they contain a lipophilic 'hydrophobic' part (the alkyl chain) and a hydrophilic 'lipophobic' (head group) which could be amino acids, polyglycerides, polysaccharides and the carboxylate groups on fatty acid chains. So, FAs could exist comfortably at an oil–water interface and can reduce the surface tension at such interfaces. This property is fundamental in all living systems and also in many foods and other manmade systems, where aqueous and fatty phases must coexist. FAs as surfactants confers surface activity that allows appropriate compounds to act as emulsifying agents, detergents, lubricants, etc. [2, 185]. The FAs composition of oils from vegetable sources varies depending on plant origin and sort. Commercially exploited seeds such as soya, rape, sunflower and linseed have been the subject of many years of breeding programmes to obtain oils with particular fatty acid patterns. In addition to breeding efforts on traditional oil crops, work is being done to domesticate alternative oil-rich plants, that may yield new potentially useful fatty acids [99]. Oleochemical surfactants including soaps are mainly made from fatty acids, fatty alcohols or fatty amines using tallow, palm oil, palm kernel oil or coconut oil as feedstock. Saponins are kind of natural soaps, which have antifungal and antibacterial activity and are produced by many plants to form part of the plant's protection against disease [184].

Unsaturated fatty acid methyl esters obtained from plant oils were converted to terminally unsaturated esters and α-olefins by metathesis with ethylene using heterogeneous rhenium or homogeneous ruthenium catalysts in the purpose of preparation of special surfactant. These esters were directly copolymerized with ethylene by an insertion-type palladium-catalyzed polymerization to functionalized polyolefins. Polyesters were synthesized by metathetical dimerization of x-unsaturated esters and subsequent polycondensation of the produced internally unsaturated dicarboxylic esters or by acidic transesterification with petrochemical diols and additional acyclic

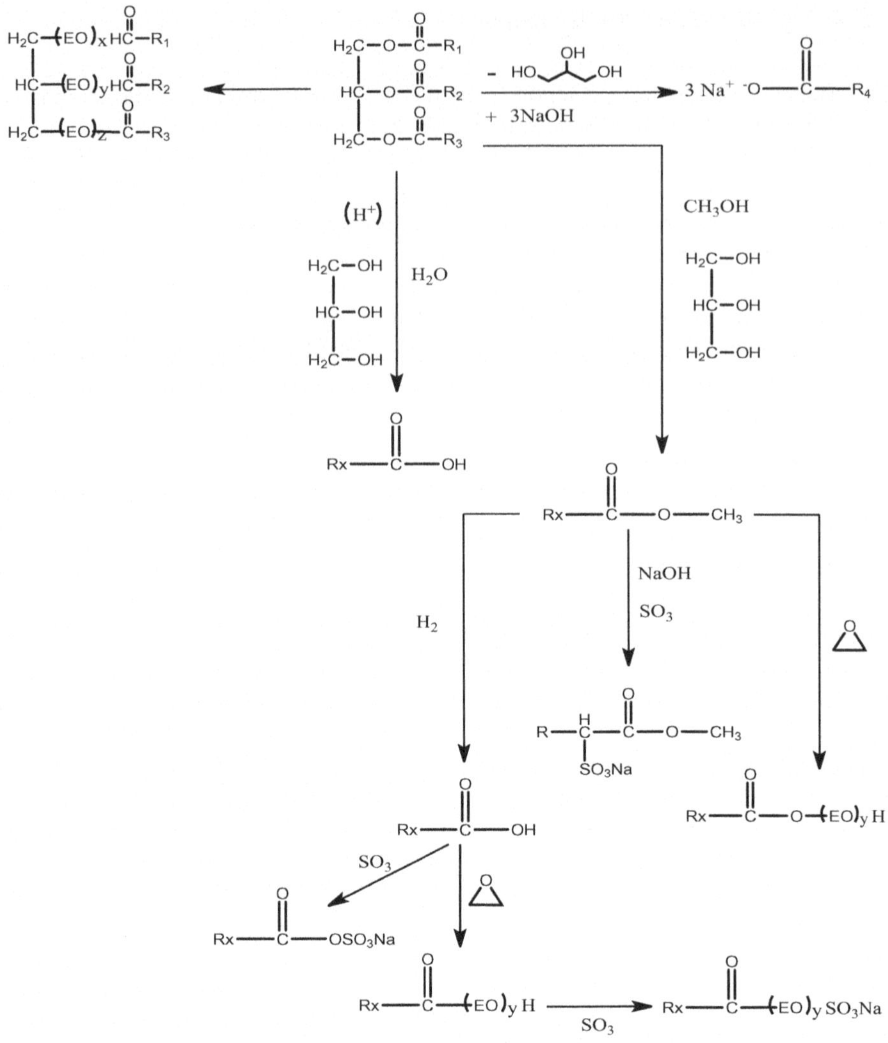

Fig. 8.2 Chemical reactions to obtain surfactants

diene metathesis polymerization. w-epoxy fatty acid methyl esters, achieved by a new method of chemoenzymatic epoxidation, were converted into polyethers with comb-structure catalyzed by aluminoxanes on the one hand and into sugar surfactants by nucleophilic ring opening with amino carbohydrates on the other hand [227, 228].

Glycerol is an organic compound produced in large amounts during industrial-scale transformations of oils and fats into soaps, fatty acids, fatty esters and, in particular, fatty acid methyl esters (used in the development of biofuels) [87].

The transformation of glycerol to different valuable derivatives by conversion of crude glycerol (CG), which is a waste by-product of biodiesel production has been rarely reported because of the CG impurity. The direct etherification of glycerol has been extensively explored to synthesize polyglycerols which are known as non-ionic surfactants that can stabilize suspensions and emulsions. Thus, polyglycerols have various biomedical applications and are widely used as cosmetic ingredients, polymer additives, nutritional additives and lubricants [110].

Glycerol has the potential to become an available and sustainable raw material, due to its low cost and intrinsic properties including having a reactivity that can be increased by transformation into glycerol carbonate by heterocyclization. The oligomerization of glycerol carbonate, assisted by the glycerol, results in the production of polyhydroxylated oligomers rich in linear carbonate groups. The polar moieties of these oligomers were supplied by glycerol and glycerol carbonate rather than ethylene oxide as in most commercial surfactants. The insertion of linear carbonate groups into the glycerol-based skeleton rendered the oligomers amphiphilic, resulting in a decrease in air/water surface tension. The polar head is constituted of homo-oligomers (oligoglycerol and/or oligocarbonate) and hetero-oligomers (oligo-(glycerol-glycerol carbonate)). Coprah oligoesters had the best surfactant properties, outperforming molecules of fossil origin. The oligo-(glycerol carbonate-glycerol ether) with relatively low molecular weights showed properties of relatively high-molecular-weight molecules, and constitute a viable 'green' alternative to ethoxylated surfactants [87].

Fully bio-sourced lauroyl oligoester surfactants, based on poly(glycerol-succinate) as polar head group with controllable sizes and topologies were prepared without solvent nor catalyst. The alkyl chain was either derived from lauric acid or from glycerol α-monolaurate. The oligoester surfactants were characterized by quantitative 13C NMR, acid values and size exclusion chromatography. Their surface activity, foamability, foam stability, wetting power, solubilization properties and biodegradability were investigated proving their potential use in a wide set of applications in replacement of petrochemical surfactants. Relative impacts of the succinic acid/glycerol monomer ratios and of the strategies of synthesis on the physico-chemical behaviour, on the biodegradability and on the stability of the lauroyl poly(glycerol-succinate) surfactants were underlined [3].

Fatty acid soaps

Usually fats and oils are hydrolyzed to glycerol and fatty acid. Fatty acids can be purified by distillation in a specific fraction before the saponification reaction. Soaps of fatty acids are subsequently produced by the neutralization with various bases, resulting in a fatty acid soap with different positively charged counterions, for example, Na, K, NH_4. Unlike fatty acids, soaps are generally water soluble and display strong surfactant properties. The solubility and surface-active properties can be tuned by the nature and combination of fatty acids, counterions and the extent of polarization. Fatty acid methyl ester salts, synthesized from renewable resources, are an example of green surfactants used in eco-friendly washing detergents because of their excellent detergent properties, biodegradability and enzyme stability [207].

Different studies on the behaviour of these compounds have been performed. In particular, the behaviour of other surfactants strengthens as the fatty acids become ionized and turn to soaps with an increasing pH. This transition, and its effect on surface-active properties, has consequently been subject to several studies. At low pH values, the predominant molecule is the undissociated fatty acid. At intermediate values (pH 4–8), undissociated acid, anionic carboxylates as well as so-called acid soaps, $(RCOO)_2H^-$, coexist in the system. At alkaline pH, carboxylate anions and acid–soap salts, $(RCOO)_2HNa$, dominate the solution and the surface layer [207].

Polyethylene glycol fatty acid esters

In industry, polyethoxylation and polypropoxylation reactions are usually performed to prepare non-ionic surfactants and polymers [55, 57, 189]. Direct ethoxylation of fatty acids and fats with conventional catalysts yields a complex mixture of mono- and diesters, as well as various polyethylene glycols as by-products, with a wide range in the number of polyethylene glycol units. Usually, the polyethoxylation and polypropoxylation reactions are usually performed using as catalyst an alkaline compound as sodium hydroxide, sodium methylate, sodium acetate, potassium hydroxide or related alkoxides [133]. In the past years, based on experiences of narrow-range ethoxylation catalysis, new catalysts have been developed that enable a direct ethoxylation of short-chain alkyl esters of fatty acids [207]. Products obtained are used as emulsifiers in food and technical applications. In general, polyoxyethylene esters of fatty acid methyl esters have been found to have good emulsifying, lubricating, dispersing and suspending power and these properties, combined with detergent and antistatic characteristics, provide a potential use in a variety of textile processing applications. Good wetting, penetrating and dispersing properties have made them useful as adjuvants in agricultural products. A comparison between fatty acid methyl ester ethoxylates and the corresponding range of alcohol ethoxylates shows that the critical micelle concentration value is somewhat higher and surface tension at this value lower for the methyl ester ethoxylates [207]. For what concern the other differences between ethoxylates methyl esters and the corresponding alcohol ethoxylates are: detergency is similar, foam is lower due to methyl group terminating the alcohol chain, cloud point is approximately 10 °C lower.

Methyl ester sulfonates

The reaction between fatty acid methyl esters and an excess of sulphur trioxide (15–30%) produce methyl ester sulfonates [68]. The reaction is carried out in a falling film reactor at about 90 °C and it is followed by neutralization with alkali, such as KOH [68].

Methyl ester sulfonates are an important class of anionic biosurfactants. Methyl ester sulfonates' biosurfactants market size may generate over USD 900 million by 2024. These products offer superior properties such as good detergency, excellent biodegradability and improved calcium hardness tolerance. Methyl ester sulfonates are environment friendly as it is low toxicity and high biodegradability characteristics. These products are cost-effective and have excellent detergent properties which

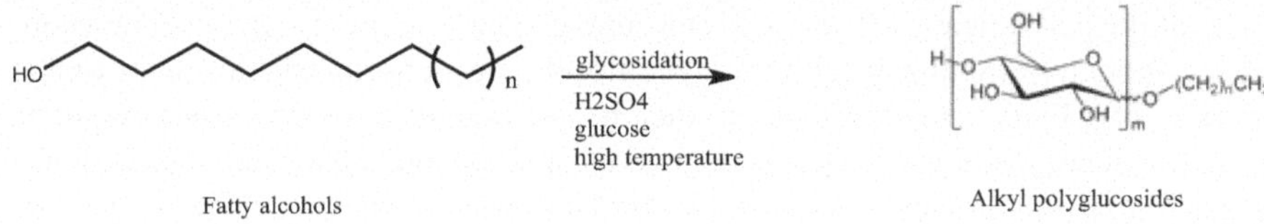

Fig. 8.3 Fatty alcohols reaction

include high absorption, solubilization and dispersing capabilities. Strong application outlook in industrial cleaners, agricultural chemicals and personal care industries should enhance product demand.

Fatty alcohols

Another important reaction of fatty acid methyl esters is the reduction. Fatty acid methyl esters can be reduced to fatty alcohol using, as catalyst, copper chromite at elevated temperature (250–300 °C) and elevated pressure (250–300 bar) [68, 114, 178, 216].

Fatty alcohols are important intermediates for surfactant production and in the world, 50% of fatty alcohols used are derived from renewable raw materials [68].

Two important classes of biosurfactants are derived from fatty alcohols: alkyl polyglucosides and fatty alcohol sulfates.

Alkyl polyglucosides

Alkyl polyglucosides, a popular class of non-ionic surfactants, are synthesized by an acid catalyzed glycosylation with glucose (Fig. 8.3).

Alkyl polyglucosides (APGs)' biosurfactants market is likely to register gains at over 4.5% by the end of 2024. The alkyl polyglucosides are non-toxic and for this property, they are ideal components in facial cleansers, shower gels, hand soaps, shampoos, oral care products and other personal care products. APGs, as a natural surfactant was produced from vegetable oils and starch, are in demand. They are claimed to possess advantages over other classes of surfactants in terms of dermatological and ocular safety, biodegradability, wettability, foam production and cleaning ability. Some APGs have been granted the status of pharmaceutical excipients. Nevertheless, it remains a challenge to persuade some formulators to replace low-cost conventional surfactants with these alternatives [156].

Fatty alcohol sulfonates

Fatty alcohol sulfates are synthesized from fatty alcohol and SO_3 in the presence of NaOH [68].

They are important surfactants used in personal care and detergent products. Their annual production capacity is 0.6 Mt.

Fatty acids

Fatty acids can be obtained from triglycerides in different ways such as acid catalysis [68] or using steam at different degrees of elevated pressure and temperature [68].

For purposes of acylation usually, the activation strategies include the formation of a fatty acid chloride or anhydride, or the formation of a primary amine for nucleophilic derivatization. Fatty acid chlorides can be generated under different conditions, including treatment with thionyl chloride, phosphorus trichloride or oxalyl chloride [68]. Fatty acid chlorides can then be used directly for acylations or can be reacted with an additional fatty acid to generate a fatty acid anhydride. Alternatively, fatty acid anhydrides can be formed directly from fatty acids via dehydration in the presence of acetic anhydride and fractional distillation [68] (Fig. 8.4).

Fatty amines

Nitrile Process is the most common route to primary and secondary fatty amines. During this process, fatty acid reacts with ammonia at high temperature (>250 °C) in the presence of a metal oxide catalyst (e.g. alumina or zinc oxide). This process results in a fatty nitrile, which can then be hydrogenated in the presence of a nickel catalyst and excess ammonia to generate primary amines in good yield. The excess ammonia in this reaction serves to suppress the formation of secondary and tertiary amines that occur as a result of alkyl amine addition to the imine intermediates. Alternatively, secondary and tertiary fatty amines can be generated directly from the reaction of fatty alcohols with alkylamines (Fig. 8.5).

Also, fatty amines represent an important class of renewable hydrophobes.

Quaternary ammonium compounds (QACs)

Quaternary ammonium compounds (QACs) are cationic surfactants that have been known for its broad spectra of antimicrobial activity, so it is commonly used as a disinfectant in medical care, food industry, detergents and glue industries. Their antimicrobial mechanisms were reported to be through the penetration of cell walls

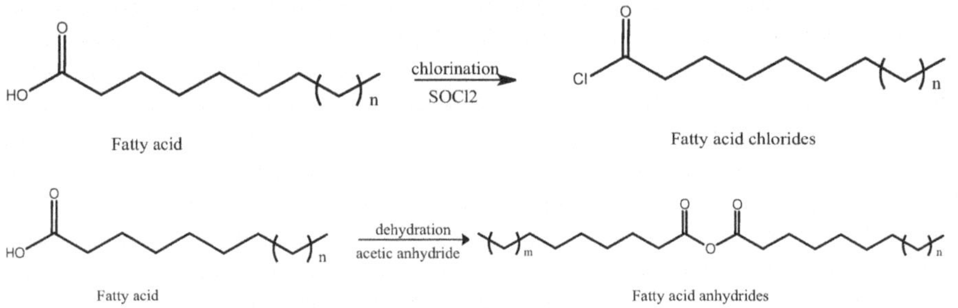

Fig. 8.4 Fatty acids reaction

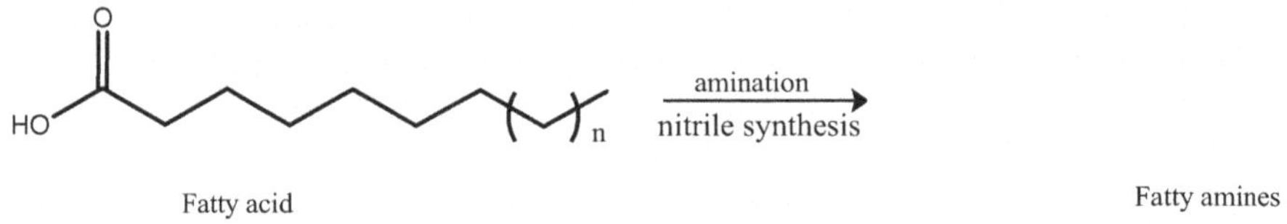

Fig. 8.5 Amination reaction

and partitioning the agents in between the core membrane and solubilize the membrane as a result and that causes lyse of the cells. The growth inhibition test with gram-positive (Staphylococcus aureus) and gram-negative (Escherichia coli and Pseudomonas aeruginosa) bacteria were used to determine the toxicity of single and mixed surfactants. Certain mixed surfactants have lower minimum inhibition concentration (MIC) as compared to the single surfactants. Besides that, it was also found that alkyl chain length and the mixing ratios of the surfactants play a significant role in determining the mixture inhibitive power [233].

A bio-surfactant was produced by Vaz et al. [222] through using Bacillus subtilis strain isolated from crude oil and it was characterized for its properties which were comparable to those of commercially available chemical surfactants. The emulsification indexes showed that the bio-surfactant possesses equal or superior capacity to form emulsions with n-hexadecane as compared to the commercial chemical surfactants. The bio-surfactant showed antimicrobial activity against Staphylococcus aureus and Escherichia coli. Nevertheless, no particular trend or special effect could be assigned to the use of commercial chemical surfactants as anti-adhesives. Suggesting that the bio-surfactant recovered from B. subtilis EG1 constitutes an interesting alternative to the commercial chemical surfactants with potential use in several industries [222].

Baphia nitida oil was extracted, characterized and used in the production of diethanolamide and epoxidized diethanolamide via transamidation reaction. Oil had C18:2 ($50.00 \pm 0.20\%$) fatty acid as the most abundant fatty acid. The epoxidized diethanolamide was synthesized by peroxyformic acid generated in situ by reacting formic acid and hydrogen peroxide with the oil of B. nitida. The formation of the diethanolamide and epoxidized diethanolamide was monitored and confirmed using FTIR and 1H NMR. The epoxidized diethanolamide showed better surface-active properties than the diethanolamide in terms of emulsion stability, and foaming power [2].

Sterols

Sterols are a more novel class of surfactant raw materials from a natural origin. It is derived either from plant or animal sources and its structure is a perhydrocyclopentanophenanthrene nucleus with a side chain in position C17, consisting of 8–10 carbon atoms. The steroid nucleus has a rather rigid character due to the condensed ring structure. The carbon side chain, on the other hand, is of a much more flexible character. The chemical structure of the carbon chain varies for the different components, as shown in Fig. 8.6. The most common sterol derived from animals (zoosterol) is cholesterol while the sterols synthesized by vegetable organisms are called phytosterols. Phytosterol is a general name for different sterol derivatives present in the unsaponifiables of different plants. The main components are usually β-sitosterol, campesterol, stigmasterol and tocopherol. A common source for these components is tall oil or sulfate soap from the pulping of wood. Other common sources for phytosterols are, e.g. maize, rapeseed, sesame, soybean, sunflower and avocado [69].

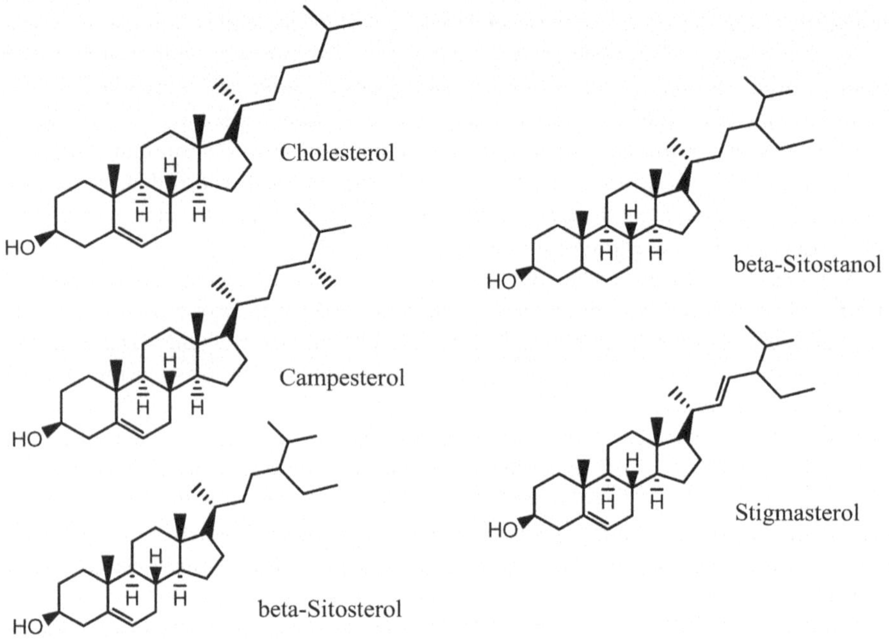

Fig. 8.6 The molecular structure of different sterols

8.2.5 Oleochemicals Surfactants Market (Source: Global Market Insights, Inc.)

The global biosurfactants market is mainly driven by the increasing demand for renewable and biodegradable products. Green products or biologically derived biosurfactants exhibit enhanced functionality as compared to chemically derived surfactants. By using natural raw materials, the products so developed are environment friendly.

The industrial biosurfactants market size was over USD 1.8 billion in 2016 and the global biosurfactants market size should witness consumption in excess of 540 kilotons by 2024 (Source: Global Market Insights, Inc.).

Oleo-surfactants are used in the world, more precisely:

- North America—U.S.,
- Europe—UK, Germany, Italy, France,
- Asia Pacific—China, India, Japan,
- Brazil.

U.S. biosurfactants market demand should register significant growth at over 4% by the end of 2024. Asia Pacific led by China, India and Japan biosurfactants market size should witness gains close to 4.5% over the forecast timeframe. The extensive presence of textile, personal care and agricultural industries along with low overall production costs and competitive economies should stimulate regional industry demand. Rising per capita disposable income and stability to purchase personal care and consumer products should drive regional industry growth. Furthermore, government initiatives such as the National Health Surveillance Agency (ANVISA) of Brazil offer necessary regulations and policies regarding manufacture, supply and import of the products in domestic and industrial applications, which should stimulate industry growth.

However, cost competitiveness with respect to conventional surfactants remains a major restraint.

8.3 Broad-Spectrum Fungicides and Plastic Wrap Film

8.3.1 Introduction

Many natural essential oils (EOs), extracted from plants, have demonstrated a broad spectrum of activity against insects and plant-pathogenic fungi acting as insecticidal, repellent, oviposition deterrent and, from a more general point of view, as 'green pesticides'. These oils have also a very long tradition of use in the protection of foods during medium and long-term storage. Different scientific researches [34, 113] indicated that some chemical constituents of these natural oils can cause, for example, an interference with the nervous system of insects inhibiting, in different cases, their reproductive activity. Another important aspect of these oils is related to their compatibility with mammal's life, as they are non-toxic. This aspect, combined with their wide availability and relatively low cost, make them suitable for the application also in other areas such as, for example, additives for plastics films used for food packaging defined, in this respect, 'active packaging'.

Some of the chemicals substances extracted from these essential oils are already used in the food industry as flavouring agents or fragrances and this should involve a relatively fast procedure for the registration of this class of 'green pesticides' as additives in packaging application such wrap films. At last, in many emerging countries a great bio-diversity of plants is present that, in perspective, can give place to a large availability of EOs with very interesting properties for the mentioned application fields.

The perspective offered by EOs is particularly attractive because, in recent years, environmental problems caused by wide use of synthetic pesticides have been a matter of concern for both scientists and public opinion. On a world basis, it has been estimated that about 2.5 million tons of pesticides are used on crops each year and the damages caused by pesticides reach around $100 billion annually [113]. The main drawbacks of the traditional pesticides reside in their high toxicity, non-biodegradable character and in the residues left in the soil and water after the use. A great impulse is then put into the research of new, highly selective and biodegradable pesticides able to solve both the long-term toxicity to mammals and to reduce the amount of pesticide used to preserve crop yields. The general framework of the Green Chemistry processes is nowadays gaining more and more relevance, which is a natural alternative to traditional synthetic pesticides as a means to reduce the negative impact on human health and the environment and such an alternative could be very probably represented by EOs. Many nature-oriented materials can be profitably employed to reduce microbial, fungi and pest populations improving food preservation and increase food production [94, 205]. As a matter of fact, essential oils can be considered today the best candidate in this field as they are safe, eco friendly and life compatible [34, 96, 113]. The global essential oils market is projected to register an estimated CAGR of 9.2%, during the forecast period, 2018–2023 and, in only the US, the EOs market is projected to reach USD 7.34 billion (13 billion on a world basis) by 2024.

8.3.2 Components of Essential Oils

It is well known from the literature that EOs are very complex mixtures of many organic compounds [34, 102, 113] and their composition can be different from plant to plant and, also, between distinct parts of the same plant [34]. A further source of variation of EOs composition, for some species of plants, could be the harvesting seasons and geographical areas of cultivation [16, 42, 62, 137].

The main constituents can be classified into different categories like in the following list in which also some examples are shown [113]:

- terpenes (hydrocarbons): myrcene, pinene, terpinene and limonene;
- terpenoids (oxygen-containing hydrocarbons);
- acyclic monoterpene alcohols: geraniol, linalool;
- monocyclic alcohols: menthol, 4-carvomenthenol, terpineol, carveol;
- aliphatic aldehydes: citral, citronellal, perillaldehyde;
- aromatic phenols: carvacrol, thymol, safrol, eugenol;
- bicyclic alcohols: verbenol;
- monocyclic ketones: menthone, pulegone, carvone;
- bicyclic monoterpenic ketones: thujone, verbenone, fenchone;
- acids: citronellic acid, cinnamic acid;
- esters: linalyl acetate.

In Fig. 8.7, chemical structures of some compounds frequently present in essential oils are reported.

An important aspect that can strongly influence the EOs composition, and ultimately also the antimicrobial activity, is the extraction method from the original matrix. For commercial purposes, the steam distillation is the most commonly used method for recovering EOs while the extraction by means of liquid carbon dioxide, at low-temperature and high-pressure conditions, resulted in more expense [144].

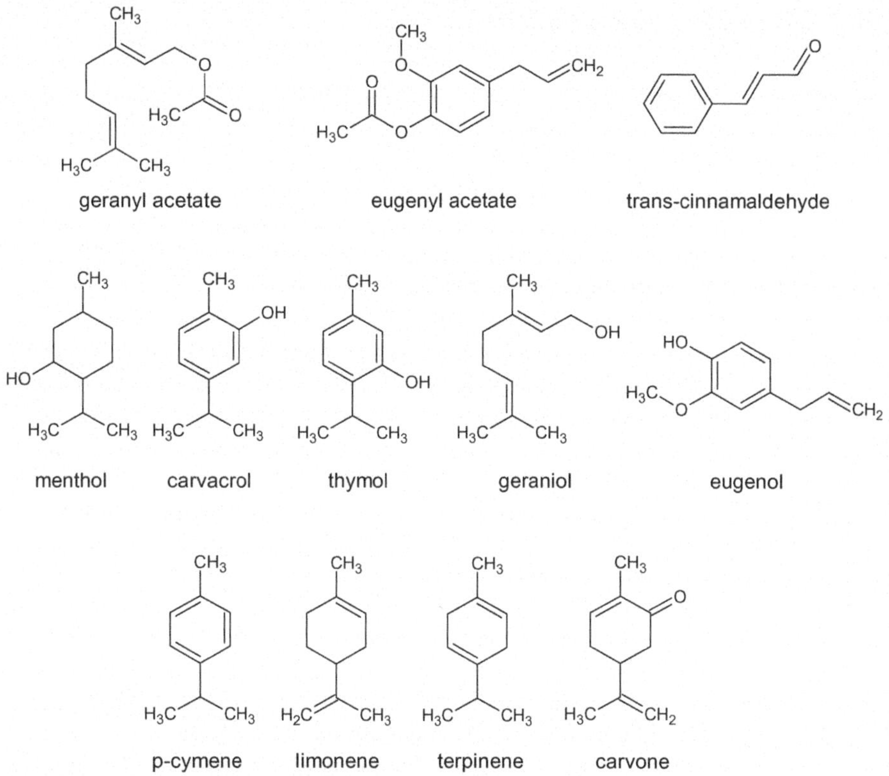

Fig. 8.7 Structure of some typical constituents of essential oils

Despite this economical consideration, CO_2-based extraction resulted in a more natural organoleptic profile, if compared with steam distillation, that may also influence the antimicrobial properties of the extracted oil. The similar difference with respect to steam distillation has been evidenced also by using solvent extraction (with hexane) that produce an oil with more pronounced antimicrobial activity [155]. Among the numerous literature publications or reviews that reported data regarding the composition of the various EOs, of particular interest resulted in the volume of Bauer et al. [23] that summarized the major components of the economically interesting EOs. For what concerns analytical techniques, suitable methods for the determination of EOs composition and components identification are gas chromatography and mass spectrometry, applied also as headspace variant [47, 97, 100, 186, 193, 231, 232]. With these techniques applied to EOs, a total number of individual components between 60 and 80 were identified [182, 195]. Notwithstanding this relatively high number of constituents, only a few major components count up to 85% of the total mass while other components are present only in traces [23].

8.3.3 Activity of Essential Oils

Although many literature reviews have been focused, in the past, on the antimicrobial properties of essential oils and their components [34], lack of extensive information is still present about the details of the action mechanism [120, 121]. Very likely, the antibacterial/antifungal activity of EOs cannot be attributed to a single specific mechanism due to a large number of different categories of chemical compounds the are present (see Chap. 2); on the contrary, it is a common opinion that several targets are located into the cells that can be considered as sites for EOs action [35, 201]. The number of chemical substances present in EOs is very high and frequently, the most abundant of them have shown a more or less pronounced antibacterial property. Moreover, as reported by many authors [1, 44, 66, 71, 120, 121], the antimicrobial or antifungal activity of essential oils can be correlated or compared with those of the individual components of the oils themselves. On the other hand, different authors examined various essential oil components and their combinations to find possible synergistic effects [58, 115, 116]. An interesting review, for example, among these property–structure relationships studies, tried to correlate the structure and molecular properties of 60 terpenoids with their antimicrobial activity [79].

Numerous literature studies have been devoted to the activity of EOs and focused on the determination of which compounds are responsible for the major antioxidant or antimicrobial effect. For example, thymol, carvacrol and eugenol are the main components responsible for the antioxidant activity of basil and thyme oils [123]. Even if the EOs activity mechanisms are not fully understood and clearly described, it seems that the antimicrobial activity is essentially due to their hydrophilicity. This characteristic enables them to be partitioned in the lipids of the bacterial cells membrane and mitochondria, disturbing the structures and rendering them more permeable [109, 200]. Terpenes, that in many cases are the major compounds of EOs, can disrupt and penetrate the lipidic structure of the bacteria cell membrane, leading to the denaturing of proteins and the destruction of the cell membrane [217]. Lambert et al. [120, 121] reported that essential oils containing a high percentage of phenolic compounds, such as eugenol, carvacrol and thymol, present stronger antibacterial properties against foods pathogens. In other cases, the action mechanism of EOs components (mainly carvacrol) against microorganisms is the channel formation through the membrane [219] by pushing apart the fatty acid chains of the phospholipids, allowing ions to leave the cytoplasm. Another study [210], reported that eugenol can induce cell wall deterioration and a high degree of cell lysis. In conclusion, not all the components of essential oils exhibit the same mechanism of action. From the point of view of the antibacterial/antifungal activity of EOs, it seems to be ascertained that phenolic components are the main responsible for the antibacterial properties [42].

As discussed before in paragraph 2, as the chemical composition of EOs depends on different factors, such as origin country and harvesting period, this type of variation could be enough large to justify the variability in the sensitivity of microorganisms and fungi to these substances. Moreover, it is a common opinion that the overall effectiveness in antimicrobial activity of EOs cannot be attributed only to the main components. EOs are complex mixtures of numerous molecules, and their biological effects are the result of a synergism of all components. Several studies have concluded, in fact, that essential oils resulted to be more effective in terms, of antimicrobial activity, with respect to an artificial mixture of the major components [73]. Essential oils and their components, commonly used as flavouring in the food industry, also present interesting antibacterial, antifungal and antioxidant properties [34].

8.3.3.1 Methods for Antimicrobial Assessment

The task of assessment of EOs, for their antibacterial and antifungal properties, is generally pursued through different conventional and widely used techniques initially developed for antibiotic testing that proceeds through a visual or an instrumental evaluation. Two fundamental basic techniques, briefly described in this paragraph, are widely used for the assessment of both antibacterial and antifungal activities of essential oils:

1. The agar diffusion method (paper disc or well),
2. The dilution method (agar or liquid broth).

Anyway, some specific properties of EOs, like volatility, high viscosity and hydrophobicity, require that some modifications must be applied to the mentioned standard assessment methods as they were initially developed for aqueous-phase antimicrobial substances. In the application of these tests on EOs, even if a proper solubilizing agent or emulsifier must be used, unequal distribution of the oil through the medium can occur affecting the results. It must be checked also if the dispersing agent can affect the growth or the differentiation of the microorganisms under assessment. Another critical issue, in the assessment test, could be the possible selective evaporation of EOs components during a long incubation period so the test condition must be carefully controlled. Usually, microorganisms derived from internationally certified collections are employed but can also be isolated from the natural environment [20, 24, 32]. All parameters which may affect the experimental results have to be strictly defined and the growth experiments must be repeated from two to seven times to ascertain reproducibility [36].

The agar diffusion method is the most widely used test for antimicrobial/antifungal activity assessment. The method is recognized by the most authors [122] as precise and reliable, even if in some cases it furnishes semi-quantitative results or, according to some authors, only qualitative [11] and with low repeatability [51]. It can, however, estimate the degree of microorganisms' growth inhibition and their morphological changes in a very simple way.

The method consists [53] in filling Petri dishes with agar broth and inoculated with microorganisms. The EO to be tested is incorporated in the system by means of a paper disc or through a hole directly in the agar medium. Critical parameters are: the diameter of the disc, the amount of EO added, the type of dispersing agent.

The effectiveness of essential oil is demonstrated by the extension of the zone of microorganism growth inhibition around the disc or well, and it is usually expressed as the diameter of this zone. The main drawbacks of the agar diffusion method are: (i) the volatile components of EOs are likely to evaporate with the dispersing solvent during the incubation time; (ii) poorly soluble EOs components do not diffuse well in the agar broth. Notwithstanding these rather serious drawbacks, this test is still the most common technique for the antibacterial and antifungal EOs activity assessment, it is easy to implement and requires only very low amounts of essential oil. It can be recommended for a preliminary screening when a large number of EOs samples have to be tested and, among them, the selection of the most active will be then submitted to more sophisticated methods [77].

The second test method considered here, consists of a sequence of dilutions (dilution method), widely employed for bacteria and fungi. The broth dilution method has recently become popular [36] also in the EOs assessment test. The test consists of putting Agar broth cultures in Petri dishes or tubes, while liquid broth cultures are cultivated in conical flasks filled with culture medium [122]. For the liquid broth in conical flasks the results, expressed as inhibitory growth index, are computed as a percent of the variation in mould's biomass compared to the control culture [122]. The inhibitory effect of essential oil in the test tubes cultures and microtiter plates is instead measured turbidimetrically [101] or with the count plate method (Kester et al. 1996). Both in agar and liquid broth, the tested microorganisms can settle onto the membrane [107], which strongly simplifies the procedure and give an accurate estimation of EO activity. As a summary, the result of this inhibition test, can be expressed in the two following ways:

- The growth inhibition index defined as the percentage ratio to the control growth culture without essential oil.
- Minimal inhibitory concentration (MIC) restraining microorganism growth or minimal lethal concentration (MLC).

In addition to the two described assessment tests, it is worth to cite at last also the turbidimetry method, also a commonly used technique, where the changes in the optical density (OD) of the growing culture is measured [122].

Fungicides

Pesticides play an important role in the agriculture sectors. Pesticides are mainly in the form of aqueous concentrates (AC) for herbicides, as emulsifiable concentrates (EC) for insecticides and as wettable powders (WP) for fungicides. Customers increasingly require safer and more convenient pesticide formulations such as suspension concentrates (SC), water-dispersible granules (WDG) and emulsions in water. Generally, the preparation of pesticide formulations will include two main components, i.e. the active ingredient(s) and inert ingredients. Active ingredient(s) is the chemical or chemicals responsible for preventing, destroying, repelling or mitigating insects, fungi, weeds, rodents, etc., whereas, inert ingredients or supplements are inactive ingredients that have no pesticide action but function as wetting agents, dispersing agents, emulsifiers, spray adjuvants, solvents, carrier/diluents, etc. At present, there is a shift from the less environmentally friendly hydrocarbon (mineral oils) to the more environmentally friendly oleochemical products as inert ingredients in pesticide formulations. This paper discusses the potential of oleochemicals derived from palm and palm kernel oils as inert ingredients in pesticide formulations [93].

Plant diseases constitute an emerging threat to global food security. Many of the currently available synthetic fungicides are highly toxic and non-biodegradable and cause extended environmental pollution and probably cause an increasing number of pathogens that develop resistance to them [134]. Surfactants (obtained by chemical processes from petrochemical and oleochemical resources) have been used as a fungicide with a noted shift towards surfactants from oleochemicals source as an environment-friendly nature and safety [72]. Vegetable oils have been used as pesticides and fungicides and continue to play an important role in integrated pest management programmes. Neem and Karanja oil are used largely for these purposes [104]. In agrochemical industries, surfactants are used as additives primarily in pesticides (herbicides, insecticides, fungicides) [239].

A new family of ultrashort antimicrobial lipopeptides which are composed of only four amino acids linked to fatty acids was reported. The activities in vitro and in planta and the modes of action of these short lipopeptides against plant-pathogenic bacteria and fungi were investigated. They act rapidly, at low micromolar concentrations, on the membranes of the microorganisms via a lytic mechanism. In vitro microscopic analysis revealed wide-scale damage to the microorganism's membrane, in addition to inhibition of pathogen growth. In planta potent antifungal activity was demonstrated on cucumber fruits and leaves infected with the pathogen Botrytis cinerea as well as on corn leaves infected with Cochliobolus heterostrophus. Similarly, treatment with the lipopeptides of Arabidopsis leaves infected with the bacterial leaf pathogen Pseudomonas syringae efficiently and rapidly reduced the number of bacteria. Importantly, in contrast to what occurred with many native lipopeptides, no toxicity was observed on the plant tissues. These data suggest that the ultrashort lipopeptides could serve as native-like antimicrobial agents economically feasible for use in plant protection [134].

A combination of palm oil based surfactant was used to stabilize the emulsion system contained essential oils for fungicide EC formulation. The best result was achieved at surfactant HLB value 10 with surfactant composition MES 20%, PDO 10% and PMO 70%, oil composition neem oil 33%, eugenol oil 33%, citronella oil 33% in oil/surfactant ratio 60/40 based on emulsions stability test for 6 h. Furthermore, this study can be developed by investigating its efficacy and residue in the environment [196].

Neem oil is a key derivative that finds application as an important pest control agent. The major active principal, azadiracthin (AZA), a ring c-seco-tetranortriterpenoid, is the most potent natural insect anti-feedant. The ever-increasing emphasis on developing environmentally benign pest control agents has brought Neem to the fore. There is a significant correlation between the stability of emulsion and HLB of surfactants suggest that the HLB requirement of emulsion varies with the chemical type of emulsifier. Creaming of emulsion could be controlled by adding viscosity-building agents like guar gum. Emulsifiable concentrate prepared from Neem seed extract was best in spreadability, stability and bioefficacy point of view [226].

Plant oils such as coconut, soybean, corn, palm, palm kernel oils and their oleochemicals are used to produce inert ingredients such as surfactants, wetting agents, dispersing agents, emulsifying agents, adjuvants, solvents and carrier/diluents for pesticide formulations. In general, the preparation of pesticide formulations will include two main components, i.e. the active ingredients and inert ingredients. Active ingredients mean the chemical or chemicals in a product responsible for the desired effect, which is capable of preventing, destroying, repelling or mitigating insects, fungi, weeds, rodents or other pests. Meanwhile, inert ingredients also called supplements are inactive ingredients that have no pesticidal action. There are various types of inert ingredients normally found in pesticide formulation, for example, emulsifiers, organic solvents, dispersing agents and wetting agents [185].

8.3.3.2 EOs as Green Pesticides

Among the various possible future uses of EOs, it is worth to mention their application in the field of traditional pesticides substitution [113]. The use of essential oils or their components in the replacement of pesticides, considering their volatility, is characterized by a limited persistence under field conditions. Moreover, several EOs have an exemption under regulatory protocols as they are usually extracted from aromatic plants, in particular, those used in perfumes and foods industry as flavours and fragrances and flavourings, respectively, and more recently in natural medicine approaches.

Even if a long tradition involved EOs in the protection of stored commodities, especially in the Mediterranean region and in Southern Asia, in the early 2000s a renewed interest emerged when insecticidal activities against a wide range of pests have been demonstrated [94, 95]. The key aspect on which this interest is based, consists in their no-toxicity to mammals, birds and fish [204] and in the moderate toxicity, with very few exceptions, of their constituents like purified terpenoids [113]. The placement of EOs under the category of 'green pesticides' is, therefore, justified and different studies reported interesting results in this perspective. For example, Kordali et al. [111], reported that oils from plants like Artemesia vulgaris, Melaleuca leucadendron, Pelargonium roseum, are very effective against various insects and fungal pathogens. Another study conducted by Tripathi et al. [211] has demonstrated that the constituents of volatile oil (Mentha species) are highly effective against Callosobruchus maculatus and Tribolium castanum, the common stored grain pests. In a more general view, EOs derived from eucalyptus and lemongrass have shown interesting properties as animal repellents, anti-feedants, insecticides, miticides, fungicides and antimicrobial products [113].

8.3.3.3 EOs as Antifungal Agents

Many literature studies appeared that are focused on the antifungal properties of essential oils or of their components. These studies [113] reported a generally positive assessment of EOs against different fungi like, for example, Botrytis cinerea, Monilinia fructicola, Rhizoctonia solani, Fusarium moniliforme and others. As an example of antifungal action of specific components present in EOs, we can cite that of thymol and carvacrol that resulted highly active against most fungal species tested [117, 146]. The action mechanism of these compounds against fungi is still unknown but the researchers believe that this activity may be very likely related to a general ability to dissolve or disrupt the integrity of cell walls and membranes (Isman et al. 2006). In Table 8.2, some examples of antifungal activity of EOs is reported. A different strategy used to test the effectiveness of EOs against fungi, consists in treating the soil with a certain amount of plant essential oil and successively check if the population of fungi and other parasites declines. Some greenhouse experiments [170] have shown that, for example, by treating the soil with 400–700 mg of oil (thymol, palmarosa and lemongrass oil) per liter of soil inhibit the growth of Ralstonia solanacearum on tomato plants. After only 7 days, the bacterial population declined to an undetectable level on 100% of plants.

8.3.4 Essential Oils in Food Applications

It is well known that food products, especially during storage, can undergo numerous physical, chemical and microbial changes and that the stability of food is a function of changes occurring in the food components, such as proteins, lipids and carbohydrates. This can be ascribed to processing and environmental factors like exposure to temperature, light, moisture, etc. The use of a protective barrier during processing, storage, and handling not only slows down the deterioration of food but may also enhance its quality. Suitable packaging can retard the deterioration phenomenon and hence, extend the shelf life of products and this is the main motivation for which, in recent years, a wide variety of packages have been developed and employed to provide desirable effects. Examples of this packaging strategy [205], are the incorporation in the polymer matrix, of oxygen, moisture and ethylene scavengers for, respectively, oxygen, moisture and ethylene sensitive foods; use of carbon dioxide in other foods; flavour imparting or scavenging chemicals; antimicrobial agents for microbiological stability of foods.

The concept of direct incorporation of these chemicals and agents directly into the packaging material is not the only possibility. Other approaches involve the placing of these active agents on the packaging layer or between this one and the food. Such strategies for obtaining desirable effects are nowadays referred to as 'active packaging,' 'interactive packaging' and 'intelligent packaging.'

Active packaging is a relatively novel concept developed to provide interaction between food and packaging material, and it aims to extend the product shelf

Table 8.2 Activity of some EOs and Eos components against different fungi with two testing methods

Fungi	Dilution method: tested oils	MIC (MLC) (ppm)	Diffusion method: tested oils
Aspergillus flavus	Carrot	2000	Laurel
	Thymol	250	Cinnamon, clove
	Cinnamaldehyde	200	Eucalyptus
	Geraniol, nerol,	500	
	citronellol	5000	
	Tea tree	500	
	Artemisia nilagirica		
Aspergillus fumigatus	Carrot	2000	Piper betle
	Thymol	125	Lemongrass
	Artemisia nilagirica	200	Palmarosa,
			peppermint
Aspergillus niger	Eucalyptus	4000	Laurel
	Tansy, elecampane	4000	Geranium
	Tea tree	4000	Oregano, thyme
	Origanum syriacum	1000	
Aspergillus parasiticus	Thyme	400	Tea tree
	Cumin	600	
	Clove	600	
	Rosemary	2000	
Candida albicans	Yarrow	2500	Mugwort
	Elecampane	2000	Rosemary
	Clove	3400–9000	Clove, coriander,
	Eucalyptus	4000	cinnamon, Marjoram
	Pine	2500–29,000	sage
	Linalool, terpineol	2500	Fir
	Chamazulene	500	Yarrow
	Polyacetylenes	500	
	Artemisia asiatica	2	
	Ocimum gratissimum	350	
	Thymol	50	
	Tea tree	2000	
	Thyme	5000	
	Cinnamon,	1000	
	cinnamaldehyde,	2000	
	Carvacrol	3100	
	Zieria sp.	3000	
	Manuka	170	
	Hyssop		
	Aniba canelilla		
Candida sp.	Aniba canelilla	360	Vicoa indica
	Black cumin	2500	Murraya exotica
	Parsley	12,500	Laurel
	Cardamom	1600	Myrtus communis
	Oregano, marjoram	5	

(continued)

Table 8.2 (continued)

Fungi	Dilution method: tested oils	MIC (MLC) (ppm)	Diffusion method: tested oils
Cryptococcus neoformans	Cinnamon, palmarosa, clove, Oregano, thyme Thymol, carvacrol Ocimum gratissimum Thymol	200 50 300 50	Mugwort Peppermint, lemongrass Piper angustifolium
Penicillium chrysogenum	Cardamom	1600	Laurel Olibanum
Saccharomyces cerevisiae	Eucalyptus Polyacetylenes Cinnamon Clove Geranium, cypress Anethol	4000 125 200 400 600 200	Rosemary Clove, coriander, cinnamon, Marjoram sage Fir

life, maintaining its nutritional quality, as well as providing microbial safety. Many researchers have reviewed the existing active products and patents [10, 67].

The intense researches and application studies in the field of active packaging have involved also the EOs [122]. The use of EOs, for food preservation purposes in food technology, is nowadays largely adopted due to the properties that EOs possess against many bacteria, fungi, insects, etc., especially when the synergistic effect of their component is exploited. Many experimental evidence are also available in the literature of EOs as antioxidants for meat, fish, fruits and vegetables [122]. In Table 8.3, different examples of packaging obtained by the addition of EOs to polymers are reported.

It has to be taken into account that the oxygen availability and food storage temperature also modifies the EO antimicrobial activity. For example, the antibacterial activity of oils from oregano and thyme against Salmonella typhimurium and Staphylococcus aureus was enhanced at low oxygen levels [158]. Therefore, the use of vacuum packaging in combination with EOs seems to be a good strategy of food preservation and an extension in the shelf life of many days (15–20 in such cases) can be obtained in comparison with the untreated fresh product [11]. The increased effectiveness of EOs at low temperatures can be explained, on the other hand, because of the higher permeability of the cell membrane at these temperatures, which allows the EOs to dissolve more easily in the lipidic bilayer [130]. The combination of EO and modified atmosphere packaging (MAP) has been widely documented over the past few years. For example, Kostaki et al. [112] evaluated the combined effect of MAP in food preservation.

Table 8.3 Examples of polymers with antimicrobial activity upon EOs addition

Polymer	Essential oil	Microorganism
Alginate	Garlic oil	Escherichia coli Salmonella typhimurium Staphylococcus aureus Bacillus cereus
Alginate	Oregano, lemongrass, cinnamon oil	Escherichia coli
Whey protein	Oregano oil Oregano, rosemary and Garlic oils	Spoilage bacteria Salmonella enteriditis Listeria monocytogenes Staphylococcus aureus Escherichia coli Lactobacillus plantarum
Chitosan	Garlic oil	Escherichia coli Listeria monocytogenes Staphylococcus aureus Bacillus cereus Salmonella typhimurium
Chitosan	Oregano oil	Escherichia coli Listeria monocytogenes
Chitosan	Cinnamon oil	Escherichia coli Listeria monocytogenes Lactobacillus plantarum Lactobacillus sakei Pseudomonas fluorescens
Chitosan	Clove oil	Escherichia coli Pseudomonas fluorescens Listeria innocua Lactobacillus acidophilus
Chitosan	Tea tree oil	Penicillium italicum Listeria monocytogenes
Chitosan	Bergamot oil	Penicillium italicum

8.3.4.1 Antimicrobial Edible Films and Coatings

An interesting approach, as already mentioned, consists in incorporating antimicrobial/antifungal agents directly into edible films or coatings in a way that they are slowly released onto the surface of the food to inhibit microbial growth [205]. Such coatings can also act as a barrier to moisture and oxygen. In particular, edible coatings have become popular in the food industry, because they have competitive costs, produce less waste being biodegradable and offer effective protection also after the package opening. Numerous literature studies appeared in the literature on this topic. For example, Hershko and Nussinovitch [84] studied the behaviour of hydrocolloid coatings on vegetative materials; Mark [136] reviewed systematically the selection of edible coatings to maximize quality and shelf life of fresh fruits and vegetables. Components of edible films and coatings can be divided into three principal categories [205]: (i) hydrocolloids (proteins and polysaccharides as cellulose and derivatives, starch, chitosan, etc.); (ii) lipids (waxes, acylglycerols and fatty acids), and (iii) composites (a combination of the two previous categories). Despite the large spectrum of available materials, the choice of materials for a film or coating depends largely on the final use of the film.

The potential applications and properties of edible films are also subjected to different reviews [51, 106, 122, 143]. Moreover, the use of a minimum amount of plasticizers in these films (like sorbitol or glycerol) may be adopted to improve mechanical properties and the possibility of incorporating active compounds (antimicrobials, antioxidants, nutraceuticals, flavours, colourants) in polymeric matrices is one of the main advantages of coatings. In particular, as commented above, the incorporation of EO into polymeric matrices gives them very interesting antimicrobial/antioxidant/antifungal properties (see Table 8.3 for some examples). The major advantage of this technology is that the diffusion rate of the antimicrobial agent can be slowed down, keeping high concentrations of the active compounds on the product surface (where the contamination is prevalent) for extended periods of time. This makes the process more effective in reducing the levels of microorganism with respect to the direct application on the product surface through spray solutions [172]. Other important issues are: the nature and amount of EOs, the EO/polymer ratio in the film and the possible interactions between the polymer and the active compounds. When the polymer itself, did not show antimicrobial activity, the effect of the EO generally increased with the EO/polymer ratio [187].

In food applications, film water vapour permeability (WVP) is another fundamental factor affecting the effectiveness of food properties preservation. Low values of WVP are usually desirable in order to minimize weight losses in the coated product and the incorporation of EO into polymeric matrices leads to an improvement in the WVP of films because of the increment in the hydrophobic character of material [122]. Less clear is the effect of EO addition on oxygen and CO_2 permeability, even if some studies [122] reported a slight decrease in oxygen permeability of the films based on alginate with lemongrass oil. In some cases, however, the incorporation of EOs into a continuous polymeric matrix of edible films, resulted in a decrease of its mechanical resistance to fracture, because of the structural discontinuities caused by the oil-dispersed phase [122]. Moreover, the use of EO can induce modifications in terms of film transparency and colour [122] and this represents a serious drawback because of the appearance of the coatings is of relevance since their commercial acceptance depends mainly on this attribute. This modification has been attributed to the migration of EOs dispersed droplets to the external surface of the film during film drying, which leads to surface irregularities. Nevertheless, observed differences in terms of colour are not significant when low concentrations of EOs are used in bioactive films [122] so a key point to be tuned is the EO concentration.

Despite the interesting properties and possibilities offered by edible packagings for improving the shelf life of the foodstuff, few industrial applications were developed [52] as the formulation of these materials still remains empirical and uneasy. Edible packagings are an integral part of food as they are located on the food surface or as thin layers between several parts within the product. Nowadays in the European Community, there is no regulation that specifies how edible packaging could be classified [13]. As in most cases, edible packagings do not provide a significative nutritional value to the coated food, it is a general opinion that they should be considered more like an additive than an ingredient.

8.3.4.2 Type of Antimicrobial Packaging

A new generation of food packaging may include materials with inherent antimicrobial properties and can be classified, in this case, as 'active packagings'. These innovative packaging technologies could play a key role in extending the shelf life of foods and reduce the risk of pathogens. Active packaging can interact with the product through the headspace between the package and the food system to realize the desired effect [31, 118, 181].

From a general point of view, active antimicrobial/antifungal packagings can be configurated in several forms such as [13]:

1. Addition of capsules containing volatile antimicrobial agents into packages.

   The most successful application of this strategy consisted in adding sachets containing, for example, oxygen of moisture absorbers or ethanol generators [89, 181].
2. Incorporation of volatile and non-volatile antimicrobial agents directly into polymeric matrix. This possibility has been commercially applied in drug and pesticide delivery, textiles, surgical implants and other biomedical devices. Few food-based applications appeared on a market scale [13]. Moreover, 'natural' antimicrobials agents have been incorporated into paper, thermoplastics and thermosets, and have been tested against a variety of microorganisms including Listeria monocytogenes, pathogenic E. coli and spoilage organisms. Among all the antimicrobials, silver-substituted zeolites are the most widely used as polymer additives for food applications, especially in Japan [31].
3. Coating or adsorbing antimicrobials onto polymer surfaces. As an applicative example of this possibility, we can cite the incorporation of fungicides into waxes

to coat fruits and vegetables or films coated with quaternary ammonium salts to wrap potatoes [199].

4. Immobilization of antimicrobials to polymers by ion or covalent bonds. Few examples of ionic and covalent immobilization of antimicrobials onto polymers or other materials have been published [13]. This type of immobilization requires indeed the presence of functional groups on both the antimicrobial and the polymer. Examples of antimicrobials with functional groups are peptides, enzymes, polyamines and organic acids.
5. Use of polymers that are inherently antimicrobial. Some polymers are inherently antimicrobial and have been used in films and coatings. Cationic polymers such as chitosan and poly-L-lysine promote cell adhesion. Another study [76] proposed charged amines that interact with negative charges on the cell membrane, causing leakage of intracellular constituents. Chitosan also has been used as a coating medium and appears to protect fresh vegetables and fruits from fungal degradation [45].

8.4 Paint and Coating

8.4.1 Introduction

A paint or coating is a formulation needed to cover materials to prevent corrosion and wear. There are four constituents in the paint:

1. Binder: it keeps the film stable on a surface;
2. Pigment: it provides the colour to the coating;
3. Solvent: a volatile liquid that dilutes the binder and evaporate once applied;
4. Additives: wetting agents, flattening agents, driers, plasticizers, emulsifiers, stabilizers, crosslinking agents.

The formulation of the four constituents strictly depends on the specific application. In the present paragraph, the main features and issues of the paint and coatings are presented, together with some economic information. Attention will be paid on bio-based constituents of paint and coatings.

Legislation related to air pollution has led the paint and coating industries to develop ways to reduce the volatile organic content (VOC) of the solvent-borne coatings without compromising the film performance and application characteristics [12, 192]. Plant oils were one of the ideal raw materials in the preparation of coating materials. Plant oil is suitable for the preparation of alkyds [150], where the oils have high molecular weight and high iodine values to enable the resins to be air-dry, and they are widely used for coatings [18]. Novel renewable green waterborne coating system based on plant oil was developed to coat wood including alkyds or polyurethane dispersions [166].

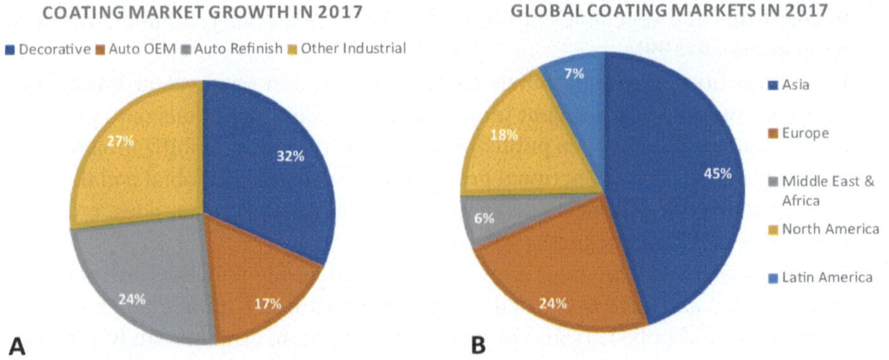

Fig. 8.8 Economic data related to 2017: **a** coating market growth; **b** global coating markets (total 152,614 million U.S. dollars)

8.4.2 *Market Overview*

The paint and coatings industry represents a 170 billion U.S. dollars market with a forecast of 210 billion US dollars to 2022 [203]. The global coatings market grew at an estimated rate of 4.2% in volume terms and 5.0% in value terms from 2016 [28]. The coating market growth in 2017 is reported in Fig. 8.8a, while the global coating markets in Fig. 8.8b [28].

As revealed from Fig. 8.8a, the main driver of the global growth rate is the production of decorative coatings [103], a fact mainly due to the high construction activity of the latest years. Auto OEM and refinish coatings are driven by automotive units build. Concerning geographic distribution, Fig. 8.8b reveals that Asia is the largest region in terms of the coating market.

8.4.3 *Paint and Coating: Main Characteristics and Issues*

Paints and coatings are commonly constituted by polymeric materials used to decorate and protect surfaces. In designing the formulation of a coating, the main properties that must be considered in common applications are: appearance, workability, durability, environmental impact and cost [138]. Balancing these five aspects is a matter of the final application: i.e. living room paints can be of high cost but has appreciable appearance and high durability, while the key point of outdoor coatings is surely workability and durability. General guidelines were drawn by Matheson [138] and Weiss [229], giving hints about the trends in coating technology, describing in detail each attribute. A summary of the considerations drawn by the authors is reported in Table 8.4.

Thus, in formulating a coating, the usual aspects to be considered are: minimizing the use of expensive components, the waste production and eventual repairs [139].

Table 8.4 Main aspects of choosing the characteristics of paint and coatings

Characteristic	Main features
Appearance	• Response to the light conditions and viewing angle and time [126] • Colour: choice of pigments or dyes
Workability	• Reproducible appearance of the colours • Avoiding over and under baking of the paint • Ease of application to the surface (i.e. adequate viscosity)
Durability	• Reduction of the repairing costs of the coated materials (i.e. wood, metal) • Resistance to photo-oxidation (using additives) • Resistance to curing • Resistance to corrosion
Environmental impact	• Volatile organic components (VOC) emissions • Solid waste: i.e. metal disposal from pigments [229]
Cost	• Dye or pigment • Formulation to special application • Molecular weight of the starting polymer • Using of bio-based raw materials

For example, the replacement of acrylic-melamine topcoats in the automotive industry by epoxy-acrylic, polyurethane, and acrylosilane coatings is an example of how customer requirements for etch resistance can lead to a technological advancement [229]. Considerable industrial and academic efforts are nowadays balancing the different attributes for individual situations, formulating complex blends of materials into a functioning coating.

8.4.4 Technological Solutions

As revealed in Table 8.1, several environmental and economic issues were highlighted. Some solutions were proposed in the literature.

The first solution deals with the reformulation of existing paint along with the use of VOC abatement or collection equipment during the application of a coating, i.e. reformulation with less expensive components and chemicals that are not classified as hazardous air pollutants. An alternative solution to the VOC emission is to use an abatement system to collect volatile organic compounds during the application and cure process, which is very expensive and high-energy demanding [229]. In any case, in selecting a solvent blend with cost-effectiveness, evaporation rates and solvent balance must be always considered. The elimination of HAPs solvents (Hazardous Air Pollutants) in a formulation is demanding. The conventional solvents based on hydrocarbons and ethylene glycol ether (HAPs) are slowly being replaced with oxygenated solvents (i.e. ketones, alcohols and propylene glycol ethers [183, 229]).

The second solution is related to the development of new 'compliant' coating technology, addresses the need for energy conservation, balances cost against necessary performance and provides compliance with environmental regulations [229]. In this case, four types of application can be possible:

- **Waterborne coatings**: coatings that utilize water as the major carrier liquid. The resins commonly used need to be high hydrophilic, i.e. acrylics, alkyds, polyesters, epoxies, polyurethanes and vinyls with common crosslinkers being melamines, aziridines, carbodiimides and glycidyl compounds.
- **Radiation curable coatings**: combinations of unsaturated monofunctional and multi-functional acrylated monomers and acrylated oligomers, characterized by the presence of epoxy, polyester or polyurethane backbone.
- **High solids coatings**: incrementing the amount (>60%vol.) of solids in a conventional solvent-borne paint, it is possible to reduce the VOC content of a coating formulation.
- **Powder coatings**: two types of powder coatings, thermoplastic and thermoset. The firsts are resins melting and flowing at elevated temperatures; the seconds provide similar melting and flow properties (75–85 °C) while changing their chemical properties upon curing (140–200 °C).

Among the classical resins used in formations, different kinds of chemicals can be used: i.e., acrylic, alkyd and polyesters, amino, epoxy resins and polyurethanes. Vegetable oils can be used as raw materials in producing most of the mentioned classes, with the advantage of starting from a renewable resource, thus obtaining environmentally friendly paints. Thus, the modern research is moving to the development of new bio-based paint and coatings [169].

8.4.5 *Bio-based Paints and Coatings*

Vegetable oils (VO) were used as the primary constituent in paints and coatings even during the days of cave paintings. VO were used as binders or additives in paint and coating technology for many centuries. They find numerous industrial applications in the field, such as plasticizers, adhesives, paints and coatings [8]. VO are non-toxic, non-depletable, non-volatile and biodegradable resources. They can be used to produce polymers that find applications in the development of paints and coatings. Several VO-based materials have been developed and are tailor-made for a huge number of applications.

The primary use of VO in coatings is as drying agent, namely highly unsaturated oils able to oligomerize or polymerize when exposed to the oxygen in air, in the presence or absence of a catalyst. This procedure leads to an overall increase in the molecular weight due to the crosslinking [128, 129]. Some of the common fatty acids found in drying oils include oleic, palmitic, myristic, stearic, linoleic, linolenic, pinolenic, ricinoleic and α-eleostearic. Therefore, only linoleic, linolenic, pinolenic, dehydrated ricinoleic and α-eleostearic acids are truly drying oils. These VO are characterized by two or more units of unsaturation separated by no more than one methylene group. Typical sources for drying oils include soybean, linseed, sunflower and dehydrated castor oil [220]. Eventual trace contaminants present in the oils can influence their acceptability in a given market, i.e. phenolic contaminants can modify the colour of the resin as it oxidizes in air.

VO find application in the formulation of different kinds of paint and coating. Natural VO are triacylglycerols of fatty acids. They contain several functionalities in their backbone that can undergo a huge number of chemical reactions useful to modify their chemical structure.

In general, VO can be directly used, without any further modification, as corrosion inhibitors and to produce drying-oil coatings. The main features of these two classes of products are reported as follows.

Corrosion inhibitors: VO and plant extracts have been used as natural corrosion inhibitors against alkaline and acid media as well as chloride ions [49, 119]. The anticorrosion activity is attributed to the presence of heterocyclic constituents such as alkaloids, cellulose, flavonoids and tannins.

True drying-oil coatings: VO have the tendency to dry forming films in their virgin forms. Drying is a characteristic of the VO, depending on the unsaturated portion, i.e. the number of carbon–carbon double bonds (iodine value or drying index). VO characterized by with high drying index are used, meaning the sum of the percentage of linoleic acid plus two times the linolenic acid content of the oil. VO with a value greater than 70 are considered drying oils. Drying or semi-drying VO are used in surface coatings. Non-drying VO can be used, instead, to incorporate suitable chemicals (e.g. hydroxyls) or modifiers (vinyls, acrylics, acrylic copolymers) in VO backbone, through chemical reactions. This fact has the aim to transform them into film formers. Drying VO can react with atmospheric oxygen to form polymeric materials with crosslinked structures [140].

VO in their virgin forms require longer drying times, and the related films do not meet the desirable physico-mechanical and corrosion resistance performance. Thus, several chemical reactions can be carried out at the functional groups present in the virgin VO, i.e. hydrogenation, transesterification, epoxidation, acrylation. The main products that can be synthesized starting from VO are listed below and summarized in Fig. 8.9.

Alkyds: oil-modified polyesters, consisting of a multi-functional acid (phthalic acid or trimellitic acid), a polyol (i.e. glycerol, trimethalolpropane or pentaerythritol) and an unsaturated fatty acid formed by a polycondensation reaction. There are three normal alkyd classes, based on the amount of unsaturated oil used in producing the alkyd resin [86]:

- *Long oil alkyds*: used for architectural paints. They contain more than 60% oil, characterized by slow drying speed and softness.
- *Medium oil alkyds*: used for architectural paints and as a co-resin in coatings, containing a VO amount of 40–60%, characterized by slow drying velocity but are slightly harder and are faster to dry than long oil alkyds.

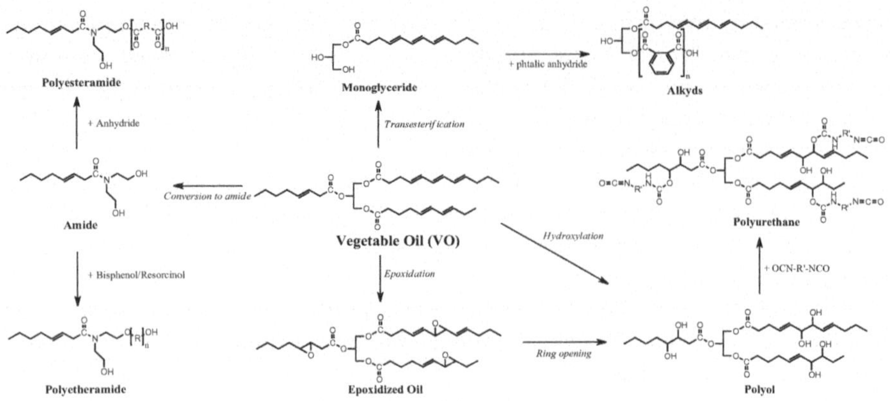

Fig. 8.9 VO chemical transformation routes to coating materials

- *Short oil alkyds*: used for rapid-dry applications, containing less than 40% VO, characterized by fast drying and hardness greater than the other two.

Alkyds can be synthesized through two different processes: fatty acid and monoglyceride process. They show good durability, gloss retention and weathering resistance but low chemical resistance in alkaline media, leading to the hydrolysis of the esters contained in the chain. They are characterized by high biodegradable nature and good compatibility with other polymers. Examples of VO used to produce alkyds are: soybean oil (used as base oil), linseed oil (used because of rapid cure), tung oil and dehydrated castor oil. Usually, alkyds show weak properties in abrasion and solvent resistance. Alkyds are available in two forms, namely *drying* or *non-drying* alkyds, that difference from the presence/absence of drying VO. The drying process is attributed to the auto-oxidation at the active groups. This procedure can be conducted either by air drying or by oven drying at elevated temperatures. Carboxylates of lead and the greener cobalt, zirconium, zinc, calcium and iron are normally used as drying accelerators. Different efforts were made to demonstrate the better properties of alkyds compared with traditional coatings: i.e. acrylated Albizia benth medium oil alkyds showed improved drying time, flexibility, adhesion, scratch, impact chemical resistance [5]; Araujo et al. demonstrated that VO-modified alkyd paints showed higher corrosion resistance in marine and industrial atmospheres [14]. Recently, alkyds were used to formulate new waterborne coatings, organic–inorganic hybrid and nanocomposite coatings with good results [8]. As an example, in 2013 Pathan and Ahmad prepared waterborne alkyd from VO to formulate anticorrosive protective coatings, including s-triazine ring [161, 162], showing good performances in alkaline media. For further details, we suggest the reading of Alam et al. review [8]. Alkyd resins are produced with the reaction of oil or fatty oil, polyol and polyacid. Alkyd resins are commonly used in coating and paint industry due to the ease of application in changing environmental conditions [92]. Blending cheap and environmental-friendly palm stearin alkyd resins with cyclohexanone–formaldehyde (CHF) resin

(70:30 w:w) resulted in desired improvement and application for coating purposes [19]. Modification of long oil palm stearin based alkyd resin was done through blending it with a commercially available ketone resin (cyclohexanone–formaldehyde) to improve its coating characteristics. The effects of blending on air drying time, hardness, adhesion, gloss, impact strength, chemical resistance and thermal behaviour of dried films of blends were investigated and compared to virgin alkyds. It was found that blending results in modifying coating characteristics of palm stearin based alkyd resins, significantly. The best result was observed for weight ratio 70:30 of alkyd and ketone resins [19].

Polyesteramides: modified alkyds obtained by the esterification reaction between VO amide diol and an acid/anhydride. They are characterized by improved properties over alkyds such as water vapour resistance, hardness, ease of drying and resistance to alkalis. The synthesis of polyesteramides follows two different steps: (i) VO amidation with diethanolamine in the presence of sodium methoxide; (ii) esterification between the hydroxyl groups of amide diols with an anhydride. These steps generally require high reaction temperatures (>120 °C) and long times (>5 h) in the presence of solvents [9]. Microwaves were recently proposed to reduce the reaction times (4 min) and temperatures [9]. The presence of ester, amide, double bonds, leads to a superior chemical resistance, thermal stability and imparts flexibility. Moreover, further chemical reactions are possible on the mentioned functional groups to give the final product chemical and physical properties tailored to the final application [240].

Polyetheramides: alternating amide and ether moieties. The synthesis of polyetheramides consists of a two-step process: (i) preparation of VO amide diols; (ii) condensation reaction between amide diols, bisphenol-A and resorcinol [8]. In this case, drying occurs by two consecutive steps: solvent evaporation and baking at elevated temperature (thermal polymerization through migration of double bonds).

Fatty acids: although fatty acid modified polyurethane dispersions are well known, fatty acid methyl esters are quite new as a raw material for coating resins. The chemical difference between these two raw materials is the carboxylic group and methyl ester group, which has no influence on the resulting polyester polyol, because both groups lead to ester bonds. However, the reaction velocity at the beginning can be accelerated by using fatty acid methyl esters, because of the better leaving group in the case of fatty acid methyl ester. Different kinetics may cause different polyester structures as well. Another reason why fatty acids have been more favoured in coating resins than fatty acid methyl esters is their fatty acid composition, which often comprises a high amount of unsaturation. High amounts of unsaturation, especially linolic (18:2) and linolenic acid (18:3), can additionally be cured by oxidation (alkyds). Fatty acids methyl esters, commercially produced for biodiesel, are often based on lower amounts of unsaturation such as rapeseed oil or palm oil. The dominant fatty acid in these fatty acid compositions is oleic acid (18:1). The lower level of unsaturation may be an advantage and lead to less yellowing [166].

Philipp et al. [166] further concluded that fatty acid methyl esters are suitable to substitute fatty acids in polyester polyurethane coatings [164]. The reaction time for polycondensation can be shortened. Furthermore, the neutralization and stirring process was optimized and DMPA was reduced to a minimum amount. It was shown that even at a lower content of emulsifier; stable dispersions and film formation are possible. It is confirmed that the lower the content of the ionic emulsifier, the higher is the chemical resistance and hydrophobicity. Regarding the dispersion stability and particle size distribution, there was no influence on the amount of hydrophobic components observed. However, the visual appearance of the coating films is improved with an increasing amount of DMPA used and even further improved with low amounts of hydrophobic components. Fatty acid methyl esters were used as monofunctional building blocks and require trifunctional building blocks to go along with them. But rapeseed oil based fatty acid methyl esters provide further possibilities for modification. Ongoing investigations are focusing on modification at the double bond and on the ester functionality. It is our objective to obtain new bifunctional monomers for polycondensation. As rapeseed fatty acid methyl ester contains mainly oleic acid, it is easier to obtain bifunctional monomers than to use triglycerides or highly unsaturated fatty acids, which lead to highly functional monomers and promote networks instead of polymer chains. In conclusion, fatty acid methyl esters offer a great opportunity to increase bio-based raw material use in coatings and to improve properties like chemical resistance and hydrophobicity at the same time [166].

Epoxides: VO coatings characterized by the presence of oxirane groups in the backbone. They can be either present in nature (i.e. Euphorbia and Cephalocroton species) or obtained by chemical reaction, the epoxidation reaction, described in detail in the next sections of this chapter. These products show high flexibility and good corrosion resistance against moisture and chemicals, due to their long hydrophobic chains [197]. Epoxy coatings can be cured with curing agents, such as acids, amines, amides and anhydrides. The formulation procedure can strongly affect the final performances of the product: i.e. films obtained blending epoxy-VO with polyvinyl chloride and polyvinyl alcohol shown high biodegradability [177], while epoxy-VO blended with polymethylmethacrylate and polystyrene are used as corrosion protective coatings [4]. Epoxy acrylate oligomer (EA) was plasticized by adding different plasticizers such as epoxidized soybean oil, glycerol and CO and cured by electron beam (EB). Incorporation of CO in epoxy acrylate diluted by 1,6-hexanediol diacrylate (HD) monomer improved the physical, chemical and mechanical properties of cured films than the other plasticizer. Sunflower-free fatty acid was epoxidized in situ under well-established conditions. The epoxidized sunflower-free fatty acids were subjected to react with aniline in sealed ampoules under an inert atmosphere at 140 °C. The produced adducts were added at different concentrations to the epoxy acrylate coatings. It was found that addition of aniline adducts to epoxy acrylate formula gives the best corrosion protection for carbon steel and compete for the commercial corrosion inhibitor efficiency with the same concentration [91].

Polyols: long aliphatic chains with various functional groups, such as hydroxyls, double bonds, active methylene groups and oxirane rings. They are present in nature (i.e. ricinoleic and lesquerolic acids) and can be obtained following different pathways: (i) hydroformylation followed by VO catalytic hydrogenation; (ii) epoxidation of VO unsaturation followed by catalytic hydrogenation or by oxirane ring-opening reaction with water, in the presence of acid catalysts; (iii) ozonolysis followed by hydrogenation. The polyol derived polyurethanes properties depend on the number, distribution, site of hydroxyls, that can cause problems in the reproducibility of the sample and imperfections, leading to a low resistance to stress and low flexibility of the polyols [8].

Polyurethanes: synthesized by polyaddition reaction between (di-or poly-) isocyanates and (di- or polyhydric) alcohols. Polyurethanes are characterized by excellent resistance to corrosion, abrasion, water and chemical resistance and high durability. VO amide diols and polyols are used as raw materials for polyurethanes synthesis. The firsts are obtained by base-catalyzed amidation of virgin VO, the latter by chemical reactions involving ester groups or unsaturation of VO backbone. These products are characterized by the presence of functional groups such as amide, acrylics, ester, vinyl, double bonds and others [6, 7], which presence improves several properties, such as adhesion, flexibility, impact resistance and scratch hardness. Moreover, they show curing or drying at ambient temperatures. Curing procedure involves three steps: (i) solvent evaporation; (ii) chemical reaction of free isocyanates of polyurethanes with moisture; (iii) auto-oxidative crosslinking.

Antimicrobial polyurethane coatings were prepared using Linseed polyol nanocomposite [LMPOL]. Nanocomposite is prepared in situ with Linseed polyol [LP] matrix as organic and Copper acetate as an inorganic constituent by 'solventless one-pot' chemical reaction. The presence of characteristic absorption bands in FTIR spectra confirmed the formation of LMPOL. TEM analysis showed the presence of nano-sized metal oxide in LMPOL. LMPOL showed good antibacterial behaviour against E. coli and S. aureus. The interactions between LMPOL and bacterial surfaces lead to good antibacterial efficacy, suggesting membrane disruption based cell death [198].

Bio-based thermoset polyurethane (PU) coatings comprised of polyesteramides were prepared from Gossypium arboreum (cottonseed) plant oil as a renewable source [142]. Oil was first converted into fatty amide that was transformed into a series of polyesteramide polyols upon esterification with different dicarboxylic acids and anhydride. Bio-based polyesteramide polyols were used to prepare thermoset PU by treating with an aromatic diisocyanate. Coatings prepared using renewable source based polyesteramides were with highly suitable for curing, mechanical/chemical performances, alkali/acid resistance and hydrophobicity. Compared to conventional PU coatings, the prepared coatings possess superior properties such as gloss, adhesion and pencil hardness. Gossypium arboreum plant oil and dicarboxylic acids were concluded to be newer eco-friendly raw materials for PU coatings and compared with standard petro-based polyols [163].

New oils: a new class of synthesis VO. For example, Lesquerella and dehydrated Lesquerella oils were studied for use in alkyd-type coatings, showing good performances in terms of drying time, flexibility and corrosion resistance. 9c,12,13 epoxy-octadecenoic acid (vernolic acid), derived from *Euphorbia lagascae* and *Vernonia galamensis*, can act as a reactive diluent in many solvent and waterborne coatings.

Linseed oil based poly (ester urethane) metallohybrids [PEUMH] was prepared from organic and inorganic precursors, respectively, in 'one-pot, multi-step' reaction.

PEUMH were characterized by spectral, physico-chemical, thermal (TGA and DSC) and morphological analyses by standard methods. The potentiality of the same as promising coating material was also evaluated. PEUMH are foreseen as prospective candidates for application as antibacterial self-sterilizing protective coatings due to oligodynamic effect of metal [198].

The Jatropha curcas Linnaeus (JCL) oil was extracted, refined and modified through epoxidation, hydroxylation and dehydration steps in order to increase the degree of unsaturation in the oil alkyl chain. The modified oil was subsequently used for alkyd resin preparation (50% oil formulation) using a two-stage alcoholysis-polyesterification method. Drying performances of white gloss paints formulated from the desaturated oil alkyd gave improved results. The blending of the improved 50% JCL oil alkyd with acrylic further enhanced the drying properties to be comparable to the commercial standards. White gloss paint formulated from the improved JCL alkyd–acrylic blends (1:3) dried hard within 2 h at an outdoor temperature of 37 °C ± 2 [151].

8.4.6 Vegetable Oils for Waterborne Paints and Coatings

Vegetable oil derivatives found applications as environment-friendly solvent-borne or waterborne coating materials, leading to improved performances and reduction or elimination of VOC emissions, due to the absence of organic solvents. VO-based coatings are normally used when penetration is desired, i.e. for wood coating.

The formulation of solvent-borne paints is not a big issue, as VO is soluble in organic solvents. On the contrary, the development of waterborne coatings is a challenging task due to the hydrophobic nature of oil chains. Several solutions to this problem were proposed.

Water-soluble resins: are rare because most resins derived from oils are insoluble in water. The real emulsions are obtained by using either a surfactant or a resin that has a surfactant-like behaviour.

VO-modified latex: VO is incorporated in the latex resin. After application, the oil portion, that lowers the glass transition temperature of the resin, of the resin crosslinks to produce a hard, durable film [220]. Acrylated oils, soybean oil-urethane-acrylic hybrids [166], epoxidized VO [197] are used in the synthesis of the latex.

Water-reducible resins: similar to the dispersions in that they are particles; however, the particle size, of spherical geometry, is generally less than 8 μm in diameter. Adding water to the resin leads to a first viscosity drop-down, followed by a further rise to a very high level. After application, water evaporates and the solvent re-dissolves leading to particles coalescence into a coating film. Even if these paints started to be used only in the 1980s, due to the difficult formulation, they can be used for many maintenance applications [220].

Radiation cures (radical UV, ionic and e-beam systems): the radical cure system is nowadays the most popular owing in part to their cost and cure speed. As a matter of fact, radical cure processes require only seconds to reach full cure, versus 20 min for ionic cure systems. VO are only recently used for radiation cure application. In general, acrylates cure through UV radical polymerization, while methacrylates, styrene and vinyl ethers react with oxygen in the air to be as useful. Epoxidized soybean oil derivatives using 2-hydroxylethyl acrylate, in the presence of a strong acid catalyst was used for UV-cure [176]. Vegetable oil acrylates from castor oil were blended with a trifunctional thiol and crosslinked via UV irradiation [26]. Due to the high polarity of epoxidized oils, they can be used for cation cure systems, characterized by less shrinkage than radical ones, and deeper curing on pigmented systems and are much slower to cure.

Natural additives in coatings for VOC abatement: VO show high potential for reducing VOC because they are a liquid at room temperature and polymerize, or at least oligomerize, with air oxidation, becoming solid with time. They can be used as additives in paints to replace solvents that can lead to VOC formation, typically the coalescent aid, an ester or ether such as 2-butoxyethanol. Recently, simple monoesters of unsaturated oils were used as potential coalescent aids [140]. These materials showed high compatibility for most commercial latex resins, with a 1:1 replacement of volatile coalescent aid.

8.5 Epoxy Plasticisers

8.5.1 Plastics and Raw Materials

Replacement of petroleum-based polymeric materials using environment-friendly materials specially from natural resources is a compelling contemporary challenge attributable to the fluctuation oil prices. Triglycerides, the primary components of vegetable oils, are an abundant, renewable and widely investigated alternative feedstock for polymeric materials. Efforts are made on a global scale to develop innovative technologies to transform these natural resources into novel monomers and polymers. Some of these technologies have already generated competitive industrial products with properties comparable to conventional petrochemical polymers. Fillers and fibres have also been incorporated into these bio-based polymer matrices to improve the physical and thermal–mechanical properties of the resulting composite materials [241].

Bio-plastics can be defined as plastic resins made entirely or partially from renewable feedstocks. Bio-based thermoplastic resins can be generally classified into three categories according to the feedstock resources of the major polymer components. The first group consists of thermoplastic polymers which are totally derived from renewable feedstock, such as poly(lactic acid) and polyhydroxyalkanoates. The second group includes polymers that are partially derived from renewable feedstock, such as polyurethanes based on plant oil derived polyols (also called oleochemical polyols or natural oil polyols) and poly(trimethylene terephthalate) synthesized using the glycerol-derived 1,3-propanediol. and the third group comprises natural polymers which are used directly for plastics, such as starch and soy protein [38].

Soybean (Sb) has been recognized as a promising plant polymer for plastic uses as it was used back in the 1930s and 1940s when Henry Ford was an enthusiastic advocate for soybean uses in domestic and industrial products. Ford Motors demonstrated a car with 14 plastic panels made of resins consisting of soybean, wheat, flax and ramie in a chemical formula. However, this 'green car' made of soybean originated plastics' project idea was suspended during World War II, and therefore the plastic car experiment. A second unit was in production at the time the war broke out, but the project was abandoned. By the end of the war, the idea of a plastic car had fallen through the cracks due to energy being directed towards war recovery efforts. The petroleum-based plastics became more economic ('http://www.thehenryford.org/research/soybeancar.aspx,' n.d.).

Triglyceride oils are expected to play a key role during the twenty-first century as enabling to synthesize polymers from renewable sources. Polymers form triglyceride oils may be prepared through various strategies depending on which one could succeed the polymerization. The presence of oil/fatty acid chain in the polymer structure improves some physical properties of polymer in terms of flexibility, adhesion, resistances of water and chemicals. Concluding that triglyceride oils as essential raw materials to be used in various applications in the future [80].

Sustainable plastics of low cost and high performance were found to be essential for meeting the demands of the growing global population with respect to health, shelter, clothing, communication, mobility, food and energy. Bio-plastics production has been demonstrated to be technologically feasible and capable of substituting plastics from petrochemical feedstock. plant oil based biopolymers and precursors are currently commercially viable and there are companies that offer derived bio-based polymers (epoxy, alkydic and polyurethanes precursors being the most frequently found) for specific applications. Plant oil based polymers were presented as a partial replacement of synthetic polymers to introduce 'green' materials in the formulation, and little more than that, the current trend is to increase the percentage of bio-based materials maintaining good overall performance and/or developing tailored special properties [128, 129].

Plant oils contain reactive functional groups in their fatty acid chains, such as hydroxyls (CO and Lesquerella oil) or epoxies (vernonia oil). These reactive groups can be used directly for polymerization, but its application tends to be limited because of the relatively low functionalities. It requires some modification prior usage as valuable monomers for polymer synthesis, which can be achieved through chemical modification of the naturally occurring reactive sites found in triglycerides (e.g. ester groups and carbon–carbon double bonds), opening synthesis routes similar to those applied for petrochemical polymers. High-performance polymers are prepared by crosslinking of these bio-monomers and incorporation of rigid components in polymer chains [241].

CO and its chemical derivatives are used as raw materials for making different types of products in chemical industries including in plastic products. There are numerous reaction possibilities in CO due to the presence of three reactive points, viz. the hydroxyl group, the double bond and the glyceride ester linkage. CO fatty acid (COFA) with a pendant hydroxyl group on each ricinoleate chain has been considered as a source of many oleochemicals after appropriate modifications. Plastic products of CO are not only soft, flexible, odourless but also completely biodegradable, which is an added advantage [236]. Several families of CO-based polymers are synthesized and used to improve thermal stabilities, flame retardancy and mechanical properties. The CO or brominated CO used as a reinforcing agent is added to epoxy resins and good toughness and flame retardancy of cured epoxy resins are obtained. Synthesis and properties of CO-based polyurethane hybrid materials have been reported [157]. Cured epoxy resin modified by brominated CO possesses good flame retardation, toughness, moderate smoke suppression and some degree of resistance to heat [235].

Mechanically strong cellulose-graft-soybean oil copolymers, which combined two natural biomasses into one was reported by Wu et al. [234]. The soybean oil based monomers with secondary amide groups (SOM1) and tertiary amide groups (SOM2) were prepared and copolymerized. Cellulose-g-P(SOM1-co-SOM2) copolymers with 0.5 wt% cellulose were prepared via atom radical transfer polymerization. Cellulose-g-P(SOM1-co-SOM2) copolymers were successfully prepared with mechanical properties superior to the linear P(SOM1-co-SOM2) copolymers [234].

Bionanocomposite films prepared with melt compounding and film blowing were evaluated for packaging applications. The nanocomposite masterbatch with PLA, chitin nanocrystals (ChNCs) and glycerol triacetate plasticizer (GTA) was melt compounded and then diluted to 1 wt% ChNCs with PLA and polybutylene adipate-co-terephthalate (PBAT) prior to film blowing. The morphological, mechanical, optical, thermal and barrier properties of the blown nanocomposite films were comparable to those of the reference material without ChNCs. Tear strength, puncture strength improved Tg increased crystallinity enhanced by the addition of the ChNCs. ChNCs had lower fungal activity and lower electrostatic attraction between the film surfaces; leading to the easy opening of the plastic bags [83].

8.5.2 *Role of Plasticisers*

The International Union of Pure and Applied Chemistry (IUPAC) council, in 1951, defined with the term plasticiser 'a substance or material incorporated in a material, to increase its flexibility, workability, or distensibility' [230]. Therefore, the plasticizers mainly act by improving the flexibility and processability of plastic by reducing its glass transition temperature (T_g) [50, 127, 202, 232]. The presence of the plasticizer allows us to ease the processability of the material in different tools (such as upon moulding, extrusion, etc.), and to optimize the applied experimental conditions. Moreover, the plasticizers enhance some physical properties of the materials, like promoting the reduction of hardness and elastic modulus, and the increase in fracture resistance.

In general, most plasticizers are liquids characterized by a very high boiling temperature and an average molecular weight included in the range between 300 and 600, with linear or cyclic carbon chains [50, 59, 127,202, 232]. Hence, the key feature of a plasticizer molecule concerns the small molecular sizes, which allow it to enter and occupy the spaces between polymeric chains, reducing the secondary forces among them. As a consequence, a plasticizer modifies the three-dimensional molecular organization of the polymers, having an effect on the secondary forces between the chains and lowering the required energy to motion.

There are many classifications related to the plasticizers; they concern: (1) the distinction between internal or external, depending on the presence or not of chemicals bonds with polymer chains; (2) the definition of primary or secondary.

An internal plasticizer has normally a bulky structure and it provides to reduce the polymer glass transition temperature and make it more flexible at room temperature (lowering the elastic modulus) [232]. It is a part of the polymer molecules, copolymerized into the polymer structure, grafted with the original polymer, thereby making the chains of the polymer more arduous to fit and compact with each other closely [141]. An external plasticizer, on the other hand, is an additive, which is not part of the polymer structure. These low-volatile molecules are added to polymers to produce swelling without chemical reaction. Even if these plasticizers can be lost by migration, evaporation or extraction, they are most widespread.

Another important classification splits the plasticizers into primary and secondary [223, 224]. If a plasticizer was soluble in the polymer at very high polymer concentration, it is said 'primary'. Thus, primary plasticizers must meet the following requests: (a) be totally soluble in the polymer; (b) not exude from the material; (c) be used alone. The secondary plasticizer, on the other side, has lower gelation capacity and compatibility, and it is inclined to exude. These plasticizers are generally used in addition to primary plasticizers to reduce costs or enhance some properties.

Over the years, several theories have been developed and proposed to shed light on the plasticizing mechanism [194]. The main ones, as resumed in Fig. 8.10, correspond to (a) the lubricity theory; (b) the gel theory; (c) the free volume theory. According to the lubricity theory, the action of the plasticizer is regarded just like a lubricant: it diffuses into the polymer, inserting between the chains and reduces the intermolecular frictions. As a result, by flexing the plastic material the macromolecules can slip over each other. The plasticizers lubricate the movement of the molecules and reduce the rigid matrix. In this way, the plasticized polymers can be depicted as parallel alternating layers of polymer and plasticizer. The gel theory was developed straight after the first one, and it can be viewed like an extension of the lubricity one. It suggests that the plasticizer act to destroy and replace the intermolecular interactions, lowering the polymer gel structure and increasing the flexibility. According to this theory, the plastic is represented as a complex three-dimensional network, where the plasticizers are bonded to polymer chains by weak secondary forces. From the definitions of these two theories, similarities and differences between them are clear. The lubricity theory treats the rigidity of the material as internal frictions between polymers and the plasticizer lubricates the layers, while the gel theory ascribes it to the attachments between polymers, and the plasticizer acts to reduce them.

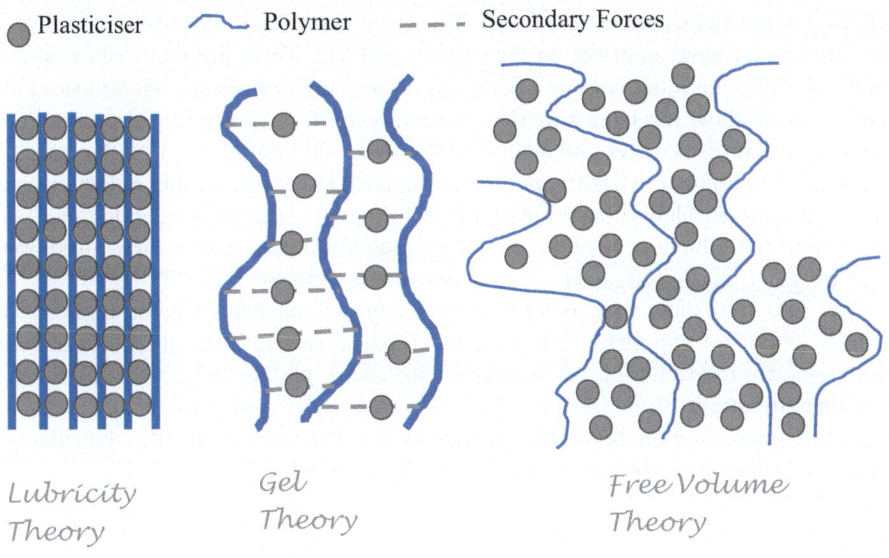

Fig. 8.10 Plasticization theories

The free volume theory, developed by Chandola and Marathe (37), is based on the definition of free volume. The latter is described as internal space available in a polymer for the movement of chains. The free volume corresponds to the difference between the observed free volume at absolute zero and the one at a selected temperature. The rigid plastics have a low free volume, while flexible resins have a higher one. Therefore, it is evident the role of plasticizers: they act to increase the free volume, and this boost is also kept by the material after the processing through cooling.

Nowadays about 1200 commercially manufactured plasticizers are available. Natural camphor and castor oil were the first plasticizers used in the 1800s. Since 1930, when the phthalates appeared as petroleum-based plasticizers, they dominated the market share for about 50 years. Since 1980, the toxicity of some phthalates began to be known and proven, and the interest towards natural and bio-based plasticizers has increased.

The plasticizers are extensively used for about 60 polymers and more than 30 commercial products. The principal plasticized polymers are based on polyvinylics (PVA, PVC, PVB), acrylics (PMMA) among petro-based polymers and polyesters (PLA and PBA), polysaccharides as biopolymers.

The compatibility and the solvation are essential for the choice of the appropriate plasticizer-polymer couple: the effective plasticization and other parameters strictly depend on it.

PVC poly(vinyl chloride) is one the largest commodity thermoplastics applied for packing materials, children toys and healthcare supplies [147]. Due to the presence of dipole on C–Cl bond, PVC is a hard and brittle plastic, and it needs the addition of plasticizers to be processed. Moreover, for soft and flexible applications, the plasticizers have an effect on the rigidity of PVC. There are many plasticizers used for PVC, the most common being based on phthalate esters. These esters of phthalic acid show the largest market share (about 80%) in the global production of plasticizers: di-isodecyl phthalate (DIDP), di-n-octyl phthalate (DNOP) and di-2-ethylhexyl phthalate (DHEP) are the widest used. However, for these plasticizers, there are some problems concerning their well-known toxicity, causing disorders in the human reproduction system and endogenous hormones, due to their migration from polymer matrix [65, 167].

Thus, six phthalates have been banned in Europe and America, for application in toys, childcare supplies, cosmetics and medical devices. For these uses, it is recommended to use bio-based alternatives instead of phthalates.

Other important plasticizers are based on carboxylic acid esters with linear or branched alcohols of medium-chain lengths (C6–C11), such as adipates, benzoates, citrates or alkane-dicarboxylic, glycols and phosphates [27].

8.5.3 Market of Plasticisers

The global market of plasticizers has been estimated at USD 12.23 billion in 2017, and it is predicted to reach USD 16.15 billion by 2022 [74, 75], with an increase of the global demand to about 9.75 million tonnes in 2024. The rising demand for plasticizers for the soft PVC is expected to drive the plasticizers market during the forecast period (a demand about 87% of the global market) [78]. The Asia Pacific area held over 59% of global demand in 2016, followed by Western Europe and North America.

As the phthalates, the most common PVC plasticizers, have been banned, because they are considered harmful for health and environment, it is expected an increase in demand for alternative phthalates. In fact, it has been recorded a loss of market share for DEHP in 2016 and a similar destine is forecast for other two important phthalates such as diisononyl phthalate (DINP) and di-isodecyl phthalate (DIDP).

In this scenario, the global bio-plasticizers market shows a noteworthy position with 1.243 billion USD in 2016 and an amount of 887 kilotons, and it is expected to reach USD 2.68 billion in 2025. Over the past, there has been an increase in demand of bio-plasticizers as an alternative to phthalates. Epoxidized soybean oil (ESBO) has dominated the relative market, with a demand of 328.8 kilotons in 2016 and a worth of USD 395.7 million, and it is projected to grow at a rate of 13.5% during the forecast period of 2016–2024 (Report review 2017).

8.5.4 Why Bio-plasticizers

As argued before, the application of phthalates is nowadays an object of debate by the international scientific community due to their toxicity issues arising from their aptitude to migration. DEHP is listed among potential carcinogenic substances, thus it has been banned in specific uses, like toys and healthcare supplies. For others, there are severe restrictions in some Countries concerning their concentration in childcare material. Therefore, it results imperative to find a valid alternative, characterized by low toxicity and limited migration.

The ideal bio-based plasticizer must lower the T_g value of polymers, be biodegradable, no toxic, no volatile and must have the minimal aptitude to migration.

Bio-plasticizers, owing to their biological nature as they are obtained from agricultural resources, seem to rise to the challenge. Many plasticizers come from different renewable and biological sources, such as cereals or oleaginous plants, or a waste of agricultural industries (Reports 2017).

Starches and cellulose, extracted from rice, potatoes and straw fibres, are natural sources for bio-based plasticizers. Sugar derivatives, such as sugar alcohols and isosorbide esters, obtained by chemical modification of glucose or mannose, are suitable renewable plasticizers. Sugar alcohols are polyols used for the plasticization of starch, in mixture with glycerol. Isosorbide esters, obtained by double esterification of isosorbide and long-chain acids, are effective plasticizers of PVC due to their structure, similar to phthalates.

From citrus fruits and sugarcane, citrate-based plasticizers are obtained. These esters of citric acid are used as plasticizers of PVC, cellulose and PLA.

Glycerol, the main by-product of biodiesel synthesis, is well known to be a suitable plasticizer for starch thanks to the presence of three hydroxyl groups.

Recently, cardanol and its derivatives have been reported as precursors of bio-plasticizers. The chemical structure of cardanol is quite similar to one of the most famous phthalates like DEHP. For this reason, it shows good miscibility with PVC and it can be used as a secondary plasticizer.

Bio-plasticizers can be also found out by natural wastes, such as tannins or derivatives of tannic acid for cement and sugar bagasse for cellulose acetate. Also, liquefied wood esters have shown good plasticizing properties.

Another important aspect that spurs the bio-plasticizers market concerns recent advances on bio-based plastics due to petroleum-raw resource depletion. As the demand for bio-plastics continuously grows, also the demand for new bio-plasticizers, with new properties and performances, and compatible additives grow in parallel. The principal biodegradable polymers are synthesized from biomass like agro-resources or obtained by microbial fermentation, such as PLA, PHA, PBSA, PHBv.

8.5.5 Vegetable Oils Based Plasticisers

Vegetable oils represent a valid resource for bio-plasticizers, because they are biodegradable, renewable and no toxic. Due to their cheap and sustainable feature, they are taking on a remarkable role in the chemical industry as raw material. They are obtained from plants and trees such as sunflowers, rapeseed, soybean. The principal components of the vegetable oils are triglycerides, made up of three fatty acids joined at a glycerol structure. The composition in terms of fatty acids, according to the number of carbons atoms in the length and of unsaturations for chain, can be variable depending on the crops, the growth conditions and the method of purification.

The reasons why oils have the potential to be applied as plasticizers have to be found in the length and complexity of the fatty acid chains: these can occupy the intermolecular space between the polymers macromolecules and in this way, they can enhance the mobility, making softer the plastic. Moreover, the presence of the esters groups, which can interact with polymer chains through secondary forces, van der Waals, for example, can aid to solve the compatibility problems, preventing the migration of plasticizer. Modification reactions of triglyceride molecules are able to further influence the compatibility of vegetable oils with the polymers.

The most important modifications of vegetable oils generally involve two main reactive sites of fatty acids: carboxyl group and double bonds. Therefore, esterification, transesterification, epoxidation and acetylation are the primary reactions suitable to the scope.

Castor and soybean oil esters are recognized as PVC plasticizers due to excellent compatibility. The HallGreen plasticizer, sold by Hallstar Company, is one of the commercially manufactured plasticizers completely based on fatty acids esters and used in PLA resins. Danisco patented a bio-plasticizer, with a bio-component of 100%, obtained by acetylation reaction of castor oil [148]. Diels–Alder adducts, derivatives by tung oil and alcohols, are reported as excellent PVC plasticizers.

8.5.6 Epoxidized Soybean Oil

Epoxidized soybean oil (ESBO) is one of the oldest bio-based plasticizer of PVC. It acts as a secondary plasticizer; notwithstanding it was initially tested as the primary one during the 1950s [132]. At that time, it was reported to have efficient plasticizing action for PVC, however, it suffered from exudation after a long time of UV-exposure, pointed out by too sticky surface of the plastics. This drawback was ascribed to the presence of remaining unsaturations in the oil, hailed from the incomplete epoxidation reaction. It was demonstrated a correlation between the amount of remaining double bonds and the ESBO permanence in PVC (lifetime) [33, 108].

Other vegetable oils have been applied to prepare epoxidized compounds. The most important, in addition to soybean, are linseed and rapeseed oil. It is evident that the double bonds content, expressed as Iodine Number (I.N., namely grams of iodine consumed by 100 grams of substance) represents a key feature for the effective plasticizing behaviour in PVC or other plastics.

On the other hand, palm oil is not advisable as plasticizers feedstock, as it has a very low double bonds content, while linseed oil, characterized by the highest I.N., is largely used for this scope.

ESBO, with its very low I.N. (<1 $gI_2/100$ g, i.e. 0.04 mol double bonds in each molecule), has been claimed as a primary plasticizer in PVC.

Beyond its plasticizer function, ESBO has revealed to be a thermal stabilizer for PVC: this dual function in PVC resin was the key to success for its wide application. PVC is sensitive to thermal degradation during processing at high temperatures. Therefore, thermal ageing, to eliminate HCl by radical free reaction, is required. ESBO has the ability to substitute labile chlorines [105], by the reaction between the epoxide ring and hydrogen chloride generated by PVC degradation. This reaction provides to make more stable PVC towards further dehydrochlorination, preserving its colour and properties.

Plasticizers are also prepared by ring-opening reaction of epoxidized vegetable oils leading to acetylated vegetable oils. Satisfactory plasticizing properties are described for these compounds.

Many companies manufactured and sold epoxy-based plasticizers in order to compete with petro-based ones: the main are Varteco, Nexdeum, Vanderputte Oleochemicals.

With commercial name Vikoflex®, PVC and PLA plasticizers are sold by Arkema, based on ESBO and epoxidized esters [60]. Novance developed Radia® plasticizers based on soybean and rapeseed oils [191]. Kalfex®, Nexo®, Resiflex® and Lankroflex® are famous examples of epoxy plasticizers, which lead the market in Southern America [27].

8.5.7 Other Epoxidized Oil

Epoxidized palm oil (EPO) was used as plasticizers in toughening and aiding the melt-processing of poly (lactic acid) (PLA) [212]. At various loadings, they were melt compounded with PLA and hot pressed into 3-mm sheets for impact strength testing. From the torque and stock temperature curves obtained during compounding, both plasticizers rapidly reduced the torque and stock temperature. While EPO readily supports the processing and toughens PLA, higher content of ESO is needed to perform correspondingly. These improvements suggest the potential of using EPO and ESO as plasticizer substitutes which shall contribute to the PLA processing industry [209].

An oligomer of a diepoxy (diglycidyl ether of bisphenol-A, DGEBA) and an aromatic diamine (MCDEA) have been used as reactive plasticizers for polycarbonate (PC). A small amount of PC chain scission occurred during this blending process, probably due to the transesterification of the PC carbonate group by the hydroxyl group of the DGEBA oligomer. The addition of DGEBA to PC was found to greatly reduce the Tg and processing temperature. Dynamic rheology measurements showed that the added epoxy can very effectively reduce the viscosity, but that the addition of epoxy also accelerated the crystallization rate of the PC [125].

A series of bio-based epoxidized plasticizers (CExEp) for soft PVC was synthesized from cardanol and various fatty acids by esterification and epoxidation of the fatty and cardanol unsaturations. The plasticizing properties of these additives for PVC films were proved considering thermal and mechanical data. They seem to be a potential bio-based alternative plasticizer to diisononyl phthalate (DINP), one of the most widespread phthalate plasticizers, with less rigid and stable films. The synergy between epoxy groups, cardanol ring and fatty chains to elaborate efficient primary plasticizers was demonstrated. Very low eco-toxicity (daphnia and algae) and reprotoxicity without significant effect on agonistic or antagonistic properties for both, female and male hormones, CEC18:1Ep plasticizer is a good fully bio-based candidate as phthalate substituent plasticizer for PVC films; revealing that using fatty cardanol plasticizer instead of DINP reduces the global toxicity and improve the environment and human health which let science consider these cardanol plasticizers in industrial applications as an alternative to DINP. Concluding the safety of applications of cardanol plasticizers in children's toys, automobile interiors, food packaging [29].

Epoxidized fatty acid methyl esters (EME) have been developed as plasticizers because they are more soluble and confers better flexibility to the plastic even at low temperatures. The main problem associated with these products lies in their content of saturated methyl esters. This fraction cannot be epoxidized and has got a very low affinity with the polymeric matrix, so that tends to migrate at the plastic surface giving not desired properties to the final products. Acetic acid and hydrogen peroxide were applied in order to obtain EME and then distillate it to remove the unreacted esters. A better bio-plasticizer was obtained by a distillation of starting biodiesel, performed at 0.5 torrs using a two equilibrium stage batch distillation apparatus, followed by the epoxidation reaction [70].

A simple chemoenzymatic method was developed for the preparation of epoxy lecithin that contains epoxy oils and phospholipids. The product was evaluated as a plasticizer in polyester laminates and compared with virgin polyester laminates. The laminates were prepared using various amounts of epoxy lecithin and evaluated for different properties. The epoxy lecithin modified laminates showed good impact strength, tensile, and chemical resistance properties. These laminates were also evaluated for Vicat softening point and water absorption. The epoxy lecithin can be used as a plasticizer in polyester laminates [173].

Castor oil (CO) based polyol esters (COPE-1, ECOPE-1, COPE-2 and ECOPE-2) were synthesized and characterized as a plant oil based plasticizer for poly(vinyl chloride) (PVC) materials. PVC materials were prepared via blending the synthesized CO-based polyol esters as the main plasticizer. Properties of the bio-plasticized PVC materials were better compared to those of commercial plasticizers (DOP and ESO) including when tested using the tensile test. Thermal stability was improved and the amount of char residual of PVC blends was reduced. Increasing the molecular weight of CO-based polyol esters enhanced the migration stability and volatility stability.

CO-based polyol esters plasticizers could be applied in food packing, children toys and medical devices as the main plasticizer [98].

CO was used as a plasticizer to improve the properties of PLA and starch which were melt compounded in a lab-scale co extruder. CO layer on starch has a positive effect on the crystallization of PLA in the ternary blend comparing to the use of CO with PLA alone. The accumulation of CO on starch greatly improves the toughness and impact strength of PLA/starch blends. The grafting content of HDI on the starch granules primarily determined the compatibility and properties of the resulted blends [236].

CO fatty acid (COFA) and furfuryl alcohol (FA) esterification by immobilized Candida antarctica Lipase B in a solvent-free system gave a plasticizer with better mechanical properties and thermal stability of a FA-COFA ester-containing ethyl cellulose (EC) film. Prepared bio-plasticizer was than the traditional plasticizer such as dibutyl phthalate (DBP) in producing good quality films [145].

Natural plasticizer was synthesized in the laboratory by esterification of rice fatty acids and polyols. The resulting product was added to PVC and natural rubber latex films (5% w/w). Mechanical tests indicated that the addition of the natural plasticizer resulted in an increase in the elongation at break and a consequent decrease in the tensile strength of the films. The plasticizing effect induced by the natural plasticizer was also confirmed by the Tg shift towards lower temperatures in the plasticized films [46].

A tung oil derived epoxidized dicarboxylic acid dimethyl ester (epoxidized-C21-DAE), was synthesized through transesterification, a Diels–Alder reaction, and epoxidation. The application of epoxidized C21-DAE as a bio-based, primary plasticizer for PVC significantly improved the thermal stability over that of DOTP and EFAME. The mechanical properties of this type of PVC were superior to those of DOTP. Migration and volatility stabilities of epoxidized-C21-DAE were much better than EFAME. Epoxidized-C21-DAE could, therefore, be fully substituted for commercial DOTP or EFAME [124].

Waste cooking oil methyl esters were structurally modified, i.e. transesterification and epoxidation reaction, to prepare epoxide waste cooking oil 2-ethylhexyl esters. Prepared epoxide via structural modification proved to be an effective alternative to dioctyl phthalate, replacing about 40% of the total plasticizer. This may lead to the eco-development of waste cooking oil materials, and makes the plasticizing process more environmentally friendly [238, 242]. Waste cooking oil and citric acid were used to synthesize acetylated-fatty acid methyl ester-citric acid ester to be used as a plasticizer for PVC. Glass transition temperature (Tg) decreased while mechanical and migration stabilities of the PVC films were obviously better than those of PVC films plasticized by ESO (epoxidized soybean oil) and as good as those of PVC films plasticized by DOP (dioctyl phthalate, 99%) [64].

Norbornyl epoxidized linseed oil was synthesized via Diels–Alder reaction of cyclopentadiene with linseed oil at high pressure and high temperature, followed by an epoxidation using hydrogen peroxide with a quaternary ammonium tetrakis(diperoxotungsto) phosphate(3−) epoxidation catalyst. Incorporation of divinyl ethers increased the curing rate and overall conversion of the epoxide groups [39].

8.5.8 Epoxidation Methods

The term 'epoxides' is usually referred to as oxiranes, organic compounds constituted by cyclic ethers with a three-membered ring, which are very reactive. For this reason, they are susceptible to ring-opening reaction, through different reactions like ketone rearrangements.

The epoxidized oils, generally, contain epoxides groups along the fatty acids chains of triglyceride, replacing the initial double bonds lost during the epoxidation reaction (Fig. 8.11).

Referring to only unsaturated vegetable oils, three epoxidation methods, to produce epoxidized vegetable oils (EVO), are currently most studied and known:

1. Epoxidation with percaboxylic acids [171, 190] (Prileschajew method), usually named as 'classical or conventional'.
2. Epoxidation with organic and inorganic hydroperoxides, which includes hydrogen peroxide epoxidation with heterogeneous catalysts [22].
3. Chemoenzymatic epoxidation [179, 206].

Nowadays, EVOs are commercially produced by the classical method via Prileschajew reaction, by using peracids, according to the reaction shown in Fig. 8.12. In this way, the unsaturated oil reacts with percaboxylic acids (or peracids) generated in situ, for safety reasons, through the acid-catalyzed peroxidation of the respective organic acids with hydrogen peroxide. The peracid formation occurs in the aqueous phase, in the presence of soluble mineral acids as catalysts (such as H_2SO_4, H_3PO_4). Subsequently, the obtained performic acid migrates into the oil-immiscible phase giving the reaction in Fig. 8.12.

Performic acid spontaneously reacts with the oil double bonds, forming epoxide groups; then it eventually comes back to the aqueous phase, and the reaction cycle restarts [190].

However, this method suffers from several disadvantages, beyond the high exothermicity: (a) low selectivity caused by side reactions; (b) very corrosive

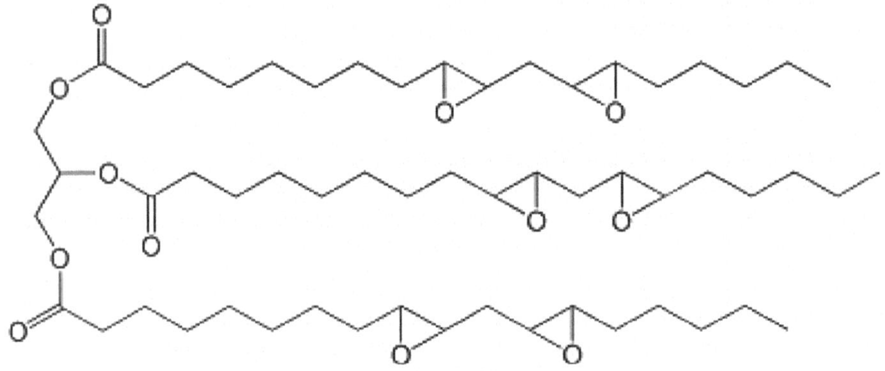

Fig. 8.11 Molecule of epoxidized triglyceride

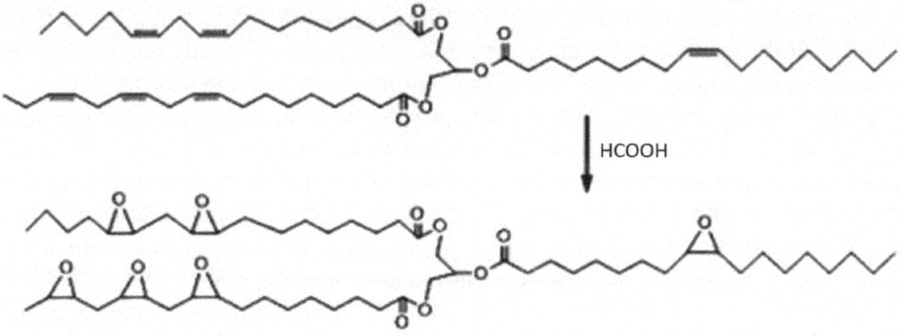

Fig. 8.12 Conventional epoxidation reaction

reaction-environment due to the presence of strong acids; (c) high instability of the peracids, which are subjected to decomposition reaction.

The soluble mineral acids promote side reactions, such as oxirane ring opening, resulting in a dramatic decrease in selectivity. Ring opening takes place through the cleavage of one of the carbon–oxygen bonds, and it is able to decrease the selectivity to oxirane. The presence of by-products, derived by side reactions, in the commercial epoxidized oils, diminishes their attractiveness as starting material for further elaboration, as the producers must perform expensive purification. The presence of acid and corrosive aqueous solutions brings problems: environmental, due to disposal of salts formed during the final neutralization of the mineral acid; technical, connected to the corrosion and difficult separation operations.

The replacement of soluble acids with acid heterogeneous catalysts has been reported as an effective answer to these issues. The use of acidic ionic exchange resins (AIER) has been described as beneficial to minimize the side reactions and improve the selectivity in the epoxidation reaction [214]. The presence of strong isolated acid sites for AIERs, located inside the pore structure, is only accessible to small molecules like hydrogen peroxide or carboxylic or percaboxylic acids, and protects the oxirane ring from reaction with proton and, therefore, from side reactions. However, the application of these catalysts on large scale is limited by their low mechanical resistance and low thermal stability. Recently, the zeolites have been reported as active and selective catalysts for the epoxidation of soybean oil with performic acid, owning the advantageous features of AIERs [212, 215].

The use of hydrogen peroxide as oxidant, to epoxidize the unsaturated oil, results in convenient and 'green', as no corrosive acids are required, and water is the only one by-product. Hydrogen peroxide is a very efficient oxidant with a very high content of active oxygen. For this reason, commercial hydrogen peroxide has been more and more frequently applied in many chemical syntheses, e.g. oxidation of propylene to propylene oxide, oxidation of amines to oximes and hydrocarbons to fatty alcohols. In homogeneous phase, the more active catalysts in epoxidation with hydrogen peroxide are salts and compounds of transition metals, such as Ti, W, Rh, Ru, Nb, Mo. However, only a few of these metals were successfully heterogenized.

The use of a heterogeneous catalyst is more desirable: apart from the beneficial recovery of the catalyst at the end of process, a suppression of side reactions could be feasible, accounting for the lack of free acidity in the reaction environment.

Many different catalytic systems for epoxidation with hydrogen peroxide have been proposed in the literature [154]. Among them, titanium silicalite (TS-1) has been praised as a milestone for the oxidation of olefins with hydrogen peroxide [208]. However, this catalyst is barely active with larger unsaturated molecules like oils and methyl esters, due to the diameter of the small pores (5.6 × 4.7 Å) of TS-1. Broad work was done to incorporate Ti(IV) in large molecular sieves pores, leading to materials such as Ti-MCM-41 and Ti-MCM-48 [180]. Recently, some niobium–silica and niobium–alumina-based solids have been described in the literature as active catalysts for the epoxidation reaction [212, 215]. A strict correlation between the surface distribution of active sites and synthesis methods has been reported. Among free-transition metals catalyst, γ-alumina-based materials are of interest. It has been shown that this oxide has been able to activate hydrogen peroxide for the oxidation, through the formation of a peroxide site [213].

Organic hydroperoxides, such as *tert*-butyl hydroperoxides (TBHP) or cumyl hydroperoxides (CHP), have been reported as alternative oxidizing agents. However, they are successfully applied in few oxidations of oils, and always in the presence of homogenous catalysts based on vanadium, titanium or molybdenum. This compromises the application on a larger scale.

The chemoenzymatic approach gained increasing interest in an alternative to the chemical one, because the use of lipases allows obtaining a very high stereoselectivity, with suppression of side reactions (Rios et al. 2011). The reaction mixture, in the chemoenzymatic epoxidation, usually consists in three phases system which involves an aqueous phase containing hydrogen peroxide, an organic phase containing solvent and oil, and a solid phase corresponding to lipase (Candida antartica Lipase B, CALB) immobilized on resin or silica (Novozym 435® or CALB-silica). The epoxidation proceeds in two steps: the first consists of the hydrolysis by the reaction of triglyceride with H_2O_2 in presence of lipase; the second is the self-epoxidation with the EVOs formation.

Although high selectivity, this route is not feasible on the industrial scale, due to various drawbacks such as the large use of solvent, high cost of lipase and rapid deactivation of the latter [48].

References

1. Adegoke GO, Iwahashi H, Komatsu Y, Obuchi K, Iwahashi Y (2000) Inhibition of food spoilage yeasts and aflatoxigenic moulds by monoterpenes of the spice Aframomum danielli. Flavor Fragr J 15:147–150. https://doi.org/10.1002/1099-1026(200005/06)15:3147::AID-FFJ8833.0.CO;2-0
2. Adewuyi A (2014) Synthesis and surface-active property of diethanolamide and epoxidised diethanolamide surfactant from the seed oil of Baphia nitida. Arab J Chem 12: 1545–1551. https://doi.org/10.1016/j.arabjc.2014.10.021

3. Agach M, Delbaere S, Marinkovic S, Estrine B, Nardello-Rataj V (2013) Synthesis, characterization, biodegradability and surfactant properties of bio-sourced lauroyl poly(glycerol-succinate) oligoesters. Colloids Surf A 419:263–273. https://doi.org/10.1016/j.colsurfa.2012.12.006
4. Ahmad S, Ashraf SM, Hasnat A, Noor A (2001) Studies on epoxidized oil and its blend with poly- styrene and polymethylmethacrylate. Ind J Chem Technol 8:176–180
5. Akintayo CO, Adebowale KO (2004) Synthesis and characterization of acrylated Albizia benth medium oil alkyds. Prog Org Coat 50:207–212. https://doi.org/10.1016/j.porgcoat.2003.09.017
6. Akram D, Sharmin E, Ahmad S (2010) Development and characterization of boron incorporated linseed oil polyurethanes. J Appl Polym Sci 116:499–508. https://doi.org/10.1002/app.31579
7. Akram D, Hakami O, Sharmin E, Ahmad S (2017) Castor and linseed oil polyurethane/TEOS hybrids as protective coatings: a synergistic approach utilising plant oil polyols, a sustainable resource. Prog Org Coat 108:1–14. https://doi.org/10.1016/j.porgcoat.2017.03.012
8. Alam M, Akram D, Sharmin E, Zafar F, Ahmad S (2014) Vegetable oil based eco-friendly coating materials: a review article. Arab J Chem 7:469–479. https://doi.org/10.1016/j.arabjc.2013.12.023
9. Alam M, Alandis NM (2011) Development of ambient cured polyesteramide coatings from linseed oil: a sustainable resource. J Polym Environ 19:391–397. https://doi.org/10.1007/s10924-011-0281-9
10. Alvarez MF (2000) Review: active food packaging. Food Sci Tech Int 6(2):97–108. https://doi.org/10.1177/108201320000600203
11. Amakura Y, Umino Y, Tsuji S, Ito H, Hatano T, Yoshida T, Tonogaia Y (2002) Constituents and their antioxidative effects in eucalyptus leaf extract used as a natural food additive. Food Chem 77:47–56. https://doi.org/10.1016/S0308-8146(01)00321-1
12. Ang DTC, Gan SN (2012) Novel approach to convert non-self drying palm stearin alkyds into environmental friendly UV curable resins. Prog Org Coat 73(4):409–414. https://doi.org/10.1016/j.porgcoat.2011.11.013
13. Appendini P, Hotchkiss JH (2002) Review of antimicrobial food packaging. Innovative Food Sci Emerg Technol 3:113–126. https://doi.org/10.1016/S1466-8564(02)00012-7
14. Araujo WS, Margarit ICP, Mattos OR, Fragata FL, de Lima-Neto P (2010) Corrosion aspects of alkyd paints modified with linseed and soy oils. Electrochim Acta 55:6204–6211. https://doi.org/10.1016/j.electacta.2010.03.088
15. Araújo K, Mahajan D, Kerr R, da Silva M (2017) Global biofuels at the crossroads: an overview of technical, policy, and investment complexities in the sustainability of biofuel development. Agriculture 7(4):32. https://doi.org/10.3390/agriculture7040032
16. Arras G, Grella GE (1992) Wild thyme, Thymus capitatus, essential oil seasonal changes and antimycotic activity. J Hortic Sci 67(2):197–202. https://doi.org/10.1080/00221589.1992.11516237
17. Atabani AE, Silitonga AS, Ong HC, Mahlia TMI, Masjuki, HH, Badruddin IA, Fayaz H (2013) Non-edible vegetable oils: a critical evaluation of oil extraction, fatty acid compositions, biodiesel production, characteristics, engine performance and emissions production. Renew Sustain Energy Rev 18:211–245. https://doi.org/10.1016/j.rser.2012.10.013
18. Ataei S, Yahya R, Gan SN (2011) Fast physical drying, high water and salt resistant coatings from non-drying vegetable oil. Prog Org Coat 72(4):703–708. https://doi.org/10.1016/j.porgcoat.2011.07.013
19. Azimi ARN, Yahya R, Gan S-N (2013) Improving coating characteristics of palm stearin alkyd by modification with ketone resin. Prog Org Coat 76(4):712–719. https://doi.org/10.1016/j.porgcoat.2013.01.001
20. Bakkali F, Averbeck S, Averbeck D Idaomar M (2008) Biological effects of essential oils—a review. Food Chem Toxicol 46:446–475. https://doi.org/10.1016/j.fct.2007.09.106
21. Bart JCJ, Palmeri N, Cavallaro S (2010) Oleochemical sources: basic science, processing and applications of oils. Biodiesel Sci Technol. https://doi.org/10.1533/9781845697761.62

22. Bartok M, Lang KL (1980) Oxirane in: the chemistry of ethers, crown ethers, hydroxyl groups and their sulfur analogues. Wiley, New York
23. Bauer K, Garbe D, Surburg H (2001) Common fragrance and flavor materials: preparation, properties and uses. Wiley-VCH, Weinheim
24. Ben Taarit M, Msaada K, Hosni K, Hammami M, Kchouk ME, Marzouk B (2009) Plant growth, essential oil yield and composition of sage (Salvia officinalis L.) fruits cultivated under salt stress conditions. Ind Crop Prod 30:333–337. https://doi.org/10.1016/j.indcrop.2009.06.001
25. Biermann U, Bornscheuer U, Meier MAR, Metzger JO, Schäfer HJ (2011) Oils and fats as renewable raw materials in chemistry. Angew Chem Int Ed 50(17):3854–3871. https://doi.org/10.1002/anie.201002767
26. Black M, Rawlins JW (2009) Thiol–ene UV-curable coatings using vegetable oil macromonomers. Eur Polym J 45:1433–1441. https://doi.org/10.1016/j.eurpolymj.2009.02.007
27. Bocqué M, Voirin C, Lapinte V, Caillol S, Robin J (2016) Petro-based and bio-based plasticizers: Chemical structures to plasticizing properties. J Polym Sci 54:11–33. https://doi.org/10.1002/pola.27917
28. Bohn D (2018) Coatings world. Global coatings market overview. https://www.coatingsworld.com/issues/2018-01-01/view_features/global-coatings-market-overview. Accessed 19 Apr 2018
29. Briou B, Caillol S, Robin J-J, Lapinte V (2019) Non-endocrine disruptor effect for cardanol based plasticizer. Ind Crops Prod 130:1–8. https://doi.org/10.1016/j.indcrop.2018.12.060
30. Britton LN (1998) Surfactants and the environment. J Surfactants Deterg 1:109–117. https://doi.org/10.1007/s11743-998-0014-6
31. Brody Z, Strupinsky A, Kline EL (2001) Active packaging for food applications. Technomic Publishing Co, Lancaster, PA
32. Brophy JJ, Davies NW, Southwell IA et al (1998) Gas chromatographic quality control for oil of Melaleuca Terpinen-4-ol type (Australian Tea Tree). J Agric Food Chem 37:1330–1335. https://doi.org/10.1021/jf00089a0271
33. Bueno-Ferrer C, Garrigos MC, Jimenez A (2010) Characterization and thermal stability of poly (vinyl chloride) plasticized with epoxidized soybean oil for food packaging. Polym Degrad Stab 95:2207–2212. https://doi.org/10.1016/j.polymdegradstab.2010.01.027
34. Burt S (2004) Essential oils: their antibacterial properties and potential applications in foods: a review. Int J Food Microbiol 94:223–253. https://doi.org/10.1016/j.ijfoodmicro.2004.03.022
35. Carson CF, Mee BJ, Riley TV (2002) Mechanism of action of Melaleuca alternifolia (tea tree) oil on Staphylococcus aureus determined by time-kill, lysis, leakage and salt tolerance assay sand electron microscopy. Antimicrob Agents Chemother 46(6):1914–1920. https://doi.org/10.1128/AAC.46.6.1914-1920.2002
36. Carson CF, Riley TV (1993) A review–antimicrobial activity of the essential oil of Melaleuca alternifolia. Lett Appl Microbiol 16:49–55. https://doi.org/10.1128/CMR.19.1.50-62.2006
37. Chandola M, Marathe S (2008) J Mol Graphics Model 26:824–826
38. Chen F, Zhang J (2012) 10.14—Processing soy protein concentrate as plastic in polymer blends. In: Matyjaszewski K, Möller M (eds) Polymer science: a comprehensive reference. Elsevier, Amsterdam, pp 249–254. https://doi.org/10.1016/B978-0-444-53349-4.00264-8
39. Chen J, Soucek MD, Simonsick WJ, Celikay RW (2002) Synthesis and photopolymerization of norbornyl epoxidized linseed oil. Polymer 43(20):5379–5389. https://doi.org/10.1016/S0032-3861(02)00404-4
40. Chu Z, Feng Y (2013) Vegetable-derived long-chain surfactants synthesized via a "green" route. Sustainable Chem Eng 1:75–79. https://doi.org/10.1021/sc300037e
41. Condon BD, Matheson KL (1994) A comparison of surfactants derived from alcohols based on petrochemical and oleochemical sources. J Am Oil Chem Soc 71:53–59. https://doi.org/10.1007/BF02541472
42. Cosentino S, Tuberoso CIG, Pisano B Satta M, Mascia V, Arzedi E, Palmas F (1999) In vitro antimicrobial activity and chemical composition of Sardinian Thymus essential oils. Lett Appl Microbiol 29:130–135. https://doi.org/10.1046/j.1472-765x.1999.00605.x

43. Cowan-Ellsberry C, Belanger S, Dorn P, Dyer S, McAvoy D, Sanderson H Versteeg D, Ferrer D, Stanton K (2014) Environmental safety of the use of major surfactant classes in North America. Environ Sci Technol 44:1893–1993. https://doi.org/10.1080/10739149.2013.803777
44. Cox SD, Gustafson JE, Mann CM, Markham JL, Liew YC, Hartland RP, Bell HC, Warmington JR, Wyllie SG (1998) Tea tree oil causes K + leakage and inhibits respiration in Escherichia coli. Lett Appl Microbiol 26:355–358. https://doi.org/10.1046/j.1472-765X.1998.00348.x
45. Cuq B, Gontard N, Guilbert S (1995) Edible films and coatings as active layers. In: Rooney ML (ed) Active food packaging. Blackie Academic and Professional, Glasgow, UK, p 111
46. da Silva MA, Vieira MGA, Maçumoto ACG, Beppu MM (2011) Polyvinylchloride (PVC) and natural rubber films plasticized with a natural polymeric plasticizer obtained through polyesterification of rice fatty acid. Polymer Test 30(5):478–484. https://doi.org/10.1016/j.polymertesting.2011.03.008
47. Daferera DJ, Ziogas BN, Polissiou MG (2000) GC-MS analysis of essential oils from some Greek aromatic plants and their fungitoxicity on Penicillium digitatum. J Agric Food Chem 48:2576–2581. https://doi.org/10.1021/jf990835x
48. Danov SM, Kazantsev OA, Esipovich AL, Belousov AS, Rogozhin AE, Kanakov EA (2017) Recent advances in the field of selective epoxidation of vegetable oils and their derivatives: a review and perspective. Catalyst Sci Tecn 7:3659–3675. https://doi.org/10.1039/C7CY00988G
49. Dar MA (2011) A review: plant extracts and oils as corrosion inhibitors in aggressive media. Ind Lubr Tribol 63:227–233. https://doi.org/10.1108/00368791111140431
50. De Groote, Devaux J, Godard P (2002) Effect of benzenesulfonamide plasticizers on the glass-transition temperature of semicrystalline polydodecamide. J Polym Sci Part B Polym Phys 40:2208–2218. https://doi.org/10.1002/polb.10281
51. Debeaufort F, Quezada-Gallo JA, Voilley A (1998) Edible films and coating: tomorrow packagings: a review. Crit Rev Food Sci 38(4):299–313. https://doi.org/10.1080/10408699891274219
52. Debeaufort F, Quezada-Gallo J, Voilley A (2010) Edible films and coatings: tomorrow's packagings: a review. Crit Rev Food Sci 38(4):299–313. https://doi.org/10.1080/10408699891274219
53. Delaquis PJ, Stanich K, Girard B, Mazza G (2002) Antimicrobial activity of individual and mixed fractions of dill, cilantro, coriander and eucalyptus essential oils. Int J Food Microbiol 74:101–109. https://doi.org/10.1016/S0168-1605(01)00734-6
54. Deleu M, Paquot M (2004) From renewable vegetables resources to microorganisms: new trends in surfactants. C R Chimie 7:641–646. https://doi.org/10.1016/j.crci.2004.04.002
55. Di Serio M, Tesser R, Russo V, Turco R, Vitiello R, Sun Y, Hreczuch W (2015) Catalysts for the ethoxylation of esters. J Surfact Deterg 18:913–918. https://doi.org/10.1007/s11743-015-1719-1
56. Di Serio M, Tesser R, Pengmei L, Santacesaria E (2008) Heterogeneous catalysts for biodiesel production. Energy Fuels 22:207–217. https://doi.org/10.1021/ef700250g
57. Di Serio M, Vairo G, Iengo P, Felippone F, Santacesaria E (1996) Kinetics of ethoxylation and propoxylation of 1- and 2-octanol catalyzed by KOH. Ind Eng Chem Res 35(11):3848–3853. https://doi.org/10.1021/ie960200c
58. Didry N, Dubreuil L, Pinkas M (1993) Antibacterial activity of thymol, carvacrol and cinnamaldehyde alone or in combination. Pharmazie 48:301–304
59. Donhowe IG, Fennema ON (1993) The effects of plasticizers on crystallinity, permeability, and mechanical properties of methylcellulose films. J Food Process Pres 17:247–257. https://doi.org/10.1111/j.1745-4549.1993.tb00729.x
60. Donnelly Z (2012) WO2012102952A1
61. Elraies KA, Tan IM, Saaid I (2009) Synthesis and performance of a new surfactant for enhanced oil recovery. Int J Petro Sci Technol 3:1–9. https://doi.org/10.1007/s11814-015-0186-8

62. Faleiro ML, Miguel MG, Ladeiro F, Venâncio F, Tavares R, Brito JC, Figueiredo AC, Barroso JG, Pedro LG (2002) Antimicrobial activity of essential oils isolated from Portuguese endemic species of Thymus. Lett Appl Microbiol 36:35–40. https://doi.org/10.1046/j.1472-765X.2003.01259.x
63. Fargione J, Hill J, Tilman D, Polasky S, Hawthorne P (2008) Land clearing and the biofuel carbon debt. Science 319(5867):1235 LP-1238. https://doi.org/10.1126/science.1152747
64. Feng G, Hu L, Ma Y, Jia P, Hu Y, Zhang M, Liu C, Zhou Y (2018) An efficient bio-based plasticizer for poly (vinyl chloride) from waste cooking oil and citric acid: Synthesis and evaluation in PVC films. J Cleaner Prod 189:334–343. https://doi.org/10.1016/j.jclepro.2018.04.085
65. Fenollar O, Sanchez-Nacher L, Garcia-Sanoguerra D, Lopez J, Balart R (2009) The effect of the curing time and temperature on final properties of flexible PVC with an epoxidized fatty acid ester as natural-based plasticizer. J Mater Sci 44:3702–3711. https://doi.org/10.1007/s10853-009-3495-7
66. Flamini G, Mastrorilli E, Cioni PL, Morelli I, Panizzi L (1999) Essential oil from Crithmum maritimum grown in Liguria: seasonal variation and antimicrobial activity. J Essent Oil Res 11:788–792. https://doi.org/10.1080/10412905.1999.9712022
67. Floros JD, Dock LL, Han JH (1997) Active packaging technologies and applications. Food Cosmet Drug Packag 20(1):10–17. https://doi.org/10.1159/000416709
68. Foley P, Kermanshahi A, Beach ES, Zimmerman JB (2012) Derivation and synthesis of renewable surfactants. Chem Soc Rev 41:1499–1518. https://doi.org/10.1039/C1CS15217C
69. Folmer BM (2003) Sterol surfactants: from synthesis to applications. Adv Colloid Interface Sci 103(2):99–119. https://doi.org/10.1016/S0001-8686(01)00100-2
70. Galli F, Nucci S, Pirola C, Bianchi C (2014) Epoxy methyl soyate as bio-plasticizer: two different preparation strategies. Chem Eng Trans 37:601–606. https://doi.org/10.3303/CET1437101
71. Gergis V, Spiliotis V, Argyriadou, Poulos C (1991) Relation between the antimicrobial activity and the chemical composition of the essential oil of Sideritis sipylea Boiss (Labiatae). Flavor Fragra J 6:93–95. https://doi.org/10.1002/ffj.2730060114
72. Geys R, Soetaert W, Bogaert IV (2014) Biotechnological opportunities in biosurfactant production. Curr Opin Biotechnol 30:66–72. https://doi.org/10.1016/j.copbio.2014.06.002
73. Gill AO, Delaquis P, Russo P, Holley R.A (2002) Evaluation of antilisterial action of cilantro oil on vacuum packed ham. Int J Food Microbiol 73:83–92. https://doi.org/10.1016/S0168-1605(01)00712-7
74. Global demand for plasticizers continue to rise (2017) Addit Polym. 10:10–11
75. Godwin AD (2017) 24 Plasticizers. https://doi.org/10.1016/B978-0-323-39040-8/00025-0
76. Goldberg S, Doyle R, Rosenberg M (1990) Mechanism of enhancement of microbial cell hydrophobicity by cationic polymers. J Bacteriol 172(10):5650–5654. https://doi.org/10.1128/jb.172.10.5650-5654.1990
77. Gomez-Estaca J, Lopez A, Lopez-Caballero ME et al (2010) Biodegradable gelatin–chitosan films incorporated with essential oils as antimicrobial agents for fish preservation. Food Microbiol 27:889–896
78. Grand Review Research, Bio-Platicizers Market and Segment Forecast to 2025 (2017)
79. Griffin SG, Wyllie SG, Markham JL, Leach, DN (1999) The role of structure and molecular properties of terpenoids in determining their antimicrobial activity. Flavo Fragra J 145:322–332. http://doi.org/10.1002/(SICI)1099-1026(199909/10)14:5322::AID-FFJ8373.0.CO;2-4
80. Güner FS, Yağcı Y, Erciyes AT (2006) Polymers from triglyceride oils. Prog Poly Sci 31(7):633–670. https://doi.org/10.1016/j.progpolymsci.2006.07.001
81. Harwood JL, Ramli US, Tang M, Quant PA, Weselake RJ, Fawcett T, Guschina IA (2013) Regulation and enhancement of lipid accumulation in oil crops: the use of metabolic control analysis for informed genetic manipulation. Eur J Lipid Sci Technol 115(11):1239–1246. https://doi.org/10.1002/ejlt.201300257
82. Heidbreder A, Höfer R, Grützmacher R, Westfechtel A, Blewett CW (1999) Oleochemical products as building blocks for polymers. Fett/Lipid 11:418–424. https://doi.org/10.1002/(SICI)1521-4133(199911)101:11418::AID-LIPI4183.0.CO;2-X

83. Herrera N, Roch H, Salaberria AM, Pino-Orellana MA, Labidi J, Fernandes SCM, Radic D, Leivad A, Oksman K (2016) Functionalized blown films of plasticized polylactic acid/chitin nanocomposite: Preparation and characterization. Mater Des 92:846–852. https://doi.org/10.1016/j.matdes.2015.12.083
84. Hershko V, Nussinovitch A (1998) The behavior of hydrocolloid coatings on vegetative materials. Biotechnol Prog 14:756–765. https://doi.org/10.1021/bp980075v
85. Hill K (2000) Fats and oils as oleochemical raw materials. Pure Appl Chem 72. https://doi.org/10.1351/pac200072071255
86. Hofland A (2012) Alkyd resins: from down and out to alive and kicking. Prog Org Coat 73:274–282. https://doi.org/10.1016/j.porgcoat.2011.01.014
87. Holmiere S, Valentin R, Maréchal P, Mouloungui Z (2017) Esters of oligo-(glycerol carbonate-glycerol): new biobased oligomeric surfactants. J Colloid Interface Sci 487:418–425. https://doi.org/10.1016/j.jcis.2016.10.072
88. Hosamani KM, Patil AS, Pattanashettar RS (2002) Acacia arabica varieties—Telia babul, Vediana and Cupressiformis seed oils: a moderate source of coronaric and cyclopropene fatty acids. Ind Crops Prod 15(2):131–137. https://doi.org/10.1016/S0926-6690(01)00103-0
89. Hoshino Z, Iijima A, Hayashi Y et al (1998). Plant freshness retentive agent. Japanese Patent 11332460
90. http://www.thehenryford.org/research/soybeancar.aspx. (n.d.)
91. Ibrahim MS, Mohamed HA, Kandile NG, Said HM, Mohamed IM (2011) Electron beam processed plasticized epoxy coatings for surface protection. Mater Chem Phys 130(1):237–242. https://doi.org/10.1016/j.matchemphys.2011.06.033
92. İşeri-Çağlar D, Baştürk E, Oktay B, Kahraman MV (2014) Preparation and evaluation of linseed oil based alkyd paints. Prog Org Coat 77(1):81–86. https://doi.org/10.1016/j.porgcoat.2013.08.005
93. Ismail A, Dzolkifli O, Ooi T, Salmiah A (1998) Pesticide formulations based on oleochemicals derived from palm and palm kernel oils. Agris Fao Org
94. Isman MB (2000) Plant essential oils for pest and disease management. Crop Protec 19:603–608. https://doi.org/10.1016/S0261-2194(00)00079-X
95. Isman MB, Machial CM (2006) Pesticides based on plant essential oils: from traditional practice to commercialization. In: Rai M, Carpinella MC (eds) Naturally occurring bioactive compounds. Elsevier, BV, pp 29–44
96. Jayasena DD, Jo C (2013) Essential oils as potential antimicrobial agents in meat and meat products: a review. Trends Food Sci Technol 34:96–108. https://doi.org/10.1016/j.tifs.2013.09.002Getrightsandcontent
97. Jerkovic I, Mastelic J, Milos M (2001) The impact of both the season of collection and drying on the volatile constituents of Origanum vulgare L. ssp. hirtum grown wild in Croatia. Int J Food Sci Technol 36:649–654. https://doi.org/10.1046/j.1365-2621.2001.00502.x
98. Jia P, Zhang M, Hu L, Feng G, Bo C, Zhou Y (2015) Synthesis and Application of Environmental Castor Oil Based Polyol Ester Plasticizers for Poly(vinyl chloride). ACS Sustain Chem Eng 3(9):2187–2193. https://doi.org/10.1021/acssuschemeng.5b00449
99. Johansson I, Svensson M (2001) Surfactants based on fatty acids and other natural hydrophobes. Curr Opin Coll Interface Sci 6(2):178–188. https://doi.org/10.1016/S1359-0294(01)00076-0
100. Juliano C, Mattana A, Usai M (2000) Composition and in vitro antimicrobial activity of the essential oil of Thymus herbabarona Loisel growing wild in Sardinia. J Essent Oil Res 12:516–522
101. Juven BJ, Kanner J, Schved F et al (1994) Factors that interact with the antibacterial action of thyme and its active constituents. J Appl Bacteriol 76:626–631. https://doi.org/10.1080/10412905.2000.9699578
102. Kalemba D, Kunicka A (2003) Antibacterial and antifungal properties of essential oils. Curr Med Chem 10:813–829. https://doi.org/10.2174/0929867033457719

103. Kalita DJ, Tarnavchyk I, Sibi M, Moser BR, Webster DC, Chisholm BJ (2018) Biobased poly(vinyl ether)s derived from soybean oil, linseed oil, and camelina oil: Synthesis, characterization, and properties of crosslinked networks and surface coatings. Prog Org Coat 125:453–462. https://doi.org/10.1016/j.porgcoat.2018.09.033
104. Karak N (2012) 3–Vegetable oils and their derivatives. In: Karak N (ed) Vegetable oil-based polymers. Woodhead Publishing, pp 54–95. https://doi.org/10.1533/9780857097149.54
105. Karmalm P, Hjertberg T, Jansson A, Dahl R (2009) Thermal stability of poly(vinyl chloride) with epoxidized soybean oil as primary plasticizer. Poly Degrad Stab 94:2275–2281. https://doi.org/10.1016/j.polymdegradstab.2009.07.019
106. Kester JJ, Fennema OR (1986) Edible films and coatings: a review. Food Technol 40(12):47–59. https://doi.org/10.1016/B978-0-12-394601-0.00009-6
107. Kim HJ, Chen F, Wu C, Wang X, Chun HY, Jin Z (2004) Evaluation of antioxidant activity of Australian Tea Tree (Melaleuca alternifolia) oil and its components. J Agric Food Chem 52:2849–2854. https://doi.org/10.1021/jf035377d
108. Kim JR, Sharma S (2012) The development and comparison of bio-thermoset plastics from epoxidized plant oils. Ind Crops Prod 36(1):485–499. https://doi.org/10.1016/j.indcrop.2011.10.036
109. Knobloch K, Pauli A, Iberl B, Weigand H, Weis N (1989) Antibacterial and antifungal properties of essential oil components. J Essent Oil Res 1:119–128. https://doi.org/10.2174/0929867033457719
110. Kong PS, Aroua MK, Daud WMAW (2016) Conversion of crude and pure glycerol into derivatives: a feasibility evaluation. Renew Sustain Energy Rev 63:533–555. https://doi.org/10.1016/j.rser.2016.05.054
111. Kordali S, Cakir A, Mavi A, Mavi A, Kilic H, Yildirim A (2005) Screening of chemical composition and antifungal activity of essential oils from three Turkish Artemisia species. J Agric Food Chem 53:1408–1416. https://doi.org/10.1021/jf048429n
112. Kostaki M, Giatrakou V, Savvaidis IN, Kontominas M(2009) Combined effect of MAP and thyme essential oil on the microbiological, chemical and sensory attributes of organically aquacultured sea bass (Dicentrarchus labrax) fillets. Food Microbiol 26:475–482. https://doi.org/10.1016/j.fm.2009.02.008
113. Koul O, Walia S, Dhaliwal GS (2008) Essential oils as green pesticides: potential and constraints. Biopestic Int 4(1):63–84
114. Kreutzer U (1984) Manufacture of fatty alcohols based on natural fats and oils. J Am Oil Chem Soc 61:343–348. https://doi.org/10.1007/BF02678792
115. Kubo I, Muroi H, Himejima M (1992) Antimicrobial activity of green tea flavor components and their combination effects. J Agric Food Chem 40:245–248. https://doi.org/10.1021/jf00014a015
116. Kubo I, Muroi H, Himejima M (1993) Antibacterial activity against Streptococcus mutans of Mate tea flavor components. J Agric Food Chem 41:107–111. https://doi.org/10.1021/jf00025a023
117. Kurita N, Miyaji M, Kurane R, Takahara Y(1981) Antifungal activity of components of essential oils. Agric Biol Chem 45:945–952. https://doi.org/10.1080/00021369.1981.10864635
118. Labuza T, Breene W (1989) Applications of active packaging for improvement of shelf-life and nutritional quality of fresh and extended shelf-life foods. J Food Process Preserv 13:189–197. https://doi.org/10.1111/j.1745-4549.1989.tb00090.x
119. Lahhit N, Bouyanzer A, Desjobert JM, Hammouti B, Salghi R, Costa J, Jama C, Bentiss F, Majidi L (2011) Fennel (Foeniculum Vulgar) essential oil as green corrosion inhibitor of carbon steel in hydrochloric acid solution. Portugal Electrochim Acta 29:127–138. https://doi.org/10.4152/pea.201102127
120. Lambert RJW, Skandamis PN, Coote P, Nychas JE (2001) A study of the minimum inhibitory concentration and mode of action of oregano essential oil, thymol and carvacrol. J Appl Microbiol 91:453–462. https://doi.org/10.1046/j.1365-2672.2001.01428.x

121. Lambert RJW (2000) Susceptibility testing: inoculum size dependency of inhibition using the Colworth MIC technique. J Appl Microbiol 89:275–279. https://doi.org/10.1046/j.1365-2672.2000.01105.x
122. Sanchez-Gonzalez L, Vargas M, Gonzalez-Martínez C, Chiralt A, Cháfer M (2011) Use of essential oils in bioactive edible coatings. Food Eng Rev 3:1–16. https://doi.org/10.1007/s12393-010-9031-3
123. Lee SJ, Umano K, Shibamoto T, Lee KG (2005) Identification of volatile components in basil (Ocimum basilicum L.) and thyme leaves (Thymus vulgaris L.) and their antioxidant properties. Food Chem 91:131–137. https://doi.org/10.1016/j.foodchem.2004.05.056
124. Li M, Li S, Xia J, Ding C, Wang M, Xu L, Yang X, Huang K (2017) Tung oil based plasticizer and auxiliary stabilizer for poly(vinyl chloride). Mater Des 122:366–375. https://doi.org/10.1016/j.matdes.2017.03.025
125. Liang GG, Cook WD, Tcharkhtchi A, Sautereau H (2011) Epoxy as a reactive plasticizer for improving polycarbonate processibility. European Poly J 47(8):1578–1588. https://doi.org/10.1016/j.eurpolymj.2011.05.005
126. Lin L, Blackman GS, Matheson RR (2000) A new approach to characterize scratch and mar resistance of automotive coatings. Prog Org Coat 40:85–91. https://doi.org/10.1016/S0300-9440(00)00113-2
127. Ljungberg N, Weeslen B (2003) Tributyl citrate oligomers as plasticizers for poly (lactic acid): thermo-mechanical film properties and aging. Polymer 44:7679–7688. https://doi.org/10.1016/j.polymer.2003.09.055
128. Lligadas G, Ronda JC, Galià M, Cádiz V (2013) Renewable polymeric materials from vegetable oils: a perspective. Mater Today 16(9):337–343. https://doi.org/10.1016/j.mattod.2013.08.016
129. Lligadas G, Ronda JC, Galià M, Cádiz V (2010) Plant Oils as Platform Chemicals for Polyurethane Synthesis: Current State-of-the-Art. Biomacromolecules 11(11):2825–2835. https://doi.org/10.1021/bm100839x
130. Lopez-Gomez A, Fernandez PS, Palop A, Periago PM, Martinez-López A, Marin-Iniesta F, Barbosa-Cánovas GV (2009) Food safety engineering: an emergent perspective. Food Eng Rev 1:84–104. https://doi.org/10.1007/s12393-009-9005-5
131. Lu Y, Larock RC (2007) New hybrid latexes from a soybean oil-based waterborne polyurethane and acrylics via emulsion polymerization. Biomacromol 8:3108–3114. https://doi.org/10.1021/bm700522z
132. Lutz JT Jr (1993) Handobook of PVC formulating. Wiley, New York
133. Maag H (1984) Fatty acid derivatives: important surfactants for household, cosmetic and industrial purpose. JAOCS 61:259–267. https://doi.org/10.1007/BF02678778
134. Makovitzki A, Viterbo A, Brotman Y, Chet I, Shai Y (2007) Inhibition of fungal and bacterial plant pathogens in vitro and in planta with ultrashort cationic lipopeptides. Appl Environ Microbiol 73(20):6629–6636. https://doi.org/10.1128/AEM.01334-07
135. Maneerat S (2005) Production of biosurfactants using substrates from renewable-resources. Songklanakarin J Sci Technol 27(3):675–683
136. Park HJ (1999) Development of advanced edible coatings for fruits. Trends Food Sci Technol 10:254–260. https://doi.org/10.1016/S0924-2244(00)00003-0
137. Marotti M, Piccaglia R, Giovanelli E, Deans SG, Eaglesham E (1994) Effects of planting time and mineral fertilization on peppermint (Mentha piperita) essential oil composition and its biological activity. Flavor Fragra J 9:125–129. https://doi.org/10.1002/ffj.2730090307
138. Matheson RR Jr (2006) Paint and coatings technology: current industrial trends. J Macromol Sci-Pol R 46:341–346. https://doi.org/10.1080/15583720600945493
139. Matheson RR (2002) 20th-to 21st-century technological challenges in soft coatings. Science 297:976–979. https://doi.org/10.1126/science.1075707
140. Meir MAR, Metzger JO, Scubert US (2007) Plant oil renewable resources as green alternatives in polymer science. Chem Soc Rev 36:1788–1802. https://doi.org/10.1039/B703294C
141. Mekonnen T, Mussone P, Khalil H, Bressler (2013) D J Mater Chem A 1:13379–13397. https://doi.org/10.1039/C3TA12555F

142. Miao S, Wang P, Su Z, Zhang S (2014) Vegetable-oil-based polymers as future polymeric biomaterials. Acta Biomater 10(4):1692–1704. https://doi.org/10.1016/j.actbio.2013.08.040
143. Morillon V, Debeaufort F, Blond G, Capelle M, Voilley A (2002) Factors affecting the moisture permeability of lipid-based edible films: a review. Crit Rev Food Sci Nutr 42(1):67–89. https://doi.org/10.1080/10408690290825466
144. Moyler D (1998) CO2 extraction and other new technologies: an update on commercial adoption. In: International Federation of Essential Oils and Aroma Trades—21st International Conference on Essential Oils and Aroma's. IFEAT, London, pp 33–39
145. Mukherjee S, Ghosh M (2017) Studies on performance evaluation of a green plasticizer made by enzymatic esterification of furfuryl alcohol and castor oil fatty acid. Carbo Poly 157:1076–1084. https://doi.org/10.1016/j.carbpol.2016.10.075
146. Muller RF, Berger B, Yegen O (1995) Chemical composition and fungi toxic properties to phytopathogenic fungi of essential oils of selected aromatic plants growing wild in Turkey. J Agric Food Chem 43:2262–2266. https://doi.org/10.1021/jf00056a055
147. New reports highlight growth opportunities for plastics additives, (2015) Addi Polym 5:10–11. https://doi.org/10.1016/S0306-3747(15)30062-2
148. Nielsen B, Sparso FV, Kristiansen JK (2001) WO2001014466
149. Nieres PD, Zelin J, Trasarti AF, Apesteguia CR (2018) Valorisation of vegetable oils by heterogeneous catalysis via metathesis reaction. Curr Opin Green Sustain Chem 10:1–5. https://doi.org/10.1016/j.cogsc.2018.02.001
150. Nosal H, Nowicki J, Warzała M, Semeniuk I, Sabura E (2016) Synthesis and characterization of alkyd resins based on Camelina sativa oil, glycerol and selected epoxidized vegetable oils as functional modifiers. Prog Org Coat 101:553–568. https://doi.org/10.1016/j.porgcoat.2016.10.003
151. Odetoye TE, Ogunniyi DS, Olatunji GA (2012) Improving Jatropha curcas Linnaeus oil alkyd drying properties. Prog Org Coat 73(4):374–381. https://doi.org/10.1016/j.porgcoat.2011.11.016
152. OECD (2014) OECD/food and agriculture organization of the United Nations, OECD-FAO Agricultural Outlook 2014. OECD Publishing. http://dx.doi.org/10.1787/agr_outlook-2014-en(,in:)
153. Orth JD, Conrad TM, Na J, Lerman JA, Nam H, Feist AM, Palsson BØ (2011) A comprehensive genome-scale reconstruction of Escherichia coli metabolism–2011. Mol Syst Biol 7:535. http://doi.org/10.1038/msb.2011.65
154. Oyama ST (2008) Mechanism in homogeneous and heterogeneous epoxidation catalysis ed. Elsevier Science
155. Packiyasothy EV, Kyle S (2002) Antimicrobial properties of some herb essential oils. Food Aust 54(9):384–387
156. Pantelic I, Cuckovic B (2014) 1—Alkyl polyglucosides: an emerging class of sugar surfactants. In: Pantelic IBT-AP (ed) Woodhead Publishing Series in Biomedicine. Woodhead Publishing, Oxford, pp 1–19. https://doi.org/10.1533/9781908818775.1
157. Park S-J, Jin F-L, Lee J-R, Shin J-S (2005) Cationic polymerization and physicochemical properties of a biobased epoxy resin initiated by thermally latent catalysts. European Poly J 41(2):231–237. https://doi.org/10.1016/j.eurpolymj.2004.09.011
158. Paster N, Menasherov M, Ravid U, Ravid U, Juven B (1995) Antifungal activity of oregano and thyme essential oils applied as fumigants against fungi attacking stored grain. J Food Prot 58(1):81–85. https://doi.org/10.4315/0362-028X-58.1.81
159. Patel M (2004) CO2 Reduction Potential from Renewable Surfactants in the EU. J Ind Ecol 7:47–62
160. Patel MK, Theiß A, Worrell E (1999) Surfactant production and use in Germany: resource requirements and CO2 emissions. Resou Conserv Recycl 25(1):1–78. https://doi.org/10.1016/S0921-3449(98)00063-9
161. Pathan S, Ahmad S (2013) S-triazine ring modified waterborne alkyd: synthesis, characterization, antibacterial, and electrochemical corrosion studies. ACS Sustain Chem 1(10):1246–1257. https://doi.org/10.1021/sc4001077

162. Pathan S, Ahmad S (2013) Synthesis, characterization and effect of s-triazine ring on physico-mechanical and electrochemical corrosion resistance performance of waterborne castor oil alkyd. J Mater Chem A 1:14227–14238. https://doi.org/10.1039/C3TA13126B
163. Patil CK, Rajput SD, Marathe RJ, Kulkarni RD, Phadnis H, Sohn D, Mahulikar PP, Gite VV (2017) Synthesis of bio-based polyurethane coatings from vegetable oil and dicarboxylic acids. Prog Org Coat 106:87–95. https://doi.org/10.1016/j.porgcoat.2016.11.024
164. Pawar MS, Kadam AS, Yemul OS (2015) Development of polyetheramide based corrosion protective polyurethane coating from mahua oil. Prog Org Coat 89:143–149. https://doi.org/10.1016/j.porgcoat.2015.08.017
165. Pfleger BF, Gossing M, Nielsen J (2015) Metabolic engineering strategies for microbial synthesis of oleochemicals. Meta Eng 29:1–11. https://doi.org/10.1016/j.ymben.2015.01.009
166. Philipp C, Eschig S (2012) Waterborne polyurethane wood coatings based on rapeseed fatty acid methyl esters. Prog Org Coat 74(4):705–711. https://doi.org/10.1016/j.porgcoat.2011.09.028
167. Pielichowski P, Swirz-Motysia B (2006) Influence of polyesterurethane plasticizer on the kinetics of poly(vinyl chloride) decomposition process. J Ther Anal Calorimetry 83:207–212. https://doi.org/10.1007/s10973-005-7007-y
168. Pletnev MY (2004) Vegetable-derived surfactants as a reply to the natural trend in the household and personal care. SOFW J 130:1–10
169. Poussard L, Lazko J, Mariage J, Raquez J-M, Dubois P (2016) Biobased waterborne polyurethanes for coating applications: how fully biobased polyols may improve the coating properties. Prog Org Coat 97:175–183. https://doi.org/10.1016/j.porgcoat.2016.04.003
170. Pradhanang PM, Momol MT, Olson SM, Jones JB(2003) Effects of plant essential oils on Ralstonia solanacearum population density and bacterial wilt incidence in tomato. Plant Dis 87:423–427. https://doi.org/10.1094/PDIS.2003.87.4.423
171. Prileschajew N (1909) Oxydation ungesättigter Verbindungen mittels organischer Superoxyde. Beritch der Deutschen Chemischen Gesellschaft 42:4811–4815. https://doi.org/10.1002/cber.190904204100
172. Quintavalla S, Vicini L (2002) Antimicrobial food packaging in food industry. Meat Sci 62:373–380. https://doi.org/10.1016/S0309-1740(02)00121-3
173. Reddy JRC, Rao BVSK, Karuna M, Kothapalli RV, Krishna AVR, Prasad RBN (2013) Lipase-mediated preparation of epoxy lecithin and its evaluation as plasticizer in polyester laminates. J App Poly Sci 127. https://doi.org/10.1002/app.37960
174. Radakovits R, Jinkerson RE, Darzins A, Posewitz MC (2010) Genetic engineering of algae for enhanced biofuel production. Eukaryot Cell 9(4):486–501. https://doi.org/10.1128/EC.00364-09
175. Rebello S, Asok AK, Mundayoor S, Shanavas J (2014) Surfactants: toxicity, remediation and green surfactants. Environ Chem Lett 12:275–287. https://doi.org/10.1007/s10311-014-0466-2
176. Rengasamy S, Mannari V (2013) Development of soy-based UV-curable acrylate oligomers and study of their film properties. Prog Org Coat 76:78–85. https://doi.org/10.1016/j.porgcoat.2012.08.012
177. Riaz U, Ashraf SM, Sharma HO (2011) Mechanical, morphological and biodegradation studies of microwave processed nanostructured blends of some bio-based oil epoxies with poly (vinyl alcohol). Polym Degrad Stab 96:33–42. https://doi.org/10.1016/j.polymdegradstab.2010.11.007
178. Rieke R, Thakur D, Roberts B, White G (1997) Fatty methyl ester hydrogenation to fatty alcohol part II: process issues. J Am Oil Chem Soc 74:333–339. https://doi.org/10.1007/s11746-997-0089-x
179. Rios LA, Echeverri DA, Franco A, (2011) Epoxidation of jatropha oil using heterogeneous catalysts suitable for the Prileschajew reaction: Acidic resins and immobilized lipase. Appl Catal. A 394:132–137. https://doi.org/10.1016/j.apcata.2010.12.033
180. Rios LA, Weckes P, Schuster H, Hoelderich WF (2005) Mesoporous and amorphous Ti–silicas on the epoxidation of vegetable oils J Catal 232:19–26. https://doi.org/10.1016/j.jcat.2005.02.011

181. Rooney Z (1995) Active food packaging. Blackie Academic and Professional, Glasgow, UK
182. Russo M, Galletti GC, Bocchini P, Carnacini A (1998) Essential oil chemical composition of wild populations of Italian oregano spice (Origanum vulgare sphirtum (Link) Ietswaart): a preliminary evaluation of their use in chemotaxonomy by cluster analysis: 1. Inflorescences. J Agric Food Chem 46:3741–3746. https://doi.org/10.1021/jf980087w
183. Russo V, Tesser R, Santacesaria E, Di Serio M (2013) Chemical and technical aspects of propene oxide production via hydrogen peroxide (HPPO process). Ind Eng Chem Res 52(3):1168–1178. https://doi.org/10.1021/ie3023862
184. Sajna KV, Höfer R, Sukumaran RK, Gottumukkala LD, Pandey A (2015) Chapter 14—White biotechnology in biosurfactants. In: Pandey A, Höfer R, Taherzadeh M, Nampoothiri KM, Larroche C (eds) Industrial biorefineries & white biotechnology. Elsevier, Amsterdam, pp 499–521. https://doi.org/10.1016/B978-0-444-63453-5.00016-1
185. Salimon J, Salih N, Yousif E (2012) Industrial development and applications of plant oils and their biobased oleochemicals. Arab J Chem 5(2):135–145. https://doi.org/10.1016/j.arabjc.2010.08.007
186. Salzer UJ (1977) The analysis of essential oils and extracts (oleoresins) from seasonings—a critical review. Crit Rev Food Sci Nutr 9(4):345–373. https://doi.org/10.1080/10408397709527239
187. Sanchez-Gonzalez L (2010) Caracterizacion y aplicacion de recubrimientos antimicrobianos a base de polisacaridos y aceites esenciales. Doctoral thesis. Universidad Politecnica de Valencia, Spain
188. Santacesaria E, Martinez Vicente G, Di Serio M, Tesser R (2012) Main technologies in biodiesel production: state of the art and future challenges. Catal Today 195:2–13. https://doi.org/10.1016/j.cattod.2012.04.057
189. Santacesaria E, Tesser R, Di Serio M (2018) Polyethoxylation and polypropoxylation reaction: kinetics, mass transfer and industrial reactor design. Chin J Chem Eng. https://doi.org/10.1016/j.cjche.2018.02.020
190. Santacesaria E, Tesser R, Di Serio M, Turco R, Russo V, Verde D (2011) A biphasic model describing soybean oil epoxidation with H2O2 in a fed-batch reactor. Chem Eng J 173:189–209. https://doi.org/10.1016/j.cej.2011.05.018
191. Santos ED, Metivier P (1996) EP0701989 B1
192. Satdive A, Mestry S, Patil D, Mhaske ST (2019) Synthesis of melamine formaldehyde cured castor oil based hydroxyl functional alkyd for coating application. Prog Org Coat 131:165–175. https://doi.org/10.1016/j.porgcoat.2019.01.044
193. Scheffer JJC, Baerheim A (1981) Zur Analytikatherische Ole Forschung 1969 bis 1979. Riechstoffe, Aromen, Kosmetica 31(2):39–41
194. Sears JK, Darby JR (1982) The technology of plasticizers. Wiley & Dpmd, New York
195. Senatore F (1996) Influence of harvesting time on yield and composition of the essential oil of a thyme (Thymus pulegioides L.) growing wild in Campania (Southern Italy). J Agric Food Chem 44:1327–1332. https://doi.org/10.1021/jf950508z
196. Septiyanti M, Meliana Y, Aiman S, Ghozali M, Triwulandari E (2018) Formulation of palm oil based surfactant for fungicide emulsifiable concentrate. In: 5th International Symposium on Innovative Bio-Production Indonesia
197. Shah MY, Ahmad S (2012) Waterborne vegetable oil epoxy coatings: preparation and characterization. Prog Org Coat 75:248–252. https://doi.org/10.1016/j.porgcoat.2012.05.001
198. Sharmin E, Akram D, Zafar F, Ashraf SM, Ahmad S (2012) Plant oil polyol based poly (ester urethane) metallohybrid coatings. Prog Org Coat 73(1):118–122. https://doi.org/10.1016/j.porgcoat.2011.09.008
199. Shetty Z, Dwelle KR (1990) Disease and sprout control in individual film wrapped potatoes. Am Potato J 67(10):705–718. https://doi.org/10.1007/BF03044017
200. Sikkema J, De Bont JAM, Poolman B (1994) Interactions of cyclic hydrocarbons with biological membranes. J Biol Chem 269(11):8022–8028
201. Skandamis PN, Nychas GJE (2001) Effect of oregano essential oil on microbiological and physico-chemical attributes of minced meat stored in air and modified atmospheres. J Appl Microbiol 91:1011–1022. https://doi.org/10.1046/j.1365-2672.2001.01467.x

202. Sperling LH (2006) Introduction to physical polymer science. Wiley, New York
203. Statista (2017) Global paint and coatings industry market value 2017–2022. https://www.statista.com/statistics/745160/global-paint-and-coatings-industry-market-value/. Accessed 19 Apr 2018
204. Stroh J, Wan MT, Isman MB et al (1998) Evaluation of the acute toxicity to juvenile Pacific salmon and rainbow trout of some plant essential oils, a formulated product, and the carrier. Bull Environ Contam Toxicol 60:923–930. https://doi.org/10.1007/s001289900716
205. Su Cha D, Chinnan MS (2010) Biopolymer-based antimicrobial packaging: a review. Crit Rev Food Sci Nutr 44:223–237. https://doi.org/10.1080/10408690490464276
206. Sun S, Ke X, Cui L, Yang G, Bi Y, Song F, Xu X (2011) Enzymatic epoxidation of Sapindus mukorossi seed oil by perstearic acid optimized using response surface methodology. Ind Crops Prod 33:676–682. http://dx.doi.org/10.1016/j.indcrop.2011.01.002
207. Svensson M (2010) Surfactants from renewable resources, Part 1: renewable hydrophobes. Wiley, pp 3–19
208. Taramasso M, Perego G, Notari B (1971) U.S. Pat. 1249079
209. Tee YB, Talib RA, Abdan K, Chin NL, Basha RK, Yunos KFM (2014) Toughening poly(lactic acid) and aiding the melt-compounding with bio-sourced plasticizers. Agri Agri Sci Procedia 2:289–295. https://doi.org/10.1016/j.aaspro.2014.11.041
210. Thoroski J, Blank G, Biliaderis C (1989) Eugenol induced inhibition of extracellular enzyme production by Bacillus cereus. J Food Prot 52(6):399–403. https://doi.org/10.4315/0362-028X-52.6.399
211. Tripathi AK, Prajapati V, Aggarwal KK et al (2000) Effects of volatile oil constituents of Mentha species against stored grain pests, Callosobrunchus maculatus and Tribolium castanum. J Med Arom Plant Sci 22:549–556
212. Turco R, Pischetola C, Di Serio M, Vitiello R, Tesser R, Santacesaria E (2017) Selective Epoxidation of Soybean Oil in the Presence of H–Y Zeolite, Ind Eng Chem Res 56:7930–7936. https://doi.org/10.1021/acs.iecr.7b01437
213. Turco R, Pischetola C, Tesser R, Andini S, Di Serio M (2016) New findings on soybean and methylester epoxidation with alumina as the catalyst RSC Advan 6:31647–31652. https://doi.org/10.1039/C6RA01780K
214. Turco R, Vitiello R, Russo V, Tesser R, Santacesaria R, Di Serio M (2013) Selective epoxidation of soybean oil with performic acid catalyzed by acidic ionic exchange resins. Green Proc Synth 2:427–434
215. Turco R, Vitiello R, Tesser R, Vergara A, Andini S, Di Serio M (2017) Niobium based catalysts for methyl oleate epoxidation reaction. Top Catal 60:1054–1061. https://doi.org/10.1007/s11244-017-0808-y
216. Turek T, Trimm DL, Cant NW (1994) The catalytic hydrogenolysis of esters to alcohols. Catal Rev Sci Eng 36:645–683. https://doi.org/10.1080/01614949408013931
217. Turina AV, Nolan MV, Zygadlo JA, Perillo MA (2006) Natural terpenes: self-assembly and membrane partitioning. Biophys Chem 122:101–113. https://doi.org/10.1016/j.bpc.2006.02.007
218. Ullmann's Encyclopedia of Industrial Chemistry (1994) A2:747–817
219. Ultee A, Kets EPW, Alberda M, Alberda M, Hoekstra FA, Smid EJ (2000) Adaptation of the food-borne pathogen Bacillus cereus to carvacrol. Arch Microbiol 174(4):233–238. https://doi.org/10.1007/s002030000199
220. Van De Mark MR, Sandefur K (2005) Vegetable oils in paint and coatings. In: Erhan SZ (ed) Industrial uses of vegetable oils. AOCS Press, Illinois, pp 149–168
221. Van De Mark MR, Sandefur K (2005) Vegetable oils in paint and coatings. Inform 16(8):478–481. https://doi.org/10.1201/9781439822388.ch8
222. Vaz DA, Gudiña EJ, Alameda EJ, Teixeira JA, Rodrigues LR (2012) Performance of a biosurfactant produced by a Bacillus subtilis strain isolated from crude oil samples as compared to commercial chemical surfactants. Colloids Surf B: Biointer 89:167–174. https://doi.org/10.1016/j.colsurfb.2011.09.009

223. Vieira MGA, da Silva MA, dos Santos LO, Beppu MM (2011) Natural-based plasticizers and biopolymer films: A review . Eur Polym J 47:254–263. http://dx.doi.org/10.1016/j.eurpolymj.2010.12.011
224. Vieira MGA, da Silva MA, Macumoto ACG, dos Santos LO, Beppu MM (2014) Synthesis and application of natural polymeric plasticizer obtained through polyesterification of rice fatty acid. Mat. Res. 17(2). http://dx.doi.org/10.1590/S1516-14392014005000017
225. Vijayakumar S, Saravanan V (2015) Biosurfactants-types, sources and applications. Res J Microbiol 10:181–192. http://dx.doi.org/10.3923/jm.2015.181.192
226. Waghmare JT, Ware AM, Momin SA (2007) Neem oil as pesticide. J Dispersion Sci Technol 28(2):323–328. https://doi.org/10.1080/01932690601062291
227. Warwel S, Brüse F, Demes C, Kunz M, Rüsch M (2001) Polymers and surfactants on the basis of renewable resources. Chemosphere 43:39–48. https://doi.org/10.1016/S0045-6535(00)00322-2
228. Warwel S, Brüse F, Demes C, Kunz M, (2004) Polymers and polymer building blocks from meadowfoam oil. Industrial Crops and Products 20(3):301–309. https://doi.org/10.1016/j.indcrop.2003.11.002
229. Weiss KD (1997) Paint and coatings: a mature industry in transition. Prog Polym Sci 22:203–245. https://doi.org/10.1016/S0079-6700(96)00019-6
230. Wilkes CE, Summers JW, Daniels CA (2005) PVC Handibook. Hanser Publishers, Munich
231. Wilkins C, Madsen JO (1991) Oregano headspace constituents. Zeitschrift fur Lebensmittel-Untersuchung und Forschung 192:214–219. https://doi.org/10.1007/BF01202881
232. Wilson AS (1995) Plasticisers: principles and practice. The Institute of Materials, London
233. Wong SP, Lim WH, Cheng SF, Chuah CH (2012) Biological properties of sodium alkyl methyl ester sulfonate/alkyltrimethylammonium bromide surfactant mixtures. Colloids Surf B: Biointerf 89:48–52. https://doi.org/10.1016/j.colsurfb.2011.08.021
234. Wu M, Zhang Y, Peng Q, Song L, Hu Z, Li Z, Wang Z (2018) Mechanically strong plant oil-derived thermoplastic polymers prepared via cellulose graft strategy. Appl Surf Sci 458:495–502. https://doi.org/10.1016/j.apsusc.2018.07.072
235. Xiao W, He P, He B (2002) Improving the toughness and flame retardancy of cured epoxy resin with brominated castor oil. J Appl Polym Sci 86(10):2530–2534. https://doi.org/10.1002/app.11097
236. Xiong Z, Zhang L, Ma S, Yang Y, Zhang C, Tang Z, Zhu J (2013) Effect of castor oil enrichment layer produced by reaction on the properties of PLA/HDI-g-starch blends. Carbo Poly 94(1):235–243. https://doi.org/10.1016/j.carbpol.2013.01.038
237. Xu X-Y, Yang H-K, Singh SP, Sharp PJ, Liu Q (2018) Genetic manipulation of non-classic oilseed plants for enhancement of their potential as a biofactory for triacylglycerol production. Engineering 4(4):523–533. https://doi.org/10.1016/j.eng.2018.07.002
238. Yang D, Peng X, Zhong L, Cao X, Chen W, Zhang X, Liu S, Sun R (2014) "Green" films from renewable resources: properties of epoxidized soybean oil plasticized ethyl cellulose films. Carbo Poly 103:198–206. https://doi.org/10.1016/j.carbpol.2013.12.043
239. Yeong SK, Idris Z, Hassan HA (2012) Palm oleochemicals in non-food applications. In: Palm oil: production, processing, characterization, and uses. AOCS Press. https://doi.org/10.1016/B978-0-9818936-9-3.50023-X
240. Zafar F, Ashraf SM, Ahmad S (2004) Air drying polyesteramide from a sustainable resource. Prog Org Coat 51(3):250–256. https://doi.org/10.1016/j.porgcoat.2004.08.003
241. Zhang C, Garrison TF, Madbouly SA, Kessler MR (2017) Recent advances in vegetable oil-based polymers and their composites. Prog Poly Sci 71:91–143. https://doi.org/10.1016/j.progpolymsci.2016.12.009
242. Zheng T, Wu Z, Xie Q, Fang J, Hu Y, Lu M, Xia F, Nie Y, Ji J (2018) Structural modification of waste cooking oil methyl esters as cleaner plasticizer to substitute toxic dioctyl phthalate. J Cleaner Prod 186:1021–1030. https://doi.org/10.1016/j.jclepro.2018.03.175

Chapter 9
Oil-Based Materials and Products

Qiang Liu, Xinfeng Xie, and Changzhu Li

Abstract Oils and fats, particularly plant based, are important raw materials for the chemical industry worldwide. Plant oil and their oleochemicals have major advantages over mineral oils in that they are renewable and readily available all around the world, often as a result of by-products from other industrial activities. Also, they are generally biodegradable, nonflammable and cause fewer medical problems and allergies to the end users. A large number of studies reported that the physicochemical nature like solubilizing and dispersing ability of the oil and fats has the hair- and skin-care effects, which have been proved to be efficient emollients in skin-care product formulations, but are also used for other applications, such as agrochemicals. In this chapter, we will systematically introduce the use and development of oils and fats as emollients in cosmetic products and as alternative solvents, dispersion media, spray auxiliaries, emulsifiers, stabilizers and additives in the agrochemicals.

Keywords Oil and fats · Oil-based material · Oleochemicals · Emollients · Cosmetic · Agrochemicals

9.1 Development of Oil-Based Cosmetic Products

Oil and fat are the basic ingredients of cream cosmetic products as they function as skin protector, smoother, and moisturizer. The terms "oil" and "fat" are often confusing. "Oil" normally refers to a lipid with short or unsaturated fatty acid chains that are liquid at room temperature, while "fat" (in the strict sense) may specifically refer to lipids that are solids at room temperature. Fatty glyceride is the main component of animal and vegetable oils. Generally, oil and fat can be hydrolyzed to produce

Q. Liu · C. Li (✉)
Central South University of Forestry and Technology and Hunan Academy of Forestry, Changsha 410004, Hunan, China
e-mail: lichangzhu2013@aliyun.com

X. Xie
School of Forest Resources and Environmental Science, Michigan Technological University, Houghton, MI 49931, USA

C. Li et al. (eds.), *Industrial Oil Plant*,
https://doi.org/10.1007/978-981-15-4920-5_9

glycerol and fatty acids. And fatty acids linked to glycerol can form monoglycerides, diglycerides, or triglycerides. Traditionally, synthetic fatty acid glycerides had been widely used in the cosmetics industry. Due to the enhancement of safety and environmental consciousness together with the improvement in oil separation and purification technologies, the yield and quality of natural oils and fats had been improved. New antioxidants had solved the problems of natural plant oil deterioration, and high-quality natural plant oils can be obtained for natural cosmetic products cost-effectively. Natural cosmetic products have become more and more favorable to general consumers nowadays. In 2007, the Journal of Soap, Perfumery & Cosmetics (SPC) published several articles about oils and fats for cosmetic uses. The supplier listed the physicochemical properties of various oils and fats in the product specifications for the formulators to select the suitable oils and fats. However, it was difficult to find the physicochemical parameters of various oils and fats.

9.1.1 Functional Assessment

The basic function of oil and fats in cosmetics is still controversial. Consumers generally believe that oil can prevent skin from drying and cracking by inhibiting moisture evaporation. It can rehydrate dry skin and hardened cuticle by restoring the cuticle to being soft and elastic, making the skin smooth, soft, and elastic. Therefore, it can be used as a moisturizer. Oil can also improve the smoothness of cosmetics, which to a large extent determines the skin feeling of skincare products, such as smoothness, lubricity, moisture retention, and breathability. Based on its smoothness and softness characters, it can also be used as skin extender, elasticity and smoothing agent. A few years ago, R Goldemburg tested 85 cosmetics emollients and proposed the concept of skin feeling parameters and found no direct relationship between molecular weight of oil and skin sensitivity [1]. Polar's researching team believes that highly unsaturated and branched oils are suitable for emollient uses [2], but John Woodruf believes that isopropyl fatty acid is particularly suitable as an emollient [3]. Hans Brand tested the lubricity and spreadability of a large number of emollients, and divided them into 3 levels: (1) protection, (2) lubricating, and (3) drying shrinkage [4].

At present, many studies have assessed the physical and chemical properties of oil through sensory test. Parente evaluated several liquid emollients, such as mineral oil, sunflower oil, squalane, decyl oleate, isopropyl myristate, and octyl dodecyl, using instrumental and sensory tests [5]. Wiechers tested their smoothness, luminescence, viscosity, smoothness and oiliness associated with the relative polarity parameters (RPI). The RPI value can be used for the polarity determination of the cuticle and the emollient, which makes it easy to select the best emollient for the skin use [6].

Studies to assess the efficacy of oils in cosmetics through sensory tests are being continued. Jan Dekker published a pamphlet that briefly describes the physical and chemical properties of various oils, explaining how the oils were obtained and applied to the appropriate locations. Steam distillation removes odors, fatty acids, peroxides, degradation products and organochlorine pesticides, as well as steroids and vitamin E. Steroids and vitamin E are then distilled and returned to the oil as natural antioxidants. The booklet also lists the grades of common natural oils, which have a permeable, spreadable, dry, and oily feel. Jan Dekker also described the relationship between iodine value and physical and chemical properties of natural oil: the higher the unsaturation degree of a fatty acid is, the higher the iodine value, and the easier it is to be oxidized. The higher the iodine value is, the easier it is to spread, and the better its permeability into the cuticle [7].

The wettability of oil in cosmetic products is mainly perceived through the senses. Natural oils used in foods like shea butter is a good moisturizer for the skin. As the triterpene in the butter resin has an anti-inflammatory function, which can stimulate the regeneration of fibroblasts, it is a good skin regeneration agent, therefore, the butter resin has the repairing and protecting function for the skin.

9.1.2 Oil and Fat Products Upgrading

In order to improve the physiological performance and market profit of natural vegetable oils, suppliers extract and concentrate the active substances in the oil and fat. AAK introduced a product called Lipex 205 [8], a modified butter resin, which improved the biological activity of unsaponified substance. The fatty acid composition of the liquid butter resin is oleic acid, stearic acid, and 8–10% linoleic acid. The advantage of Lipex 205 is that it does not crystallize when used as a lipstick, moisturizer, and other high-fat products. Signaline, one of the ISP-Vincience products, is a patented active substance, mainly consisting of 1,2 glycerol diester produced by enzymatic hydrolysis of olive oil and 10% fatty alcohols extracted from the wood wax in Simmonds from California. The efficacy experimental results showed that Signaline could gradually activate the caloric-dependent enzyme, improve the cell function and skin differences, enhance the generation of collagen, fiber, and keratin, enrich the skin fibroblasts, improve the skin's toughness and elasticity, and reduce wrinkles [9]. Olive oil is an important starting material for cosmetic ingredients. Fancor extracted unsaponified parts from olive oil as part of the Silk Olea Oil product [10]. Non-saponifying substances in natural oils are usually the main components of active substances, and their properties are improved by enrichment. For example, the activity-based concept of pomegranate ABS is believed to improve skin hydration and enhance skin protection. Atrium-combined octanoic, decanoic, succinate triglycerides with sesame oil, Triticum vuigare germ oil, and acetate prevents skin dehydration and inflammation.

9.1.3 ω-3 and ω-6 Fatty Acids

Linoleic acid is a ω-3 fatty acid, a tissue material of ceramide, and a signaling substance for skin. Linoleic acid and α-linoleic acid (ω-3) are essential fatty acids for the skin, and they are usually supplied from food. But the oil can penetrate the skin surface, and the topical swirl can directly affect the lipid layer of the middle skin to improve the skin's quality. The ω-6 linoleic acid is particularly important, which is mainly stored in the nerve amide and cholesterol as a barrier to keep the skin water. The linoleic acid can also repair dry skin and prevent the skin's fatty acid levels from decreasing due to surfactants, resulting in reduced water loss in the deeper layers of the skin.

The important function of ω-3 fatty acids is to prevent skin inflammation and sunburn. In general, dietetics suggest that the proportion of nutrition for adults should be ω-3: ω-6 = 1:2, while the natural proportion for healthy skin is 1:7. AAK deduced the three grease compounds with the best proportion of ω-3: ω-6. Lipex ω-3:6 the fatty acid in the product mainly comes from the seeds of camelina camelata; Lipex ω-EPO products contained fatty acid from poppy seed oil, and the principal component is linoleic acid and ω-linoleic acid. Lipex ω-passiflora produces fatty acids from the herb oil. The advantage of these oils is that they are highly antioxidant [11].

AAK uses 5% or more of the ω-fatty acid in its formula to improve dry wrinkled skin, such as inherited allergic dermatitis and chronic dry skin. DSM is known as a producer of vitamins, nutrients, and cosmetic active compounds. The product, Ropufa, which is Borago Officinalis seed oil, contains ω-6 fatty acids extracted from poppies [12]. Jan Dekker also mentions that polyunsaturated fatty acids are as important as vitamins, mineral salts, and proteins, which are not only for skin liposomes, cell membranes, but also synthetic substances of hormones and prostaglandins. Prunus oil, which is extracted from the seeds of the Ente tree, is a natural spice with good penetration to keep the skin smooth and dry. Its fatty acid composition is similar to that of Prunus, which is about 70% oleic acid and 20% linoleic acid. The high content of vitamin E makes it antioxidant [13]. Carapa guaianensis seed oil, produced in the Amazon region, is often applied to the skin by the locals to repel mosquitoes. It can prevent skin inflammation and lubricate the skin. Free acids are usually extracted from ripe fruits, and natural vitamin E is added to improve antioxidant properties.

9.1.4 Vegetable Oils

Vegetable oils come mainly from plant fruits and seeds, also from the leaves, stems, roots and flowers of plants. Brazilian nut oil from the Amazon is often used locally as cooking oil and soap. The tree producing Brazilian nut is up to 30 m and the seed weight is 2.5 kg. Brazilian nuts are the most famous resource in the Amazon and an important source for antioxidants. Its protein is rich in sulfur, amino acids, and methionine, especially glutamine, glutamate, and arginine. Chemyunion introduced the local collection and utilization of Brazil nuts and the character of nut oil. He also introduced Caryocar brasiliense fruit oil, which mainly comes from the Pequi tree grown in the central and southern parts of the United States and contains 60% edible oil rich in vitamin E and protein. It is also an important food and nutrient for the local people. The pulp is rich in carbohydrates, trace elements, fatty acids, amino acids, and volatile substances [14]. The Sachainchi oil in the Peru area is also rich in ω-3, ω-6, and ω-9 unsaturated fatty acid. Calophyllum tacam ahaca oil is extracted from the seed of wild Calophllum inophynum trees along the Madagascar coast which is used as a pure crude oil by CAC Company. It contains about 92% neutral oil, 6.4% sugar and lipids and 1.6% phospholipid. Previous experiments have shown that it can improve skin toughness, elasticity, and smoothness, as well as antibacterial properties.

Indian mango pulp is also a natural vegetable oil, which generally includes the cutting out distillates at 20–40.5 °C and waxes containing fatty acids and fatty alcohols above 40.5 °C. The detailed composition of mango fruit was introduced by Jan Dekker. Other types of plant pulp include cocoa seed pulp, African cheese pulp, and avocado oil, some of which have been hydrogenated to improve the viscosity. Zenitech also offered a variety of vegetable oils, including raspberry oil, cranberry oil, goose pear oil, Brazilian palm oil, lemon oil, lime oil, watermelon oil and blackcurrant oil, all of which come from high-purity cold-pressed vegetable oil and royal jelly. There is an extraordinary sense of skin comfort and skin protection [15]. Natural concentrated vegetable oil is suitable for various formulations. When the content of Compritol 888 is in the range of 1–3%, the stability and viscosity of emulsifier can be improved and the use time of facial creams can be prolonged; when the content is between 4 and 8%, anhydrous gel can be formed; at 10–20%, the gel can be formed with other esters; the gel can be formed with mineral oil and silicone rubber as long as the content reaches 15% [16].

The main issues of natural vegetable oils are their readiness to be oxidized and stench. Traditional antioxidants are not certified natural organic products, but vitamin E is an important natural antioxidant vegetable oil, a complex mixture of α, β, γ, δ tocopherol and trienol. They play an important role in anti-aging and providing skin barrier function, and can effectively delay the service life of cosmetics. Tocm ix L70 is also a natural vegetable oil. It is rich in tocopherol and antioxidant, and low in prooxidant. Therefore, Jan Dekker produced antioxidant Phytox LT and Danox R-3204. The application of modern extraction technologies can effectively remove the odor of plant oil and improve the purity of vegetable oil [17].

9.2 Development of Oil-Based Material of Agrochemicals

The mechanism and the latest research progress of various oil-based additives were summarized. Since the 1990s, people have paid special attention to the environmental impacts of agrochemical formulations. Therefore, highly selective and safe agrochemicals have become the mainstream direction of agrochemicals development, and more opportunities for green agrochemical products have been arising. However, poor compatibility between high-efficiency agrochemicals and the solvent and auxiliary agents which cannot achieve the environmental expectations, which requires green solvent and auxiliary agents. If the traditional developmental pattern cannot be changed in the process of agrochemical preparation, it is difficult to form a new agrochemical product to meet the environmental requirements nowadays. Oil and fats are primary products and derivatives of plants and petroleum. In this part, the application research progress of oil and fat in the field of agrochemical auxiliaries in recent years is reviewed.

9.2.1 Application in Emulsions and Emulsions in Water

Solvent is an important component in emulsions and water emulsions of agrochemicals. The solvent mainly plays a role in dissolving and diluting the active components. The better solvent should include enough solubility for the active components, non-reactive to the active ingredients, low toxicity to humans and animals, safe to the crops, cost-effective, high volatility, safe to the environment, etc. At present, the common solvents are either aromatic or non-aromatic. Aromatics will be limited or gradually banned due to toxicity and causing environmental problems. Oil solvents are a kind of non-aromatic hydrocarbon solvent, and have been used as a substitute for the harmful aromatic hydrocarbon solvents and polar solvents. Oil solvents are a better choice for the future development of environmentally friendly emulsions and water emulsions.

Common plant oils include turpentine, vegetable oils, modified vegetable oils, and plant essential oils. Turpentine is a colorless or pale yellow liquid with a solubility between solvent oil and benzene. Pine oil is extracted by the solvent or steam distillation of branches and leaves of pine trees. It has strong emulsification, wettability, and permeability for washing, paint, and oil solvents. Vegetable oils are mainly corn oil, rapeseed oil, and soybean oil. The main component of vegetable oil is fatty acid triglyceride, which is composed of saturated or unsaturated fatty acid glycerol of C14–C18. Vegetable oils are safe, but the complex components, low solubility and stability are their major disadvantages. The modified vegetable oil is a vegetable oil that forms methyl ester by chemical or enzymatic interesterification. The advantage of modified vegetable oils is the enhanced safety and dissolution performance, and its synergetic effect on most agrochemicals. The disadvantage of modified vegetable oils is the complexity of their composition. For example, poor stability due to the nature of vegetable oil species; high condensation point due to methyl esterification or methylation of vegetable oil. Plant essential oils are a kind of volatile oily liquid produced from secondary metabolite of plants. Most of them are photo-reactive. Plant essential oils have excellent biological activity, excellent solubility, and high safety.

Except vegetable oils, the non-aromatic solvents also include mineral oils, such as kerosene, liquid paraffin and xylene esters. Kerosene is a mixture of alkanes, olefins, cycloalkanes, and aromatic hydrocarbons. Its main component is saturated hydrocarbons, and its disadvantages are highly flammable. Liquid paraffin is a positive alkane of C10–C17 extracted from light oil in the intermediate fraction of petroleum. Dimethyl ester (DBE), commonly known as dimethyl nitrite, has a certain solubility for most of the active components and is an ideal substitute for aromatic solvents.

In recent years, numerous studies have been conducted to promote the development of oil and fat solvents. In the 1970s, the Shenyang Chemical Research Institute in China developed 25% of DDT emulsifiable oil prepared from light diesel as a solvent. And the institute of agricultural sciences of Fujian province in China has studied the preparation of methylamino abamectin benzoate emulsifiable oil from pine tree oil, palm oil, auxiliary solvent, and active agent. The Institute has developed three types of auxiliary solvents using vegetable oil instead of xylene as a solvent. Liu [18] studied the application of biodiesel as a solvent for the processing of herbicide emulsifiable concentrate, and successfully prepared emulsifiable concentrated spermifluoxaline and enoxanone. The results showed that the sediments of three kinds of emulsifiable oils prepared by biodiesel increased significantly. Fu et al. [19] studied that the methylamino abamectin benzoate emulsifiable oil prepared with xylene as a solvent was harmful to the green cabbage Brassica Campestris, but the methylamino abamectin benzoate emulsifiable oil prepared with two solvents: refined pine oil + palm oil and refined pine oil had no obvious damage to the vegetable plants of Brassica Campestris. Zhang [20] prepared high-efficiency cypermethrin emulsion with biodiesel as a solvent and studied the effects of different methyl oleate dosage on the physical properties and efficacy of the cypermethrin emulsion dilution solution. Wang [21] used biodiesel as a solvent to prepare 15% of the highly effective Haloxyfop-methyl emulsion in water, and 3% of OP-10P was selected as an emulsifier. After the hot storage test, there was no oil precipitation occurred in the sample, and the particle size of the sample met the requirements for water emulsion.

There were few agrochemical emulsifiable products prepared with oil and fat solvents in the market, because they have less ability to dissolve most of the active components of agrochemicals. They are also difficult to match with existing emulsifiers, reducing their versatility. Moreover, the use of edible oil also has the problem of competing with food resources, which limited their further development.

9.2.2 Application in Oil Suspension Agent

Oil suspension is a highly dispersed and stable suspension system with oil base as the medium. The oil base includes vegetable oils or mineral oils. As a diluent, the oil itself is required to have good adhesion and spreading properties to waxy or smooth surfaces of leaves. Therefore, better oil suspension should have advantages of good adhesion and spreading properties, and strong resistance to rainwater. In order to make the solvent highly dispersed, it is necessary not only to require high flash point, low toxicity, and low volatility, but also to require sufficient viscosity of the dispersion medium. At present, the relevant research and experience are rare, thus the suitable oil base can only be selected as a medium based on the properties of agrochemical varieties and preparations.

Oil base screening is important for an oil suspension preparation. Generally, oils bases that were widely used include vegetable oils (such as soybean oil, corn oil, rapeseed oil, cottonseed oil, castor oil, coconut oil, palm oil, turpentine oil, vegetable oil concentrate, sunflower oil, etc.), vegetable oil esters (rapeseed oil methyl ester, soybean oil methyl palm oil methyl ester, etc.), and mineral oils (paraffin, methyl naphthalene advanced aliphatic oil, white oil, etc.), pine fat fatty acid ester, terpine-butyl ester, coconut oil heptyl ester, triglyceride, and its mixed solvent. Among them, the methylated vegetable oils (methyl oleates) with low viscosity are favored by various agrochemical research institutions because of their significant efficacy, low cost, non-toxic, and environmental-friendly characteristics.

Chou et al. [22] prepared 20% oil suspension of fenaminstrobin and tebuconazole using methyl soybean oil as an oil base. It was proved in indoor tests that the product has advantages in keeping biological activity. It also has excellent rainwater erosion resistance and spreading performance. Liu [23] successfully obtained a 5% oil suspension of bacillus subtilis using vegetable oil. In the preparation of 15% oil miscible flowable concentrate (OF), Liang et al. [24] compared methyl oleate, soybean oil and its mixture, which had poor solubility to chlorantraniliprole-emamectin benzoate as a dispersion medium, which had no adverse effect on the thermal stability of the active ingredients. The methyl oleate was selected as the oil base according to the parameters of fluidity, emulsifying dispersion, and suspension. Li et al. [25] used soybean oil to prepare 25% oxadiargyl oil miscible flowable concentrate. Zhao [26] prepared 20% thiamethoxam oil flowable concentrate using fatty acid methyl ester.

Among the products of oil suspensions, the most widely prepared and used is the nicosulfuron, and researchers also used different oil bases to prepare nicosulfuron series oil suspensions. The 4% oil flowable concentrate in the water of nicosulfuron was prepared using methyl oleate as an oil base, and its dispersion in water was studied [27]. Yang [28] used the engine oil as the dispersing medium in the preparation of 6% nicosulfuron oil flowable concentrate. And blending soybean oil and dioctyl phthalate with a volume ratio of 6:1 as the dispersion medium when preparing at 4% suspension of nicosulfuron was studied. Liu [29] chose biodiesel as the dispersing medium and also prepared a 4% suspension with superior performance.

9.2.3 Applications of Oils in Spray Auxiliaries

The spray auxiliaries are agrochemical auxiliaries which are directly added to the spray tank or the agrochemical box before the mixture is sprayed, and they can improve the physicochemical properties of the liquid agrochemical after being uniformly mixed. They are also generally called bucket additives. Oil-based additives are an important type of spray auxiliaries. Oil-based spray auxiliaries are classified into vegetable oil emulsifiers, methyl esterified vegetable oil emulsions, mineral oil additives, and synthetic ester emulsions according to the source of the raw material. The raw materials of the vegetable oil spray auxiliaries include soybean oil, rapeseed oil, cottonseed oil, turpentine oil, etc. The methyl esterified vegetable oil is divided into different types of fatty acid methyl esters according to the fatty acid carbon chain properties, such as saturated bonds, unsaturated bonds, and a number of carbon atoms. Mineral oil spray auxiliaries are divided into liquid paraffin oil, white oil, engine oil, diesel oil, etc. Oil-based spray auxiliaries have been widely used in herbicides, among them the vegetable oil emulsions and methyl esterified vegetable oil emulsions are the most widely used. Oil-based spray auxiliaries are also used in insecticides including mineral oils. Oil emulsifier's synergist is less used in the fungicide. The main functions of oil-based spray additives are: (1) Reducing evaporation; (2) Increasing the deposition of active ingredients to reduce the loss of volatilization and drift; (3) Enhancing the permeability by improving the physical and chemical properties of the wax layer of plant leaf and increasing pests absorption and transmission of agrochemicals; (4) Biodegradable and safe to the crops; and (5) Forming films on the foliage to improve the ability of agrochemicals to withstand rain and improve agrochemical utilization.

(1) Vegetable oil-based spray additives, the major raw materials of vegetable oil-based spray auxiliaries are derived from soybean oil and cottonseed oil. Since oils are not miscible with water, it is necessary to add emulsifiers to prepare emulsions for agrochemical spray. This kind of auxiliaries is widely applied in herbicides, and the related technology is also well developed. When preparing an herbicide spray, the amount of 0.5–1.0% vegetable oil-based auxiliary agent is added in an amount of the spray amount. Moreover, it has been proved by practice that under the condition of drought, the addition of vegetable oil-based synergist to the herbicide spray solution can overcome the adverse environmental factors, such as high temperature and dryness. In addition, it can stabilize the drug effect without damaging the crops.

Yang et al. [30] compared the activities of three kinds of adjuvants such as ammonium sulfate, azone, and vegetable oils on the control of barnyard grass with ammonium phosphate and Amaranthus retroflexus. The experimental results showed that the synergistic effect of vegetable oil was greater than that of azone. Ryckaert et al. [31] concluded that the vegetable oil adjuvant could significantly increase the deposition of propiconazole on the top and bottom surfaces of the leaves, which significantly increased the efficacy. The Northwest Agriculture and Forestry University in China studied the effects of pepper seed oil emulsion, cottonseed oil emulsion, and penny oil emulsion on the control of fenvalerate against aphids. It was believed that the addition of oils could improve the wet spreading ability of insecticides. It helped the insecticides to penetrate the insect cuticle and inhibited the activities of certain detoxifying enzymes of the insects. Du et al. [32] studied the synergistic effect of cottonseed oil, sunflower oil, corn oil, soybean oil, peanut oil, and sesame oil on acetamiprid.

The experimental results showed that cottonseed oil, sunflower oil, and soybean oil had the best synergistic effect among them.

(2) Methyl esterified vegetable oil based spray additives, the methyl esterified vegetable oil based spray auxiliaries are a refined variety of vegetable oil-based spray auxiliaries. Vegetable oils have a better affinity with crops after methylation. Studies have shown that its enhancement effect is more obvious than that of surfactants on the market, mineral oils, and vegetable oils, and it is safe for the crops. The raw materials of these additives include methylated cottonseed oil, rapeseed oil, soybean oil, sunflower oil, coconut oil, pine nut oil, refined oil, or crude oil. The methyl esterified vegetable oil is like the vegetable oil which is incompatible with water. Therefore, it is also necessary to add emulsifiers to prepare emulsions and to form an agrochemical spray aid in the form of emulsions.

Guo et al. [33] conducted a special study on the synergistic effect of methyl esterified vegetable oil on sulcotrione. The experimental results showed that esterified vegetable oil can significantly increase the activity of sulcotrione. It was recommended to use methyl ester vegetable oil with the amount of 0.5–1% to the spray. Teng [34] explored the effect of methyl esterified vegetable oil on the activity of chlorflufenate against soybean weeds. The experimental results showed that the use of methyl esterified vegetable oil at 0.5% of the spray volume could significantly improve the activity of chlorflufenate. Lu [35] used methyl potting method to determine the effects of eight kinds of methyl esterified vegetable oil additives, including methyl esterified rapeseed oil, cottonseed oil, peanut oil, corn oil, soybean oil, sunflower oil, linseed oil, and castor oil, on the control of herbicides and amaranthus retroflexus by four common stem and leaf herbicides of sulcotrione, atrazine, nicosulfuron, and quizalofop-lung. The effect of methyl esterified rapeseed oil, cottonseed oil, and peanut oil on herbicides was better, while the effects of methyl esterified linseed oil and castor oil on herbicides were poor.

(3) Mineral oil-based spray additives: Mineral oils have been used as agrochemical for more than 100 years and are widely used around the world. Using mineral oils as a synergist has a long history. The 1970s is the extensive application era of mineral oil emulsions. The mineral oils used as synergists mainly include paraffin oil, engine oil, diesel oil, kerosene, etc. Emulsions generally need to be added and are evenly dispersed in water with very small particles to form a stable opaque emulsion, which can be used as a spray aid. At present, mineral oil emulsions are generally composed of 80–87% mineral oil and 13–20% nonionic emulsifier, both of them have oil adhesion, resistance to volatility, permeability, and surfactants to reduce surface tension.

The addition of mineral oil into insecticides has a good synergistic effect. In 1995, Xing et al. [36] studied the synergistic effect of oil emulsion mixed with cypermethrin. The results showed that the combination of NO. 32 oil emulsion mixed with cypermethrin could increase the knockdown rate of fenvalerate to extend the period of validity. The acute toxicity analysis of toxic chlorpyrifos and avermectin in agricultural mineral oil was studied; the results showed that mineral oil did not increase toxic after mixing with chlorpyrifos and avermectin. The toxic compounding agent could be used as an active agent for agrochemical development. The synergistic effect of mineral oil on the control of certain insecticides against thrips and cotton leaf moths was explored. The results showed that the mineral oil with suitable physical and chemical properties could double the effect of certain agrochemical sprays on the control of thrips and cotton moths while prolonged its residual effect. Liu et al. [37] studied the plant-derived agrochemical matrine, azadirachtin and mineral oil to prevent small green leaf empoasca vitis and tarsonemidmites. When diluted 500 times in water, both 0.3% matrine and mineral oil complex solution and 0.3% azadirachtin and mineral oil complex solution have good effects on the pests and diseases in the autumn.

9.2.4 Applications of Oils in Emulsifiers

Emulsifiers are important surface-active agents that are essential in liquid agrochemical formulations, such as emulsifiable concentrates, microemulsions, and water emulsions. Some oils can also be used as agrochemical emulsifiers after modification. Castor oil polyoxyethylene ethers are a kind of commonly used emulsifiers, which are ethylene oxide modified castor oil. EL and BY series are a general term for castor oil polyoxyethylene ether in the agrochemical industry. The products have different specifications depending on the EO addition, such as the commonly used products of EL-20 and EL-40. Hydrogenated castor oil polyoxyethylene ether (HEL) is produced from hydrogenated castor oil and has a lower iodine value. It is also can be used as an agrochemical emulsifier. Researchers are constantly exploring the use of other oils and ethylene oxide in addition to preparing new emulsifiers, such as soybean oil polyoxyethylene ether, vegetable oil polyoxyethylene ether, rice bran oil polyoxyethylene ether, bitter buckwheat Ren oil polyoxyethylene ether, rosin acid polyoxyethylene ether, etc.

9.2.5 Applications of Oils and Fats in Stabilizers and Synergists

The active ingredients of an agrochemical preparation may be decomposed or deteriorated in physical properties during storage and transportation. Moreover, the degradation and decomposition processes of various agrochemicals and preparations are quite complicated as the stability mechanisms are not the same and many are not completely clear. The addition of an appropriate stabilizer during the preparation of the agrochemical can prevent the decomposition of the active ingredient. Some fats and oil derivatives can be added to agrochemical preparations as stabilizers. Related studies mainly involve two aspects. One is synthetic fatty acid polyoxyalkylene ester phosphates, such as oleic acid, linoleic acid, and ricinoleic acid. Its derivatives can be used in organophosphorus insecticides and bactericides of organophosphorus in emulsifiable concentrates. The other is epoxidized vegetable oils. Several vegetable oil epoxides are commonly used, including soybean oil, linseed oil, rapeseed oil, and cottonseed oils. Arsenic epoxide has also been used as stabilizers in emulsifiable concentrates.

Some vegetable oils can be used as synergists of insecticides. The synergistic effect of sesame oil and deltamethrin on the control of cotton bollworm was studied, the results showed that by mixing sesame oil and deltamethrin at 1:1 ratio, and the amount of deltamethrin could be reduced by a half to achieve a similar effect as straight deltamethrin. According to the analysis, the main synergistic mechanism is that sesamin in sesame oil contains methylene dioxyphenyl group, which inhibits the multifunctional oxidase in the worms [38]. Edible vegetable oil also can increase the control of avermectin EC against Plutella xylostella, and cottonseed oil could be used as a synergist of monocrotophos to prevent blind piles. Experimental results showed obvious synergistic effects on pyrethrum ester, carbamate, and organophosphorus agrochemicals [39].

9.3 Outlook

At present, the cosmetic market of vegetable oil is keeping growing. A large number of new multifunctional cosmetics components are based on vegetable oils, and their hydrogenation products and fatty acid esters are widely used in the cosmetics industry. With the increasing market demand of people to "return to nature and advocate green", as well as the improvement of plant extraction and purification techniques, the application of natural vegetable oils has gradually become a hot spot. At the same time, with the improvement of cosmetic efficacy evaluation technology, people's understanding of skin metabolism process, as well as the integration of personal care products industry, we believe there will be more and better oils and fats to be used in cosmetics that will also give more new, unique and special functions and characteristics to personal care products to bring people more beauty and happiness.

The world is facing problems of increasing population, deteriorating environment, food and energy resources shortage. Agrochemicals play a pivotal role in food production. But petroleum-based agrochemicals and their additives as chemical substances not only need to consume energy, but also pose a great threat to our environment. Oils and fat have the advantages of being environmentally friendly, biodegradable, strong affinity with targets, readily available, relatively low cost, and using less dosage of additives when processed, which should be widely promoted in the processing and use of agrochemicals.

References

1. Goldemberg RL, Consuelo P, Rosa DL (1971) Correlation of skin feel of emollients to their chemical structure. J Soc Cosmet Chem 22:635–654
2. Polar P, Kairo MTK, Moore D et al (2005) Comparison of water, oils and emulsifiable adjuvant oils as formulating agents for *Metarhizium anisopliae* for use in control of *Boophilus microplus*. Mycopathologia 160:151–157
3. Woodruff J (1999) Safe in the sun with less irritation. Deterg Cosmet 2
4. Brand HM, Pieter RR (1992) Diesters and their use in waxes. European Patent Office, EP0389269
5. Parente ME, Gámbaro A, Solana G (2010) Study of sensory properties of emollients used in cosmetics and their correlation with physicochemical properties. J Cosmet Sci 27:354–354
6. Wiechers JW, Musee N (2010) Engineered inorganic nanoparticles and cosmetics: facts, issues, knowledge gaps and challenges. J Biomed Nanotechnol 6:408
7. Dekker J, Budzelaar PHM, Boersma J et al (1984) The nature of the reformatsky reagent. Crystal structure of (brznch2coo-t-bu.thf)2. Organometallics 3:15–20
8. Lodén M, Andersson AC (2010) Effect of topically applied lipids on surfactant-irritated skin. Br J Dermatol 134:215–220
9. Lee JH, Park M J, Lee JS et al (2004) Functional Cloning of xusil gene and investigations of its involvement in FGF signaline during the early development of *Xenopus laevis*. Korean J Lab Anim Sci
10. Altıok E, Bayçın D, Bayraktar O et al (2008) Isolation of polyphenols from the extracts of olive leaves (*Olea Europaea* L.) by adsorption on silk fibroin. Sep Purif Technol 62:342–348
11. Warren R (2011) Wipe article comprising lotion composition comprising omega-6 fatty acid and method of improving skin barrier function. United States Patent, US2011159061
12. Alonso DL, Maroto FG (2000) Plants as 'chemical factories' for the production of polyunsaturated fatty acids. Biotechnol Adv 18:481–497
13. Raynal J, Mourgues J, Conte T (1991) Pectic substances from plum fruits (*Prunus Domestica* L.) c.v. d'ente. franctional extraction and general characteristics. Revista De Administración Pública 30:221–238
14. Amaral LF, Moriel P, Foglio MA et al (2014) Evaluation of the cytotoxicity and phototoxicity of Caryocar Brasiliense supercritical carbon dioxide extract. BMC Complement Altern Med 14:450
15. Zenitech (2006) Zenitech develops new series of products for personal care. Focus Surfactants 8:4
16. Barthelemy P, Laforêt JP, Farah N et al (1999) Compritol® 888 ATO: an innovative hot-melt coating agent for prolonged-release drug formulations. Eur J Pharm Biopharm 47:87–90
17. Dekker J (1977) Four compartment nested container for cosmetics or the like. United States Patent, USD244256 S
18. Liu YQ (2009) Study on the application potential of biodiesel as a solvent for stem and leaf herbicide emulsion. Doctoral dissertation, Shandong Agricultural University
19. Fu JW, Liu LM, Zou HJ (2011) Effect of safety to *Brassica campestris* of methylamino abamectin benzoate mixed with plant oil substitute for xylene solvent. Chin Agric Sci Bull 27:307–311
20. Zhang Y (2011) The development of the environmentally friendly and efficient cypermethrin emulsion in water. Doctoral dissertation, Shandong Agricultural University
21. Wang XX (2011) Screening Study of the high-efficiency haloxyfop-methyl emulsion in water and its stability. Doctoral dissertation, Shandong Agricultural University
22. Chou JY, Yu L, Dan ZG et al (2012) The formulation and bioactivity of Fenaminstrobin. Tebuconazole 20% OD. Agrochemicals 51:801–804
23. Liu Y (2012) Preparation of bacillus subtilis 5% oil miscible flowable concentrate (OF). Anhui Chem Ind 38:57–58
24. Liang JJ, Guan LT, Xu LZ (2012) Research on chlorantraniliprole emamectin benzoate 15% OF. Modern Agrochem 5:20–22

25. Li T, Dai Q, Wu XJ et al (2012) Formulation study of 25% oxadiargyl oil miscible flowable concentrate (OF). Anhui Chem Ind 38:59–62
26. Zhao K, Zhao ZY, Chi GT (2012) Preparation of 20% thianmethoxam oil flowable concentrate. Anhui Agric Sci 6:3345–3346
27. Zhai LL, Lu FS, Xia H et al (2012) The dispersibility of nicosulfuron 4% oil flowable concentrate in water. Agrochemicals 51:651–653
28. Yang JH, Jing RJ (2009) Development of the 6% nicosulfuron oil flowable concentrate. Guizhou Agric Sci 37:74–75
29. Liu YQ, Zhu BY, Du YC (2008) Preparation of 4% nicosulfuron of formulation with biodiesel as dispersing medium. Agrochem Res Appl 1:20–22
30. Yang FY, Ni HW (2009) Compare with the effect of three adjuvants to glufosinate inhibiting weeds. Agrochemicals 48:221–223
31. Ryckaert B, Spanoghe P, Heremans B et al (2008) Possibilities to use tank-mix adjuvants for better fungicide spreading on triticale ears. J Agric Food Chem 56:8041
32. Du XL, Ren AZ, Xing GY et al (2011) Study on the affection of plant oil to pesticide. J Liaocheng Univ Nat Sci Ed. 24:34–38
33. Guo MX, Meng XQ, Peng L et al (2008) Chemistry control of invasive weed *Flaveria bidentis* (L.) Kuntze. Ecology Environ 17:1184–1189
34. Teng CH (2008) Control emeacy of ethoxyfen-ethyl on weeds of soybean. Soybean Sci 27:301–305
35. Lu M (2005) Studies on the enhancement effect of methylated vegetable oil adjuvants on soybean. Doctoral dissertation, Shandong Agricultural University
36. Xing LG, Lai JC (1995) Preliminary report on synergistic effect of a new mineral oil emulsion mixed with cypermethrin on the control of pine caterpillars. Forest Pest Dis. 1:21–22
37. Liu SD (2010) Study on the efficacy of two complex liquid of matrine, azadirachtin and mineral oil against *Empoasca Vitis* and *Tarsonemidmites*. Chin Agric Sci Bull 26:238–242
38. Cui TS, Zhang ZX, Wu CL (1985) Synergistic effect of sesame oil on cotton bollworm control. Agrochemicals 5:53–54
39. Han YZ, Zhou QX, Zhang HY (1986) Preliminary study on synergistic effect of vegetable oil mixed with long-acting phosphorus. Agrochemicals 6:25–30

Chapter 10
Comprehensive Utilization of Processed Residues of Industrial Oil Plants

Jiangshan Ma, Bing Pi, and Peiwang Li

Abstract As kinds of renewable resources, industrial oil plants have been widely exploited for industrial applications. With the increasing development of the processing of the industrial oil plants, a large amount of the processed residues accounted for 70% of industrial oil plants were produced. It is critical to promote the economic sustainability of the industrial oil plant industry by improving the utilization value of the resources of the processed residues. In this chapter, we firstly introduced the different functions of the processed residues. Then, the advances of comprehensive utilization of processed residues from industrial oil plants were summarized. Furthermore, the prospects and the challenges of the utilization of the processed residues from industrial oil plants were discussed.

Keywords Industrial oil plants · Processed residues · Function · Comprehensive utilization · Value

10.1 Introduction

Industrial oil plants are a category of oil plants that produce oil-based energy, chemical and material products and are a major part of forestry resources. Since the fruits of some industrial oil plants are toxic or some are inedible, they are mainly used as raw materials in industrial productions, such as biodiesel, bio-lubricants, biofuels, and chemical products by corresponding conversion technologies. In addition, their pruned or thinned stems can also be used directly for the transformation of solid or gaseous fuels [1, 2]. On the one hand, as the plant grows, solar energy is captured efficiently, biomass raw materials are synthesized and oxygen is released,

J. Ma (✉) · B. Pi · P. Li
Key Laboratory of State Forestry and Grassland Administration on Utilization Science for Southern Woody Oilseed, Hunan Academy of Forestry, Changsha 410004, Hunan, China
e-mail: jiangshanma@hnu.edu.cn

J. Ma
Central South University of Forestry and Technology, Changsha 410004, Hunan, China

P. Li
State Key Laboratory of Utilization of Woody Oil Resource, Changsha 410004, Hunan, China

C. Li et al. (eds.), *Industrial Oil Plant*,
https://doi.org/10.1007/978-981-15-4920-5_10

which absorbs greenhouse gas like carbon dioxide and improves the ecological environment with properties of renewability and environment friendliness [3]. On the other hand, the comprehensive development and utilization of raw materials from oil plants have wide application prospects such as co-producing of energy, chemicals, pharmaceuticals, fodder and fertilizer products to realize the replacement of petrochemical products [4]. Meanwhile, the produced products can reduce greenhouse gas emissions and particulate matter releases to improve the ecological environment. The common problem in economy and society of China is how to protect the ecological environment and realize the sustainable supply of energy resources. Full use of non-cultivated land resources to promote industrial oil plants can be a new way to solve the problem of sustainable economic and social development in China. Therefore, it is of great significance to develop the industrial oil plants industry to safeguard China's energy security, alleviate ecological and environmental crisis, and increase farmers' income [5, 6].

As the green and the poverty alleviation industry, industrial oil plants have attracted a lot of interest [7–9]. The regional distribution of industrial oil plants mainly located in mountainous areas, and considering the particularity of oil products, the Central Committee of the Party and the State Council attach great importance to the development of industrial woody oil plants, which has been integrated into the national food and oil security strategy with policies fully guaranteed. It has great development potentials. The industrial oil plants species such as Camellia oleifera fruit shell, walnut, tung tree, Litsea cubeba, are widely distributed and cultivated in China (about 120 million Mu) [5]. The oil plant resource is abundant, amounting to the great potential of industrial oil plants. Industrial oil plants can not only be processed into high-quality edible vegetable oil, but also partially replace some petrochemical resources with industrial raw materials, and the demand in the market is huge. It is suggested that the annual demand in 2014 for vegetable oil in China reached 34.7 million tons (27 million tons for food and 7.7 million tons for industrial use), and the output value was over 200 billion yuan [10].

Despite the bright future of industrial woody oil plants, there are constraints in the development of industrialization of industrial oil plants. At present, the industry is still in the initial stage of development, lacking neither comprehensive design on industrial chain nor integrative management concepts on R&D. Now, the processing enterprises have large quantities but are mini-type distributed with an extensive mode of development and structural overcapacity. With regard to the processing technology, there are a number of vital problems to solve in terms of slow efficiency on fresh fruits picking, shortage of exclusive drying and storing technology, insufficient utilization of oil plants resource, limited production type, mostly manufacturing primary products, inconsequent products structure and absence of high value-added iconic end-products. Moreover, many of the bioactivity substances in oil plants are not effectively utilized [3, 11, 12].

10.2 Processed Residues of Industrial Oil Plants

In recent years, with the rapid development of the processing industrial woody plants for tung oil, Camellia oleifera, and Litsea cubeba, a large amount of fruit shells and dregs of the processed residues were produced. The processed residue accounts for 70% of woody oil plant [13]. For a long time, most of these resources have been regarded as waste, which is not only a waste of resources, but also causes great pollution to the environment. Therefore, it is important to promote the quality and efficiency of woody oil plants processing industry by integrating the utilization of high-valued processed residues. At present, the processed by-products of woody oil plants are 90 million tons, with an average annual growth rate of 3.9%, including more than 10 million tons of husks, and over 80 million tons of meals, dregs, oil stocks, and soap stocks. The comprehensive utilization rate of these by-products was less than 20% [14]. The by-products including carbohydrate, protein, fat, and other effective substances, are raw materials of food, medicine, healthcare products, energy, and chemical products [15, 16]. The low rate of comprehensive utilization leads to a large number of resources being discarded. With the continuous development of the woody oil plant processing industry, the surplus of processing residues has been increased obviously. How to efficiently make use of these residues is an urgent problem to be solved on the way forward in the woody oil plants processing industry, and is also an important way to sustain development and improve quality and efficiency of the whole woody oil industry [17, 18].

10.3 Processed Residues of Camellia Oleifera Fruit Shell

Camellia oleifera fruit shell is one of the main by-products of the oil processing of Camellia oleifera fruit, accounting for as much as 60% of the weight of a whole fresh fruit [19]. According to an estimated yield of Camellia oleifera oil, supposing 2.5 million tons in 2020, there will be about 1.5 million tons of Camellia oleifera fruit shell to be produced, if it is used effectively, it will inevitably bring about huge economic effects [20]. Camellia oleifera fruit shell is rich in active substances such as tea saponin, tannin, and pentosan, but due to the complexity of the composition, the husks of the fruits are difficult to be directly used, most of them are burned as fuel or discarded straightway. The main applications of Camellia oleifera fruit shell are listed as follows: (1) Camellia oleifera fruit shell can be used to prepare activated carbon because of the high levels of lignin. Sun et al. studied two methods of preparing activated carbon from Camellia oleifera fruit shell, and obtained the activated carbon with the excellent performance [21]. (2) Saponin is a natural nonionic surfactant with good dispersing, foaming, wetting, and emulsifying functions. It can be widely used in paper, leather, and paint industries. Besides, saponin has analgesic, anti-inflammatory, and other pharmacological functions that are often applied to cosmetics and pharmaceutical industries. Based on the diversified applications of saponin, most of the researches on saponin mainly focuses on the extraction and purification of it. For example, Yang et al. have further studied on the improvement of the method of extracting and refining saponin [22]. (3) Preparation of xylose. Xylose is an important industrial raw material that has a wide utility in leather, coating, paint, and plastic. But there are very few studies on the preparation of xylose from Camellia oleifera fruit shell. (4) Preparation of furfural. The tea-oil fruit husk is rich in furfural due to its abundance of the aldehyde group, a kind of good raw material for the preparation of furfural [23]. (5) Extraction of tannins. Tannins, also known as Danning in China, have significant effects in inhibiting the cardiovascular disease, anti-aging, anti-mutation, and anti-cataract. It can effectively prevent and control the further deterioration of tumor, thus much attention has been attracted to the pharmaceutical industry [24].

10.4 The Comprehensive Utilization of Camellia Oleifera Fruit Shell

The Camellia oleifera fruit shell accounts for 50–60% of the weight of a whole fresh Camellia oleifera fruit in general. For each unit of Camellia oleifera fruit shell that may produce 100 kg of tea oil, there will be 36 kg of tannin extracts, 32 kg of furfural, 60 kg of activated carbon, and 60 kg of potassium carbonate being extracted; and meanwhile, 6.4 kg of glacial acetic acid and 25.6 kg of sodium acetate can be derived. In the meantime, for every 100 kg of tea oil produced, the 90 kg of saponin, 20 kg of crude tea oil, and 200 kg of high-quality feedstuff can be extracted [25–27].

10.4.1 Production of Furfural and Xylitol

Furfural is a colorless and transparent oily chemical product widely used in rubber, synthetic resin, paint, medicine, pesticide, and foundry industries, which is an important chemical material. Furfural produced by tea-oil husk can be obtained by means of hydrolysis of polypentene. The theoretical content is 18.16–19.37%, close to or exceeds the content of major raw materials like corncob (9.00%) and cottonseed hull (7.50%), which are mainly used in the current production of furfural. Hydrolysis of polypentose sugars can also produce xylose as it is hydrogenated to form the xylitol. Xylitol is a sweet substance with rich nutritional value, easy to be absorbed by the human body, completely metabolized, and no stimulation on insulin. It is an ideal sweetener for diabetic patients, and is also an important industrial raw material widely used in national defense, leather, plastics, paints, coatings, etc. The yield of xylitol from Camellia oleifera fruit shell is about 12–18%. In the process of hydrolyzing polypentanose to produce furfural or xylitol, some other industrial products such as glucose, ethanol, acetic acid, propionic acid, formic acid and sodium acetate can also be produced [28]. In general, 1.2–1.3 ton of crystalline sodium acetate can be recovered from each ton of furfural that is produced.

10.4.2 Production of Tannin Extract

The Camellia oleifera fruit shell contains 9.23% tannin, which can be extracted by the water extraction method [25]. Tannin is the main raw material for the tanning industry and can also be used as a flotation agent in the mineral industry. The residues after extracting tannin can be used to make furfural or fertilizer.

10.4.3 Production of Activated Carbon

Activated carbon is a porous adsorbent widely used in decolorization, deodorization, and impurity separation of food, medicine, chemical, environmental metallurgy, and oil refining industries. The Camellia oleifera fruit shell contains a large amount of lignin and it has a unique physical structure, which is a good material for producing activated carbon [26, 29]. Camellia oleifera fruit shell can be pyrolyzed (carbonized, activated) to produce activated carbon with greater activity and adsorption capacity, and the comprehensive performance is satisfied. The quality indicators such as activity, yield, raw material consumption, and production cost are close to or better than other fruits shells or lignin materials. The activated carbon produced by Camellia oleifera fruit shell is mainly via ways of gas activation and zinc oxide activation method. The zinc oxide activation method is more commonly used since the effect is better, and the yield rate of the finished product is about 10–15%.

10.4.4 Production of Nutrient Medium

The Camellia oleifera fruit shell contains various chemical components that are beneficial for the cultivation of edible fungi such as mushrooms, oyster mushrooms, and phoenix mushrooms [28, 30]. The edible fungi cultured by tea-oil husks are close to or superior to cotton husks, straws, and sawdust in the perspective of forms and ingredients [31, 32]. The suitable content ratio of Camellia oleifera fruit shell for the cultivation of shiitake mushrooms is about 40–50%, the yield rate is slightly higher than that of the pure Fagaceae sawdust as the amino acid content is increased by 50%. By reducing the cost of 16.7–20.8% for per ton nutrient medium, 900 kg weight of fresh mushrooms (90 kg of dried mushrooms) can be produced, which is worth ￥2700 yuan and saves 1 m^3 wood for per ton Camellia oleifera fruit shell.

10.4.5 Utilization of Theasaponin from the Processed Camellia Oleifera Residues

Theasaponin is a high-performance nonionic natural surfactant agent. It has distinctive surfactivities such as emulsification, dispersion, wetting, decontamination, foaming, and foam stabilization. The deep processed products of theasaponin that have been developed and applied to the industry include tea saponin paraffin wax emulsifying agent, bubble stabilizer and foam stabilizer for the aerated concrete industry, theasaponin and shampoo, pretreatment agent for textile printing and dyeing, foam stabilization agents in beer production, anti-friction agent in machinery industry, etc. In the meanwhile, theasaponin also has a medicinal effect on inhibiting tumor and liver fibrosis. Theasaponin, however, has a destructive effect on an animal's erythrocyte, resulting in Hemolysis. The activity level is measured by the maximum dilution that brings about hemolysis, as is noted, the haemolytic index. The hemolytic property of theasaponin is much lower than that of Sasanqua saponin, but is equivalent to tea leaf saponin and Camellia saponin. The hemolysis index of Sasanqua saponin is at 1,000,000 while the hemolysis index of theasaponin, tea leaf saponin, and Camellia saponin is at 100,000. Theasaponin produces hemolysis effects only on erythrocyte (including red blood cells of nucleated fish blood, chicken blood and non-nucleated human blood), but has little effect on leucocyte. Therefore, theasaponin can cause a toxic effect on fish, but not on shrimps. Its hemolysis mechanism is believed to be an outcome induced by theasaponin that shall cause changes in the permeability of cytomembrane containing cholesterol. Initially, it destroys the cell membrane, and then leads to cytoplasmic extravasation, and ultimately disintegrates the entire red blood cell. The premise of the dissolubility is that theasaponin should be in direct contact with blood; therefore it is nontoxic when taken orally by people and livestock.

China's tea-oil cakes dregs resources are abundant and have great potential for industrial utilization. The application of theasaponin is of great practical significance and has broad prospects. By adopting certain reasonable extraction and purification process, the theasaponin with high output and good purity can be obtained. The extracted and purified theasaponin can be used in industrial and agricultural fields, and the Camellia oleifera cake dregs can be turned into something useful, alluded as "Trash to Treasure" with good economic value. In the meantime, since theasaponin is a product from nature, it has no adverse effects on plant production and is easily degraded in the open environment. Theasaponin is nontoxic to humans or livestock, and has good environmental compatibility. The characteristics of theasaponin lay a good foundation for its broad application and development prospects. At present, the extraction of theasaponin from tea-oil cake dregs now faces the following problems: the purity rate of industrialized theasaponin is not high, the purification technology needs to be improved, the extraction cost needs to be reduced to a reasonable level; the basic theory of structure–function relationship and activity mechanism of theasaponin have not been studied in detail. In this regard, it is necessary to put more efforts into researches and provide a reliable theoretical basis for the comprehensive utilization of theasaponin.

10.4.6 The Efficient Extraction of Theasaponin from the Processed Camellia Oleifera Oil Residues

The extraction and purification methods of theasaponin in Camellia oleifera fruit shell dregs include water extraction, organic solvent, physical auxiliary method, macroporous resin adsorption, etc. The water extraction is the earliest method for extracting theasaponin, which is characterized as easily dissolved in hot water, and then extracted from hot water as extractant. Since the extractant is water, there is no pollution and the cost is low, and it is convenient to obtain. However, the long-duration of hot water immersion can cause hydrolysis and deterioration of theasaponin with a large number of dissolved impurities that can gelatinize the starch and colloidize the protein, making it difficult to separate the impurities and increase the difficulties of subsequent purification. On this account, the extracted theasaponin has a low recovery rate and purity, deep colored, and poor quality of water-soluble impurities, and the product is mostly slurry, and the purification becomes difficult. In order to obtain high-purity theasaponin, it is extracted on the basis of water extraction in combination with other purification processes [32].

Compared with the water extraction method, organic solvent method reacts faster, and obtains product with higher purity, and is convenient for separation and concentration. The extracted theasaponin can be used as additive; however, the production requirements and costs on process and equipment are much higher than the method of water extraction [33]. Nowadays, the intense solvent extracting and n-butanol leaching method is a general method for extracting theasaponin. It is generally extracted by reflow and concentration with methanol or dilute ethanol. The concentrated product is suspended in water, then extracted and defatted with diethyl ether and ethyl acetate, and then extracted with n-butanol to ultimately obtain the total theasaponin. The n-butanol extraction can obtain theasaponin with higher purity, so it is often used for the purification of theasaponin.

Ultrasound is used under the following principle that ultrasonic waves acting on liquids can create cavitation effects, and strong pressure causes destruction on biological cytoderm. The overall crushing process is completed in an instant. The vibration generated by the ultrasonic waves enhances the release, diffusion, and dissolution of the intracellular substances but the biological activity remains unchanged when the extracted material is re-destructed. It also improves the crushing rate and the extraction rate at the same time, which has been widely used for theasaponin extraction [34].

10.5 High-Valued Utilization of Processed Tung Oil Residues

Tung meal is the main by-product of tung oil extraction. Its protein content can reach between 36% and 45%, and the content of extracted amino acid and essential amino acid can reach 93.8% and 45.4%, respectively, which shows the excellent resource of plant protein for industrial development and utilization [35]. China's tung oil production ranks first in the world, and the annual output of Tung meal is 500,000 tons. As Tung oil meal contains albumin, protamine, neodymidine, and saponin, toxic to humans and animals, the current utilization of tung meal occupies less than 20%. Most of them were abandoned, not only failing to take full advantage of the resource value, but also causing great pressure on the environment protection [36].

In addition to the high content of amino acids, Tung meal also contains trace elements required by various plants such as 2–7% nitrogen, 1–3% phosphorus, and 1–2% potassium. It's also a plant fertilizer matrix that can be developed and utilized. As a fertilizer, Tung meal can not only improve the yield of crops, but also improve the physical and chemical properties of the soil and enhance its moisturizing and water-retaining ability. The residual toxic substances left in the tung meal can kill pests. In addition, since the crude protein, crude fat, sugar, and fiber content levels are high, tung meal can be used as feedstuff. Although it has a certain degree of toxicity, the tung oil meal can be used for feed production after detoxification treatment. The Sichuan Grain Storage Research Institute and other organizations use the solvent method, ammonia treatment method and microbial fermentation method to carry out detoxification experiment of tung meal. In addition, some other studies have shown that the toxic substances in tung meal are poisonous to insects such as cotton weevil and cotton boll weevil, possibly to be developed for use as pesticides [37].

10.6 Processed Flaxseed Oil Residues

China is one of the major producers of flaxseed in the world, and its planting area ranks third. However, for a long time, a large amount of flaxseed cake dregs left as feedstuff or thrown-outs have caused a great wastage of resources. In view of the fact that it contains a variety of active ingredients, if flaxseed dregs are fully utilized, the health care and medicinal products with higher economic and nutritional values should be obtained therefrom. It will greatly increase the economic added value and bring about huge economic and social benefits as well.

10.6.1 Protein Value of Processed Flaxseed Oil Residues

The protein content of flaxseed meal is 32–49%, mainly composed of albumin and globulin. This is a good protein resource, containing all the essential amino acids required by the human body, and has excellent properties for industrial processing. With the further development of research on flax protein in recent years, it has been found that flax protein has certain preventive effects on certain diseases of heart, kidney, and cancer. Flax protein is a protein with properties of high branched-chain amino acids, low aromatic amino acids, and high Fisher ratio, all of which may have special physiological functions for the patients with special needs, like the malnourished patients with cancer, burns, trauma, or hepatitis. Therefore, flaxseed protein can be proportioned and formulated as a functional food for the above patients. Since the lysine/arginine ratio is a key factor leading to blood cholesterol and atherosclerosis, as that ratio in flaxseed protein is low, the impact on blood is minimal. Flaxseed is also a very good source of lysine, glutamine, and histidine. These three amino acids are known to have a strong effect on the human body's immune function. The cysteine and methionine levels of flaxseed proteins increase the human body's antioxidant levels, potentially stabilizing DNA during cell division and reducing the risk of colon cancer formation. Flaxseed protein also affects blood glucose level. Due to its combination with vegetable gums, it stimulates the secretion of insulin, following a response that reduces hyperglycemia. The combination of flaxseed protein and soluble polysaccharide can significantly reduce the effects of luminal in the colon, thereby preventing the production of ammonia that causes tumor growth.

Flaxseed protein has good properties of water retention, oil retention, foaming, emulsifying, and gelling, so it can be processed as excellent food additives. For example, by adding 3% of flaxseed protein in the canned fish sauce it can make the fish sauce creamy and smooth but eliminate all undesirable flavors; flaxseed protein can significantly reduces the loss of meat ingredients when baking, the higher the protein concentrates, the better it effects; by mixing starch with flaxseed protein, a stable complex that delays the aging time of the starch and maintains the stability of the remaining components can be formed.

10.6.2 Dietary Fiber Utilization of Processed Flaxseed Oil Residues

Dietary fiber in flaxseed cake dregs accounts for about 45.4% in dry weight. The dreg is rich in both soluble and insoluble dietary fiber, an ideal natural green health food. The main component of soluble dietary fiber is flax gum, which consists of acidic polysaccharides as a major component, and neutral polysaccharides. Acidic polysaccharide is composed of arabinose, xylose, and galactose. The methods for extracting dietary fiber from flaxseed meal include solvent extraction, enzymatic hydrolysis, microwave-assisted extraction, etc. In some cases, in order to achieve a better extraction effect, these methods are adopted cooperatively. The solvent extraction method generally uses hot water, acid, alkali, ethanol, etc., as solvents, and mostly use hot water. The water extraction method is safe, nontoxic, but time-consuming and inefficient, and acid-alkali extraction is inclined to destroy the spatial structure and the activity of the flaxseed gum. A main principle of the enzymatic hydrolysis method is to use an enzyme capable of degrading the components of the plant's cytoderm, and therefore the active ingredient could be sufficiently released. The enzymatic extraction method asks for mild extracting conditions. It is energy-saving, the obtained dietary fiber has a high purity rate, but the price of the enzyme itself is high, thus the extraction cost is relatively high. Microwave-assisted extraction can significantly shorten the extracting duration by water extraction method, the extraction rate is obviously increased, but the energy consumption is high, which is not conducive to industrialization and expansion of production. At present, the most promising method for preparing soluble dietary fiber is membrane separation, but this method cannot produce insoluble dietary fiber flax gum as the dietary fiber.

10.6.3 Extraction and Utilization of Lignin from Processed Flaxseed Oil Residues

Lignin is a phytoestrogen, a natural phenolic compound synthesized from bimolecular phenylpropanoid. Lignin is widely distributed in many plants and in a variety of foods, especially among the whole grains and beans, with the highest content in flaxseed (0.7–2%), which is 75–800 times higher than other foods. The main lignans in flaxseed are secoisolariciresinol (SECO) and its secoisolariciresinol diglucoside (SDG). In recent years, researchers in Canada, Australia, the United Kingdom, and the United States have done a lot of work on the research and development of the medicinal efficacy of flaxseed cake dregs. High-purified, standardized lignan extracts that conform to the curative requirements have been developed and adopted in various foods. Unilever Co., Ltd. and Elcom Biotech Co., Ltd. have applied for patents on the utilization of lignan in food, medicine, and cosmetics. Take another example, Golden Flax 4U and Natunola of Canada have developed lignan products in the form of natural flaxseed husk; a variety of products such as nutritionally functional super-emulsions and whole milk beverages containing lignans also have been appearing in the market. In addition, some components for health care contained in lignans have particular utility in the production of functional foods with properties of anti-inflammatory or anti-aging.

At present, the main methods for the extraction of lignin from flaxseed cake dregs are the following: traditional ethanol-water extraction, microwave-assisted extraction, enzymatic extraction, high-pressure, and low-polarity water extraction. Traditional ethanol extraction has the disadvantages of long time-consumption, low recovery rate, and environmental pollution. Enzymatic extraction has advantages of good selectivity, but the cost is high and the yield is low. Microwave-assisted extraction has the disadvantages that the extraction yield is low for each time and is not suitable for amplification for the process research. The high-pressure and low-polarity water extraction method requires high-performance equipment and that is not easy to achieve. In sum, choosing a safe, fast, and economical extraction process has a significant research value.

10.7 Comprehensive Utilization of Processed Woody Oil Plant Residues in Our Research Group

Aiming at the technical bottlenecks, namely low utilization rate of the industrial woody oil plants, backwardness of the processing technology and equipment, weakness of unitary and low value-added products, the research team groups the woody oil plant residues into remaining oil, proteins, saponins, and fibers, and looks to achieve high-valued utilization among all resources. The whole-chain structured layout and integrated design are carried out from three dimensions as the basic research innovation, the key technology breakthrough, and the regional integrated demonstration. By breaking through with a number of steering technologies and focusing on creating core technologies of independent intellectual property rights, the all-resources utilization rate of processed woody oil plant residues is targeted from 20 to 80%. The overall value of the industry aims to be increased by 30%.

First, by developing poly-generation technology of the non-oil component inverted to fertilizer and energy, the utilization rate of oil residue resources can be increased by over 20%. The production technology of bio-organic fertilizer with natural active source has been innovated, and multifunctional bio-organic fertilizer that is eligible for fertilizer feeding, soil improving, and pest controlling has been produced. Four bacteria strains, which are capable of antagonizing plant pathogens and easily degrading the crude protein of tung tree meals into water-soluble amino acids, are isolated and screened with independent intellectual property rights. By means of strain purification, refinement, rejuvenation, and cultivation expansion, a bacteria agent for rapid fermentation and maturing can be prepared (Fig. 10.1). Single-factor test and response surface method are adopted to optimize the solid-state fermentation conditions of the combined strains. The optimal processing parameters of fermentation duration, fermentation temperature, material proportion, etc., have been determined to produce a multifunctional bio-organic fertilizer that is competitive in fertilizing, soil improving, and pest controlling. The total content of free amino is 43.9 times higher than that before fermentation. The product meets the requirements of the bio-organic fertilizer standard (NY 884-2012), and three invention patents have been applied.

Second, the technology of preparing briquette fuel by coupling, drying, and molding the oil plants and forest processing residues has been developed. It is made clear that three oil plants processed residues (husk, dregs, and bio-asphalt, additive amount

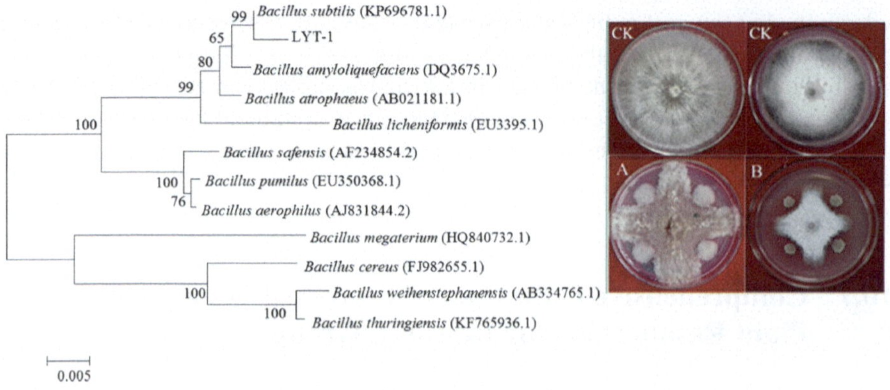

Fig. 10.1 Phylogenetic analysis of the strain based on 16S rRNA sequence and its antagonistic effect against the pathogenic fungi

$\leq$ 15%) can promote the producing mechanism of hydrogen bond and "solid bridge" in the process of ligneous biomass compound modulating. A drying reaction device for stabilizing the fluidization of ligneous biomass bed under the conditions of low oxygen and none-fluidization additives has been invented. A preparation technology of biomass compound briquette with wide material adaptability where steps of materials such as multi-modulating, fluidized bed drying, and molding are integrated and linked has been structured. High-quality solid-formed granule fuel has been developed with the extrusion of energy consumption reduced by 18% but water absorption, calorific value, and Meyer intensity of the formed granule increased by 20%, 16%, and 15%, respectively (Fig. 10.2) [38].

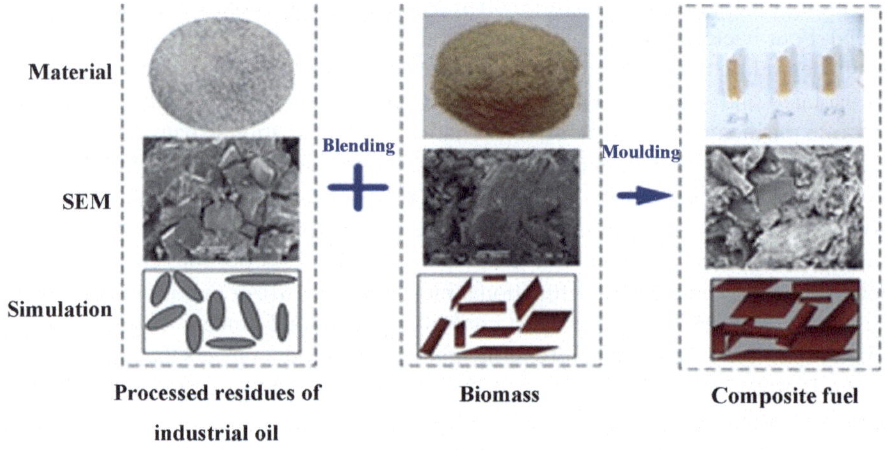

Fig. 10.2 Mechanism of the composite fuel production using processed residues of industrial oil

10.8 Future Perspective and Suggestion

Although the high value-added utilization of woody oil plant residues has attracted an increasing amount of attention over the past 10 years, the comprehensive utilization of processed residues of woody oil plants confronts the following challenges:

(1) Low rate of comprehensive utilization and a severe wastage of resources

At present, the comprehensive utilization rate of processed residues of industrial woody oil plants is only at 20%. The residue's ingredients like carbohydrate, protein, fat, and other bioactive substances are not fully and efficiently employed, resulting in a complete waste of the precious resources and inevitable pollution to the environment.

(2) Backwardness of key technology and equipment for the comprehensive utilization

The technological innovation system based on enterprises has not been fully established, and the research and development of core technologies and equipment lag behind the world's advanced level. Currently, the equipment of the deep processing industry for by-products of woody oil plant have poor stability and high energy consumption; the promotion of new advanced technology spreads slowly, and the key equipment technology still relies on imports from abroad; the existing technical equipment has low performance, low automation, low-end manufacturing level, and lacks integrated and automated production; last but not the least, low level of industrialization of by-product processing technology is the main problem that restrains the comprehensive utilization of processed residues of woody oil plants.

(3) Absence of standardization construction and management

At present, the standards that are set in connection with the processing of woody oil plant residues are scarce and the standardization construction is greatly insufficient. The entries of standards are simple and technical indicators are scanty: the standard coverage is limited and the range for usage is narrow. The standard upgrading cannot keep up with the needs of market development, which is inadaptable to the transition of the processed by-products from quantity to quality-oriented. Due to the absence of corresponding standardization management, some products relating to woody oil residues do not have unified standards for the process and ingredients at the time of producing, leading to unstable product quality and the suspicions on some products from the customers. Today, most products with comprehensive utilization do not have national standards or industry criteria, let alone an effective back-up from the perspectives of primary standard, method standard, and management standard for product standard.

(4) Demand for policy support

The demand for policy support includes fiscal service, legal service, scientific and technological service, etc. At present, there is no substantial policy application or effective and reasonable policy incentives for these processed by-products. As a result, the related processing enterprises are not very active and enthusiastic about the comprehensive utilization of woody oil plant residues, which consequently downsizes the possibility of research and development on new technology and equipment, and hinders the progress of comprehensive utilization of the processed by-products of ligneous oil plant residues.

References

1. Aguilar FX, Song N, Shifley S (2011) Review of consumption trends and public policies promoting woody biomass as an energy feedstock in the US. Biomass Bioenerg 35(8):3708–3718
2. Kinoshita T, Ohki T, Yamagata Y (2010) Woody biomass supply potential for thermal power plants in Japan. Appl Energy 87(9):2923–2927
3. Tan X, Ma L, Li F et al (2012) Industrialization development strategy of woody grain and oil in China. Nonwood For Res 30(1):1–5
4. Chen YX, Mao ZQ, Wu ZB et al (2006) Comprehensive exploitation and utilization of Jatropha oil plants. China Oils Fats 31(3):63–65
5. Zhang H, Pang X, Liu T (2006) Woody oil plant resources in China and the present situation of exploitation and utilization. Biomass Chem Eng S1
6. Chao M, Xing Y, Guangdong W (2009) Present status of application and problems on woody oil plant in China. Chin Agri Sci Bulletin 24
7. Tao W (2005) A Survey of the woody plant resources for biomass fuel oil in China. Sci Technol Rev 5
8. Grebner DL, Perez-Verdin G, Henderson JE et al (2009) Bioenergy from woody biomass, potential for economic development, and the need for extension. J Exten 47(6):8
9. Huang HJ, Ramaswamy S, Al-Dajani W et al (2009) Effect of biomass species and plant size on cellulosic ethanol: a comparative process and economic analysis. Biomass Bioenerg 33(2):234–246
10. Purba HJ, Sinaga BM, Novianti T et al (2018) The impact of changes in external factors on the world vegetable oil market. Int J Econ Finan Issu 8(6):176
11. Jin-shuan FAN (2008) Current situation and prospect of woody oil production development in China. Nonwood For Res 2
12. Zhongwu W (2012) Analysis and strategies on woody oil industry development in China. Forest Res Manage 1
13. Yoshida T, Suzuki H (2010) Current status of woody biomass utilization in ASEAN countries. Biomass 113–123
14. Jian L, Yue-jun Z, Guang L (2001) The woody oil plants in north China. Quart Forest By-Prod Special China 1
15. Arato C, Pye EK, Gjennestad G (2005) The lignol approach to biorefining of woody biomass to produce ethanol and chemicals. Appl Biochem Biotechnol 123(1–3):871–882
16. Taylor EL (2009) Co-products and by-products of woody biorefinery processing. In: Transition to a bio economy: the role of extension in energy, farm foundation conference (Little Rock, AR)
17. Amidon TE, Wood CD, Shupe AM et al (2008) Biorefinery: conversion of woody biomass to chemicals, energy and materials. J Biobased Mater Bioenergy 2(2):100–120
18. Wu J, Sperow M, Wang J (2010) Economic feasibility of a woody biomass-based ethanol plant in central Appalachia. J Agri Res Econo 522–544
19. Zhong X, Zhang B, Zhu J (2007) The comprehensive utilization of camellia oleifera seed. Sci Technol Cereals, Oils Foods 2
20. Luo X, Zhu W (2010) Processing technology of oil-tea camellia seed oil and comprehensive utilization of camellia oleifera resources. China Oils Fats 9

21. Sun K, Jiang JC, He Y et al (2010) Preparation of activated carbon from Camellia oleifera shell and its mesoporous structure modification. Chem Indus For Prod 30(6):24–28
22. Yang K, Huang M (2002) Research on the extraction and purification of tea saponin. Chem Indus For Prod 2001–2002
23. Zhang L, He Y, Zhu Y et al (2018) Camellia oleifera shell as an alternative feedstock for furfural production using a high surface acidity solid acid catalyst. Biores Technol 249:536–541
24. Xu Z, Li X, Feng S et al (2016) Characteristics and bioactivities of different molecular weight polysaccharides from camellia seed cake. Int J Biol Macromol 91:1025–1032
25. Xu M, Chen JH, Wang YM et al (2009) Preliminary report of extraction and analysis of tannins from Camellia oleifera fruit shell. Chem Indus For Prod 29(Supplement):187–191
26. Jin X (2012) Bioactivities of water-soluble polysaccharides from fruit shell of Camellia oleifera Abel: antitumor and antioxidant activities. Carbohyd Polym 87(3):2198–2201
27. Zhang J, Wang Z, Du M et al (2015) Physicochemical properties of camellia nut shell and its thermal degradation characteristics. BioResources 10(1):647–659
28. Xie Y, Ge S, Jiang S et al (2018) Study on biomolecules in extractives of Camellia oleifera fruit shell by GC–MS. Saudi J Biolog Sci 25(2):234–236
29. Kang S, Jian-chun J, Dan-dan C (2011) Preparation of activated carbon with highly developed mesoporous structure from Camellia oleifera shell through water vapor gasification and phosphoric acid modification. Biomass Bioenerg 35(8):3643–3647
30. Zhang J, Ying Y, Li X et al (2018) Evaluation of three kinds of nutshell with respect to utilization as culture media. BioResources 13(4):7508–7518
31. Hu J, Chang S, Peng K et al (2015) Bio-susceptibility of shells of Camellia oleifera Abel. fruits to fungi and termites. Int Biodeter Biodegra 104:219–223
32. Hu JL, Nie SP, Huang DF et al (2012) Extraction of saponin from Camellia oleifera cake and evaluation of its antioxidant activity. Int J Food Sci Technol 47(8):1676–1687
33. Shi G, Wang H, Hua R et al (2011) Technical study on ethanol-ammonia water extraction of tea saponin from Camellia cakes. Mod Chem Ind 31(4):37–40
34. Guo H, Deng Z, Peng Y et al (2008) Study on extraction of tea saponin by microwave and light wave. Sci Technol Food Ind 11
35. Zheng X, Zhang J, Wu F et al (2016) Extraction and composition characterisation of amino acids from tung meal. Nat Prod Res 30(7):849–852
36. Li-xia ZAN (2011) Exploitation and utilization of Aleurites fordii Hemsl. Ani Husban Feed Sci 6
37. Zan L, Song W, Liu J et al (2018) Study on toxicity of water extract from tung meal on aboveground and underground pests. Nonwood For Res 1:13
38. Cao L, Yuan X, Li H, Li C, Xiao Z, Jiang L, Huang B, Xiao Z, Chen X, Wang H, Zeng G (2015) Complementary effects of torrefaction and co-pelletization: energy consumption and characteristics of pellets. Bioresour Technol 185:254–262

The manufacturer's authorised representative in the EU is Springer Nature Customer Service Centre GmbH, Europaplatz 3, 69115 Heidelberg, Germany. If you have any concerns regarding our products, please contact ProductSafety@springernature.com

Printed and bound by CPI Group (UK) Ltd, Croydon, CR0 4YY
07/10/2026
02229171-0001